MW01025732

INTRODUCTION TO

Scientific Programming and Simulation Using R

INTRODUCTION TO

Scientific Programming and Simulation Using R

Owen Jones, Robert Maillardet,
and Andrew Robinson

CRC Press is an imprint of the
Taylor & Francis Group, an **informa** business
A CHAPMAN & HALL BOOK

Chapman & Hall/CRC
Taylor & Francis Group
6000 Broken Sound Parkway NW, Suite 300
Boca Raton, FL 33487-2742

Printed in the United States of America on acid-free paper
10 9 8 7 6 5

International Standard Book Number-13: 978-1-4200-6872-6 (Hardcover)

Visit the Taylor & Francis Web site at
http://www.taylorandfrancis.com

and the CRC Press Web site at
http://www.crcpress.com

Preface

This book has two principal aims: to teach scientific programming and to introduce stochastic modelling. Stochastic modelling in particular, and mathematical modelling in general, are intimately linked to scientific programming because the numerical techniques of scientific programming enable the practical application of mathematical models to real-world problems. In the context of stochastic modelling, simulation is the numerical technique that enables us to analyse otherwise intractable models.

Simulation is also the best way we know of developing statistical intuition.

This book assumes that users have completed or are currently undertaking a first year university level calculus course. The material is suitable for first and second year science/engineering/commerce students and masters level students in applied disciplines. No prior knowledge of programming or probability is assumed.

It is possible to use the book for a first course on probability, with an emphasis on applications facilitated by simulation. Modern applied probability and statistics are numerically intensive, and we give an approach that integrates programming and probability right from the start.

We chose the programming language R because of its programming features. We do not describe statistical techniques as implemented in R (though many of them are admittedly quite remarkable), but rather show how to turn algorithms into code. Our intended audience is those who want to make tools, not just use them.

Complementing the book is a package, `spuRs`, containing most of the code and data we use. Instructions for installing it are given in the first chapter. In the back of the book we also provide an index of the programs developed in the text and a glossary of R commands.

Course structure options

This book has grown out of the notes prepared for a first year course consisting of 36 lectures, 12 one-hour tutorials, and 12 two-hour lab classes. However it now contains more material than would fit in such a course, which permits

its use for a variety of course structures, for example to meet prerequisite requirements for follow-on subjects. We found the lab classes to be particularly important pedagogically, as students learn programming through their own experimentation. Instructors may straightforwardly compile lab classes by drawing on the numerous examples and exercises in the text, and these are supplemented by the programming projects contained in Chapter 22, which are based on assignments we gave our students.

Core content The following chapters contain our core material for a course on scientific programming and simulation.

Part I: Core knowledge of R and programming concepts. Chapters 1–6.

Part II: Thinking about mathematics from a numerical point of view: applying Part I concepts to root finding and numerical integration. Chapters 9–11.

Part III: Essentials of probability, random variables, and expectation required to understand simulation. Chapters 13–15 plus the uniform distribution.

Part IV: Stochastic modelling and simulation: random number generation, Monte-Carlo integration, case studies and projects. Chapters 18.1–18.2, 19, 21.1–21.2 and 22.

Additional stochastic material The core outlined above only uses discrete random variables, and for estimation only uses the concept of a sample average converging to a mean. Chapters 16 and 17 add continuous random variables, the Central Limit Theorem and confidence intervals. Chapters 18.3–18.5 and 20 then look at simulating continuous random variables and variance reduction. With some familiarity of continuous random variables the remaining case studies, Chapter 21.3–21.4, become accessible.

Note that some of the projects in Chapter 22 use continuous random variables, but can be easily modified to use discrete random variables instead.

Additional programming and numerical material For the core material basic plotting of output is sufficient, but for those wanting to produce more professional graphics we provide Chapter 7. Chapter 8, on further programming, acts as a bridge to more specialised texts, for those who wish to pursue programming more deeply.

Chapter 12 deals with univariate and multivariate optimisation. Sections 12.3–12.7 on multivariate optimisation, are harder than the rest of the book, and require a basic familiarity with vector calculus. This material is self-contained, with the exception of Example 17.1.2, which uses the `optim` function. However, if you are prepared to use `optim` as a black box, then this example is also quite accessible without reading the multivariate optimisation sections.

Chapter outlines

1: Setting up. Here we describe how to obtain and install R, and the package `spuRs` which complements the book.

2: R as a calculating environment. This chapter shows you how to use R to do arithmetic calculations; create and manipulate variables, vectors, and matrices; work with logical expressions; call and get help on built-in R functions; and to understand the workspace.

3: Basic programming. This chapter introduces a set of basic programming structures that are the building blocks of many programs. Some structures are common to numerous programming languages, for example `if`, `for` and `while` statements. Others, such as vector-based programming, are more specialised, but are arguably just as important for efficient R coding.

4: Input and output. This chapter describes some of the infrastructure that R provides for importing data for subsequent analysis, and for displaying and saving results of that analysis. More details on the construction of graphics are available in Chapter 7, and we provide more information about importing data in Chapter 6.

5: Programming with functions. This chapter extends Chapter 3 to include user-defined functions. We cover the creation of functions, the rules that they must follow, and how they relate to the environments from which they are called. We also present some tips on the construction of efficient functions, with especial reference to how they are treated in R.

6: Sophisticated data structures. In this chapter we study R's more sophisticated data structures—lists and dataframes—which simplify data representation, manipulation, and analysis. The dataframe is like a matrix but extended to allow for different data modes in different columns, and the list is a general data storage object that can house pretty much any other kind of R object. We also introduce the factor, which is used to represent categorical objects.

7: Better graphics. This chapter provides a deeper exposition of the graphical capabilities of R, building on Chapter 4. We explain the individual pieces that make up the default plot. We discuss the graphics parameters that are used to fine-tune individual graphs and multiple graphics on a page. We show how to save graphical objects in various formats. Finally, we demonstrate some graphical tools for the presentation of multivariate data (lattice graphs), and 3D-graphics.

8: Further programming. This chapter briefly mentions some more advanced aspects of programming in R. We introduce the management of and interaction with packages. We present details about how R arranges the objects that we create within the workspace, and within functions that we are running. We provide further suggestions for debugging your own functions. Finally, we present some of the infrastructure that R provides for object-oriented programming, and for executing code that has been compiled from another computer language, for example, C.

9: Numerical accuracy and program efficiency. In this chapter we consider technical details about how computers operate, and their ramifications for programming practice, particularly within R. We look at how computers represent numbers, and the effect that this has on the accuracy of computation results. We also discuss the time it takes to perform a computation, and programming techniques for speeding things up. Finally we consider the effects of memory limitations on computation efficiency.

10: Root-finding. This chapter presents a suite of different techniques for finding roots. We cover fixed-point iteration, the Newton-Raphson method, the secant method, and the bisection method.

11: Numerical integration. This chapter introduces numerical integration. The problem with integration is that often a closed form of the antiderivative is not available. Under such circumstances we can try to approximate the integral using computational methods. We cover the trapezoidal rule, Simpson's rule, and adaptive quadrature.

12: Optimisation. This chapter covers the problem of finding the maximum or minimum of a possibly multivariate function. We introduce the Newton method and the golden-section method in the context of a univariate function, and steepest ascent/descent and Newton's method for multivariate functions. We then provide some further information about the optimisation tools that are available in R.

13: Probability. In this chapter we introduce mathematical probability, which allows us to describe and think about uncertainty in a precise fashion. We cover the probability axioms and conditional probability. We also cover the Law of Total Probability, which can be used to decompose complicated probabilities into simpler ones that are easier to compute, and Bayes' theorem, which is used to manipulate conditional probabilities in very useful ways.

14: Random variables. In this chapter we introduce the concept of a random variable. We define discrete and continuous random variables and consider various ways of describing their distributions, including the distribution function, probability mass function, and probability density function. We define expectation, variance, independence, and covariance. We also consider transformations of random variables and derive the Weak Law of Large Numbers.

15: Discrete random variables. In this chapter we study some of the most important discrete random variables, and summarise the R functions relating to them. We cover the Bernoulli, binomial, geometric, negative binomial, and the Poisson distribution.

16: Continuous random variables. This chapter presents the theory, applications of, and R representations of, a number of continuous random variables.

We cover the uniform, exponential, Weibull, gamma, normal, χ^2, and t distributions.

17: Parameter estimation. This chapter covers point and interval estimation. We introduce the Central Limit Theorem, normal approximations, asymptotic confidence intervals and Monte-Carlo confidence intervals.

18: Simulation. In this chapter we simulate uniformly distributed random variables and discrete random variables, and describe the inversion and rejection methods for simulating continuous random variables. We also cover several techniques for simulating normal random variables.

19: Monte-Carlo integration. This chapter covers simulation-based approaches to integration. We cover the hit-and-miss method, and the more efficient Monte-Carlo integration method. We also give some comparative results on the convergence rate of these two techniques compared with the trapezoid and Simpson's rule, which we covered in Chapter 11.

20: Variance reduction. This chapter introduces several sampling-based innovations to the problem of estimation. We cover antithetic sampling, control variates, and importance sampling. These techniques can vastly increase the efficiency of simulation exercises when judiciously applied.

21: Case studies. In this chapter we present three case studies, on epidemics, inventory, and seed dispersal (including an application of object-oriented coding). These are extended examples intended to demonstrate some of our simulation techniques.

22: Student projects. This chapter presents a suite of problems that can be tackled by students. They are less involved than the case studies in the preceding chapter, but more substantial than the exercises that we have included in each chapter.

Bibliography/further reading

For those wishing to further their study of scientific programming and simulation, here are some texts that the authors have found useful.

The R language

W.N. Venables and B.D. Ripley, *S Programming.* Springer, 2000.

W.N. Venables and B.D. Ripley, *Modern Applied Statistics with S, Fourth Edition.* Springer, 2002.

J.M. Chambers and T.J. Hastie (Editors), *Statistical Models in S.* Brooks/Cole, 1992.

J. Maindonald and J. Braun, *Data Analysis and Graphics Using R: An Example-Based Approach, Second Edition.* Cambridge University Press, 2006.

Scientific programming/numerical techniques

W. Cheney and D. Kincaid, *Numerical Mathematics And Computing, Sixth*

Edition. Brooks/Cole, 2008.
M.T. Heath, *Scientific Computing: An Introductory Survey, Second Edition.* McGraw-Hill, 2002.
W.H. Press, S.A. Teukolsky, W.T. Vetterling and B.P. Flannery, *Numerical Recipes, 3rd Edition: The Art of Scientific Computing.* Cambridge University Press, 2007.
C.B. Moler, *Numerical Computing with Matlab*, Society for Industrial Mathematics, 2004.

Stochastic modelling and simulation
A.M. Law and W.D. Kelton, *Simulation Modeling and Analysis, Third Edition.* McGraw-Hill, 1999.
M. Pidd, *Computer Simulation in Management Science, Fifth Edition.* Wiley, 2004.
S.M. Ross, *Applied Probability Models with Optimization Applications.* Dover, 1992.
D.L. Minh, *Applied Probability Models.* Brooks/Cole, 2001.

Caveat computator

R is under constant review. The core programmers schedule a major release and a minor release every year. Releases involve many changes and additions, most of which are small, but some of which are large. However, there is no guarantee of total backward compatibility, so new releases can break code that makes assumptions about how the environment should work.

For example, while we were writing this book, the upgrade from version 2.7.1 to 2.8.0. changed the default behaviour of `var`, to return an `NA` where previously it returned an error, if any of the input were `NA`. Happily, we had time to rewrite the material that presumed that an error would be returned.

We conclude that R changes, and we note that this book was written for version 2.8.0. The spuRs package will include a list of errata.

Thanks

Much of this book is based on a course given by the first two authors at the University of Melbourne. The course was developed over many years, and we owe much to previous lecturers for its fundamental structure, in particular Steve Carnie and Chuck Miller. We are also indebted to our proof readers and reviewers: Gad Abraham, Paul Blackwell, Steve Carnie, Alan Jones, David Rolls, and especially Phil Spector. Olga Borovkova helped with some of the coding, and we thank John Maindonald for drawing our attention to `playwith`.

We would like to acknowledge the dedication and the phenomenal achievement of the community that has provided the tools that we used to produce this book. We are especially grateful to R-core, to the LaTeX community, the GNU community, and to Friedrich Leisch for `Sweave`.

Of course we could not have written the book without the support of our partners, Charlotte, Deborah, and Grace, or the bewilderment of our offspring, Indigo, Simone, André, and Felix.

ODJ
RJM
APR

October 2008

Contents

Preface v

I Programming 1

1 Setting up 3

1.1 Installing R 3
1.2 Starting R 3
1.3 Working directory 4
1.4 Writing scripts 5
1.5 Help 5
1.6 Supporting material 5

2 R as a calculating environment 11

2.1 Arithmetic 11
2.2 Variables 12
2.3 Functions 13
2.4 Vectors 15
2.5 Missing data 18
2.6 Expressions and assignments 19
2.7 Logical expressions 20
2.8 Matrices 23
2.9 The workspace 25
2.10 Exercises 25

3 Basic programming **29**

3.1 Introduction 29
3.2 Branching with `if` 31
3.3 Looping with `for` 33
3.4 Looping with `while` 36
3.5 Vector-based programming 38
3.6 Program flow 39
3.7 Basic debugging 41
3.8 Good programming habits 42
3.9 Exercises 43

4 I/O: Input and Output **49**

4.1 Text 49
4.2 Input from a file 51
4.3 Input from the keyboard 53
4.4 Output to a file 55
4.5 Plotting 56
4.6 Exercises 58

5 Programming with functions **63**

5.1 Functions 63
5.2 Scope and its consequences 68
5.3 Optional arguments and default values 70
5.4 Vector-based programming using functions 70
5.5 Recursive programming 74
5.6 Debugging functions 76
5.7 Exercises 78

6 Sophisticated data structures **85**

6.1 Factors 85
6.2 Dataframes 88
6.3 Lists 94
6.4 The `apply` family 98
6.5 Exercises 105

7 Better graphics **109**

7.1 Introduction 109

7.2 Graphics parameters: `par` 111

7.3 Graphical augmentation 113

7.4 Mathematical typesetting 114

7.5 Permanence 118

7.6 Grouped graphs: `lattice` 119

7.7 3D-plots 123

7.8 Exercises 124

8 Pointers to further programming techniques **127**

8.1 Packages 127

8.2 Frames and environments 132

8.3 Debugging again 134

8.4 Object-oriented programming: S3 137

8.5 Object-oriented programming: S4 141

8.6 Compiled code 144

8.7 Further reading 146

8.8 Exercises 146

II Numerical techniques **149**

9 Numerical accuracy and program efficiency **151**

9.1 Machine representation of numbers 151

9.2 Significant digits 154

9.3 Time 156

9.4 Loops versus vectors 158

9.5 Memory 160

9.6 Caveat 161

9.7 Exercises 162

10 Root-finding **167**

10.1 Introduction 167

10.2 Fixed-point iteration 168

10.3 The Newton-Raphson method 173

10.4 The secant method 176

10.5 The bisection method 178

10.6 Exercises 181

11 Numerical integration **187**

11.1 Trapezoidal rule 187

11.2 Simpson's rule 189

11.3 Adaptive quadrature 194

11.4 Exercises 198

12 Optimisation **201**

12.1 Newton's method for optimisation 202

12.2 The golden-section method 204

12.3 Multivariate optimisation 207

12.4 Steepest ascent 209

12.5 Newton's method in higher dimensions 213

12.6 Optimisation in R and the wider world 218

12.7 A curve fitting example 219

12.8 Exercises 220

III Probability and statistics **225**

13 Probability **227**

13.1 The probability axioms 227

13.2 Conditional probability 230

13.3 Independence 232

13.4 The Law of Total Probability 233

13.5 Bayes' theorem 234

13.6 Exercises 235

14 Random variables **241**

14.1 Definition and distribution function 241
14.2 Discrete and continuous random variables 242
14.3 Empirical cdf's and histograms 245
14.4 Expectation and finite approximations 246
14.5 Transformations 251
14.6 Variance and standard deviation 256
14.7 The Weak Law of Large Numbers 257
14.8 Exercises 261

15 Discrete random variables **267**

15.1 Discrete random variables in R 267
15.2 Bernoulli distribution 268
15.3 Binomial distribution 268
15.4 Geometric distribution 270
15.5 Negative binomial distribution 273
15.6 Poisson distribution 274
15.7 Exercises 277

16 Continuous random variables **281**

16.1 Continuous random variables in R 281
16.2 Uniform distribution 282
16.3 Lifetime models: exponential and Weibull 282
16.4 The Poisson process and the gamma distribution 287
16.5 Sampling distributions: normal, χ^2, and t 292
16.6 Exercises 297

17 Parameter Estimation **303**

17.1 Point Estimation 303
17.2 The Central Limit Theorem 309
17.3 Confidence intervals 314
17.4 Monte-Carlo confidence intervals 321
17.5 Exercises 322

IV Simulation **329**

18 Simulation **331**

18.1 Simulating iid uniform samples 331
18.2 Simulating discrete random variables 333
18.3 Inversion method for continuous rv 338
18.4 Rejection method for continuous rv 339
18.5 Simulating normals 345
18.6 Exercises 348

19 Monte-Carlo integration **355**

19.1 Hit-and-miss method 355
19.2 (Improved) Monte-Carlo integration 358
19.3 Exercises 360

20 Variance reduction **363**

20.1 Antithetic sampling 363
20.2 Importance sampling 367
20.3 Control variates 372
20.4 Exercises 374

21 Case studies **377**

21.1 Introduction 377
21.2 Epidemics 378
21.3 Inventory 390
21.4 Seed dispersal 405

22 Student projects **421**

22.1 The level of a dam 421
22.2 Roulette 425
22.3 Buffon's needle and cross 428
22.4 Insurance risk 430
22.5 Squash 433
22.6 Stock prices 438

Glossary of R commands 441

Programs and functions developed in the text 447

Index 449

PART I

Programming

CHAPTER 1

Setting up

In this chapter we show you how to obtain and install R, ensure R can find your data and program files, choose an editor to help write R scripts, and access the extensive help resources and manuals available in R. We also tell you how to install the `spuRs` package, which complements this book and gives access to most of the code examples and functions developed in the book.

R is an implementation of a functional programming language called S. It has been developed and is maintained by a core of statistical programmers, with the support of a large community of users. Unlike S-plus, the other currently available implementation of S, R is free. It is most widely used for statistical computing and graphics, but is a fully functional programming language well suited to scientific programming in general.

1.1 Installing R

Versions of R are available for a wide variety of computing platforms including various variants of Unix, Windows, and MacOS.

You can download R from one of the many mirror sites of the Comprehensive R Archive Network (CRAN), for example `http://cran.ms.unimelb.edu.au/`. In the first instance it will be sufficient to obtain the base distribution. Advice on downloading and installing R is available from the FAQs provided on the CRAN site.

1.2 Starting R

The Windows R implementation is called `Rgui.exe` (short for R graphical user interface). The MacOS R implementation is called `R.app`. In UNIX you start an R session simply by entering the command `R` (we are assuming that your *path* includes the R binaries).

When R starts it loads some infrastructure and provides you with a prompt:

```
>
```

This prompt is the fundamental entry point for communicating with R. We can type expressions at the prompt; R evaluates the expressions, and returns output.

```
> 1 + 1
```

```
[1] 2
```

R is object oriented, meaning that we can create objects that persist within an R session, and manipulate these objects by name. For example,

```
> x <- 1 + 1
> x
```

```
[1] 2
```

When you are finished using R, you quit with the command `q()`. R asks if you would like to save your workspace, which amounts to all the objects that you have created. See Section 2.9 for more information about the workspace.

1.3 Working directory

When you run R, it nominates one of the directories on your hard drive as a *working directory*, which is where it looks for user-written programs and data files. You can determine the current working directory using the command `getwd()`. The first thing you should do when you start an R session is to make sure that the working directory is the right one for your needs. You can do this using the command `setwd("dir")`, where `dir` is the directory address. Alternatively, if you are using `Rgui.exe` in Windows, then there is a menu command for changing the working directory.

For example, if you had a USB drive mounted as drive E and you wanted to save your solutions to the Chapter 2 exercises in the directory `E:\spuRs\ch2`, you would type `setwd("E:/spuRs/ch2")`. Note that R uses the UNIX convention of forward slashes `/` in directory and file addresses; `.` refers to the current directory and `..` refers to the parent directory.

```
> getwd()
```

```
[1] "/home/andrewr/0.svn/1.research/spuRs/trunk/manuscript/chapters"
```

```
> setwd("../scripts")
> getwd()
```

```
[1] "/home/andrewr/0.svn/1.research/spuRs/trunk/manuscript/scripts"
```

On Windows you can set R to automatically start up in your preferred working directory by right clicking on the program shortcut, choosing properties, and completing the 'Start in' field. On MacOS you can set the initial working directory using the Preferences menu.

1.4 Writing scripts

Although we can type and evaluate all possible R expressions at the prompt, it is much more convenient to write *scripts*, which simply comprise collections of R expressions that we intend R to evaluate sequentially. We will use the terms *program* and *code* synonymously with script.

To write programs you will need a text editor (as distinguished from a word processor). The Windows R implementation has a built-in text editor, but you might also like to try Tinn-R,[1] which is available from `http://www.sciviews.org/Tinn-R/`. For more advanced users, emacs and Xemacs also work very well with R, and we particularly recommend the Emacs Speaks Statistics (ESS) package for these applications.

1.5 Help

This book does not cover all the features of R, and even the features it does cover are not dealt with in full generality. To find out more about an R command or function `x`, you can type `help(x)` or just `?x`. If you cannot remember the exact name of the command or function you are interested in, then `help.search("x")` will search the titles, names, aliases, and keyword entries of the available help files for the phrase `x`.

For a useful HTML help interface, type `help.start()`. This will allow you to search for help and also provides links to a number of manuals, in particular the highly recommended 'An Introduction to R.'

A short glossary of commands is included at the end of the book. For further documentation, a good place to start is the CRAN network, which gives references and links to online resources provided by the R community. Some references to more advanced material are given in Chapter 8.

Of course reading the help system, R manuals, and this book will start you on the way to understanding R and its applications. But to properly understand how things really work, there is no substitute for trying them out for yourself: learn through play.

1.6 Supporting material

We give examples of R usage and programs throughout the book. So that you do not have to retype all of these yourself, we have made the longer programs and all of datasets that we use available in an online archive, distributed using the same CRAN network that distributes R. In fact, the archive has

[1] At the time of writing we recommend Version 1.17.2.4, which is easier to set up than the latest version.

the added functionality of what is called a *package*. This means that it can be *loaded* within R, in which case some of our functions and datasets will be directly available, in the same way that built-in functions and datasets are available.

We describe how to obtain, install, and load the archive below. When successfully installed, you will have a new directory called `spuRs`, within which is a subdirectory `resources`, which contains the material from the book. You will see that `spuRs` contains a number of other subdirectories: these are required to provide the package functionality and can be safely ignored. The `resources` directory contains two subdirectories: `scripts`, which contains program code; and `data`, which contains the datasets.

When the package is installed and then loaded in R, you get direct access to some of our functions and datasets. To obtain a list of these, type `?spuRs` once the package has been loaded. To use the dataset called `x`, in addition to loading the package you need to type `data(x)`, at which point it becomes available as an object called `x`. You can also get the code for function `f` just by typing `f` in R.

Within the text, when giving the code for a program `prog.r`, if it is included in the archive it will begin with the comment line

```
# spuRs/resources/scripts/prog.r
```

The code for a function `f` that is available as part of the package will begin with the line

```
# loadable spuRs function
```

Note that the code for `f` will also be available as the file

```
spuRs/resources/scripts/f.r
```

within the `spuRs` archive.

1.6.1 Installing and loading the package when you have write privileges

In order for the following approaches to succeed, your computer needs access to the Internet. If your computer is behind a firewall, then further steps may be required; consult your network manual or local support.

If your computer is appropriately configured, then you may be able to install the archive in a single step, from within R. The key facility is that you need to be able to write to the R application directory. That is, you need to be able to save files in the directory that R was installed in. Start R as usual, then try:

```
> install.packages("spuRs")
```

If no errors or warnings ensue, then any time you want access to the objects in the package, type

```
> library(spuRs)
```

If installation was successful, then a compressed version of the archive will have been downloaded to a temporary location. This location is reported in case you wish to move the compressed version, as otherwise it will be deleted when R quits. The decompressed archive is saved in the directory `spuRs` in the `library` subdirectory of the R application directory. That is, within the R application directory, you will find our program code in the subdirectory `library/spuRs/resources/scripts` and datasets in the subdirectory `library/spuRs/resources/data`.

If the process failed, then either downloading or installation failed. If downloading failed, then there may be a problem with the network; try again in a short time. If it fails again, then consult your local support.

1.6.2 Installing and loading the package with limited write privileges

This section covers the necessary steps to obtain and install the archive even if the user has limited write access to the computer.

Preparation Create a directory to download the archive to, in a convenient area of your hard drive. You may elect to delete this directory when you are done, so its location is not very important. For example, we might create a directory called:

```
D:\temporary
```

Now, create a directory to house the archive, in a convenient area of your hard drive. You will probably want to keep this directory. For example, we might create a directory called:

```
D:\library
```

Note that our examples assume that we have write access to the `D:\` drive. If we do not have write access, then we would create the directories elsewhere. Make a note of the locations.

Start R normally, then check that you have created the directories successfully An easy way to do this is using the `list.files` function, to list the contents of the directories.

```
> list.files("D:/library")

character(0)

> list.files("D:/temporary")
```

```
character(0)
```

These will give a warning if the directory cannot be found. As we have noted earlier, within R we use the forward slash for addresses, regardless of the operating system.

Download Having created the directories and verified that R can find them, we proceed to downloading the archive. This is performed using the `download.packages` function.

```
> info <- download.packages("spuRs", destdir = "D:/temporary")
```

You will be asked to select a CRAN mirror for use in your session. A closer mirror will be slightly faster. R will provide you information about the URL and the size of the archive. Notice that the `download.packages` command gives output, which we have saved in the object called `info`.

The compressed archive has been saved to `D:\temporary`.

Install Next we install the package to get direct access from within R to many of the functions and datasets. Here is where the `info` object becomes useful.

```
> info

     [,1]       [,2]
[1,] "spuRs"     "D:/temporary/spuRs_1.0.0.zip"
```

Note that the second element of `info` is the address of the archive. We can easily install the archive now via:

```
> install.packages(info[1,2], repos = NULL, lib = "D:/library")

package 'spuRs' successfully unpacked and MD5 sums checked
updating HTML package descriptions
```

R will have created a subdirectory `spuRs` within `library`, containing the archive. That is, you will find our program code in the subdirectory `library/spuRs/resources/scripts` and datasets in the subdirectory `library/spuRs/resources/data`.

At this point we can use the `library` command to load the `spuRs` package to our session. We have to include the directory to which the archive was installed in the `lib.loc` argument.

```
> library(spurs, lib.loc = "D:/library")
```

It is also useful to add this directory to the R's list of recognised libraries, so that the various help tools are aware of it. The second of the following three expressions is necessary; the others show its effect.

```
> .libPaths()

[1] "C:/PROGRA~1/R/R/library"

> .libPaths("D:/library")
> .libPaths()

[1] "D:/library"                "C:/PROGRA~1/R/R/library"
```

Now when we invoke a help search, or use **help.start**, R knows to look in the local library `D:\library` as well as in the usual places.

CHAPTER 2

R as a calculating environment

You can use R as a powerful calculator for a wide range of numerical computations. Using R in this way can be useful in its own right, but can also help you to create and test code fragments that you wish to build into your R programs, and help you to learn about new R functions as you meet them.

This chapter shows you how to use R to do arithmetic calculations; create and manipulate variables, vectors, and matrices; work with logical expressions; call and get help on inbuilt R functions; and to understand the workspace that contains all the associated objects R creates along the way.

Throughout this book examples of typed R input will appear in the `Slanted Typewriter` font and R output and code in plain `Typewriter`. The right angle bracket `>` is the R input prompt. In R you can separate commands using a newline/return or a semicolon, though the latter usually leads to a lack of clarity in programming and is thus discouraged. If you type return before a command is finished then R displays the `+` prompt, rather than the usual `>`, and waits for you to complete the command.

R provides a very rich computing environment, so to avoid overwhelming the reader we will introduce different objects and functions as their need arises, rather than all at once.

2.1 Arithmetic

R uses the usual symbols for addition `+`, subtraction `-`, multiplication `*`, division `/`, and exponentiation `^`. Parentheses `( )` can be used to specify the order of operations. R also provides `%%` for taking the modulus and `%/%` for integer division.

```
> (1 + 1/100)^100
[1] 2.704814
> 17%%5
[1] 2
> 17%/%5
```

```
[1] 3
```

The [1] that prefixes the output indicates (somewhat redundantly) that this is item 1 in a vector of output. R calculates to a high precision, but by default only displays 7 significant digits. You can change the display to `x` digits using `options(digits = x)`. (Though displaying `x` digits does not guarantee accuracy to `x` digits, as we will see in Chapter 9.)

R has a number of built-in functions, for example `sin(x)`, `cos(x)`, `tan(x)`, (all in radians), `exp(x)`, `log(x)`, and `sqrt(x)`. Some special constants such as `pi` are also predefined.

```
> exp(1)

[1] 2.718282

> options(digits = 16)
> exp(1)

[1] 2.718281828459045

> pi

[1] 3.141592653589793

> sin(pi/6)

[1] 0.5
```

The functions `floor(x)` and `ceiling(x)` round down and up respectively, to the nearest integer.

2.2 Variables

A variable is like a folder with a name on the front. You can place something inside the folder, look at it, replace it with something else, but the name on the front of the folder stays the same.

To assign a value to a variable we use the assignment command `<-`. Variables are created the first time you assign a value to them. You can give a variable any name made up of letters, numbers, and `.` or `_`, provided it starts with a letter, or `.` then a letter. Note that names are case sensitive.

To display the value of a variable `x` on the screen we just type `x`. This is in fact shorthand for `print(x)`. Later we will see that in some situations we have to use the longer format, or its near equivalent `show(x)`, for example when writing scripts or printing results inside a loop.

```
> x <- 100
> x
```

```
[1] 100

> (1 + 1/x)^x

[1] 2.704814

> x <- 200
> (1 + 1/x)^x

[1] 2.711517
```

We can also show the outcome of an assignment by surrounding it with parentheses, as follows.

```
> (y <- (1 + 1/x)^x)

[1] 2.711517
```

When assigning a value to a variable, the expression on the right-hand side is evaluated first, then that value is placed in the variable on the left-hand side. It is thus possible (and quite common) to have the same variable appearing on the right- and left-hand sides.

```
> n <- 1
> n <- n + 1
> n

[1] 2
```

In common with most programming languages, R allows the use of `=` for variable assignment, as well as `<-`. We prefer the latter, because there is no possibility of confusion with mathematical equality. For example, we understand the assignment `n <- n + 1` by thinking of `n` as the name of a data location in the computer memory, whose contents change as the assignment is processed. Contrast this with the usual mathematical interpretation of $n = n + 1$, where the variable n is thought of as having the same value on both sides (so this equation has no finite solution).

A good programming practice is to use informative names for your variables to improve readability.

2.3 Functions

In mathematics a function takes one or more arguments (or inputs) and produces one or more outputs (or return values). Functions in R work in an analogous way.

To call or invoke a built-in (or user-defined) function in R you write the name of the function followed by its argument values enclosed in parentheses and separated by commas. We illustrate with the `seq` function, which produces arithmetic sequences:

```
> seq(from = 1, to = 9, by = 2)

[1] 1 3 5 7 9
```

Some arguments are optional, and have predefined default values, for example, if we omit `by` then R uses `by = 1`:

```
> seq(from = 1, to = 9)

[1] 1 2 3 4 5 6 7 8 9
```

To find out about default values and alternative usages of the built-in function `fname`, you can access the built-in help by typing `help(fname)` or `?fname`.

Every function has a default order for the arguments. If you provide arguments in this order, then they do not need to be named, but you can choose to give the arguments out of order provided you give them names in the format `argument_name = expression`.

```
> seq(1, 9, 2)

[1] 1 3 5 7 9

> seq(to = 9, from = 1)

[1] 1 2 3 4 5 6 7 8 9

> seq(by = -2, 9, 1)

[1] 9 7 5 3 1
```

Each argument value is given by an expression, which can be a constant, variable, another function call, or an algebraic combination of these.

```
> x <- 9
> seq(1, x, x/3)

[1] 1 4 7
```

In R functions can have a variable number of arguments, including no arguments at all. A function call always needs the parentheses, even if no arguments are required. If you just type the name of the function, then R types out the function 'object', which is simply the program defining the function itself. Try typing `demo` and then `demo()` to see the difference. (Then type `demo(graphics)` to see a good demonstration of some of R's graphics capabilities.)

Generally, when we describe functions, we will only describe the most important or commonly used options. For complete definitions you should use the built-in help facility.

2.4 Vectors

A vector is an indexed list of variables. You can think of a vector as a drawer in a filing cabinet: the drawer has a name on the outside and within it are files labelled sequentially $1, 2, 3, \ldots$ from the front. Each file is a simple variable whose name is made up from the name of the vector and the number of the label/index: the name of the `i`-th element of vector `x` is `x[i]`.

Like variables, vectors are created the first time you assign values to them. In fact, a simple variable is just a vector with length 1 (also called atomic). To create vectors of length greater than 1, we use functions that produce vector-valued output. There are many of these, but the three basic functions for constructing vectors are `c(...)` (combine); `seq(from, to, by)` (sequence); and `rep(x, times)` (repeat).

```
> (x <- seq(1, 20, by = 2))

 [1]  1  3  5  7  9 11 13 15 17 19

> (y <- rep(3, 4))

[1] 3 3 3 3

> (z <- c(y, x))

 [1]  3  3  3  3  1  3  5  7  9 11 13 15 17 19
```

The functions `seq(from, to, by = 1)` and `seq(from, to, by = -1)` are used all the time and so R provides the shorthand `from:to`. Note that `:` takes precedence over algebraic operators such as `+` and `-`, so to get the sequence from 1 to $n+1$, you need to use `1:(n+1)` and not `1:n+1`, which produces the sequence $2, 3, \ldots, n, n+1$.

To refer to element `i` of vector `x`, we use `x[i]`. If `i` is a vector of positive integers, then `x[i]` is the corresponding subvector of `x`. If the elements of `i` are negative, then the corresponding values are *omitted*.

```
> (x <- 100:110)

 [1] 100 101 102 103 104 105 106 107 108 109 110

> i <- c(1, 3, 2)
> x[i]

[1] 100 102 101

> j <- c(-1, -2, -3)
> x[j]

[1] 103 104 105 106 107 108 109 110
```

It is possible to have a vector with no elements. The function `length(x)` gives the number of elements of `x`.

```
> x <- c()
> length(x)

[1] 0
```

Algebraic operations on vectors act on each element separately, that is elementwise.

```
> x <- c(1, 2, 3)
> y <- c(4, 5, 6)
> x * y

[1]  4 10 18

> x + y

[1] 5 7 9

> y^x

[1]   4  25 216
```

When you apply an algebraic expression to two vectors of unequal length, R automatically repeats the shorter vector until it has something the same length as the longer vector.

```
> c(1, 2, 3, 4) + c(1, 2)

[1] 2 4 4 6

> (1:10)^c(1, 2)

 [1]   1   4   3  16   5  36   7  64   9 100
```

This happens even when the shorter vector is of length 1, allowing the shorthand notation:

```
> 2 + c(1, 2, 3)

[1] 3 4 5

> 2 * c(1, 2, 3)

[1] 2 4 6

> (1:10)^2

 [1]   1   4   9  16  25  36  49  64  81 100
```

R will still duplicate the shorter vector even if it cannot match the longer vector with a whole number of multiples, but in this case it will produce a warning.

```
> c(1,2,3) + c(1,2)

[1] 3 4 4
Warning message:
In c(1,2,3) + c(1, 2) :
  longer object length is not a multiple of shorter object length
```

A useful set of functions taking vector arguments are `sum(...)`, `prod(...)`, `max(...)`, `min(...)`, `sqrt(...)`, `sort(x)`, `mean(x)`, and `var(x)`. Note that functions applied to a vector may be defined to act elementwise or may act on the whole vector input to return a result:

```
> sqrt(1:6)

[1] 1.000000 1.414214 1.732051 2.000000 2.236068 2.449490

> mean(1:6)

[1] 3.5

> sort(c(5, 1, 3, 4, 2))

[1] 1 2 3 4 5
```

2.4.1 Example: mean and variance

```
> x <- c(1.2, 0.9, 0.8, 1, 1.2)
> x.mean <- sum(x)/length(x)
> x.mean - mean(x)

[1] 0

> x.var <- sum((x - x.mean)^2)/(length(x) - 1)
> x.var - var(x)

[1] 0
```

2.4.2 Example: simple numerical integration

```
> dt <- 0.005
> t <- seq(0, pi/6, by = dt)
> ft <- cos(t)
> (I <- sum(ft) * dt)
```

```
[1] 0.5015487

> I - sin(pi/6)

[1] 0.001548651
```

In this example note that `t` is a vector, so `ft` is also a vector, where `ft[i]` equals `cos(t[i])`.

To plot one vector against another, we use the function `plot(x, y, type)`. When using plot, `x` and `y` must be vectors of the same length. The optional argument `type` is a graphical parameter used to control the appearance of the plot: `"p"` for points (the default); `"l"` for lines; `"o"` for points *over* lines; etc.

2.4.3 Example: exponential limit

```
> x <- seq(10, 200, by = 10)
> y <- (1 + 1/x)^x
> exp(1) - y

 [1] 0.124539368 0.064984123 0.043963053 0.033217990 0.026693799
 [6] 0.022311689 0.019165457 0.016796888 0.014949367 0.013467999
[11] 0.012253747 0.011240338 0.010381747 0.009645015 0.009005917
[16] 0.008446252 0.007952077 0.007512533 0.007119034 0.006764706

> plot(x, y)
```

The output is given in Figure 2.1.

2.5 Missing data

In real experiments it is often the case, for one reason or another, that certain observations are missing. Depending on the statistical analysis involved, missing data can be ignored or invented (a process called imputation).

R represents missing observations through the data value `NA`. They can be mixed in with all other kinds of data. It is easiest to think of `NA` values as place holders for data that should have been there, but for some reason, are not. We can detect missing values using `is.na`.

```
> a <- NA                  # assign NA to variable A
> is.na(a)                 # is it missing?

[1] TRUE

> a <- c(11,NA,13)         # now try a vector
> is.na(a)                 # is it missing?
```

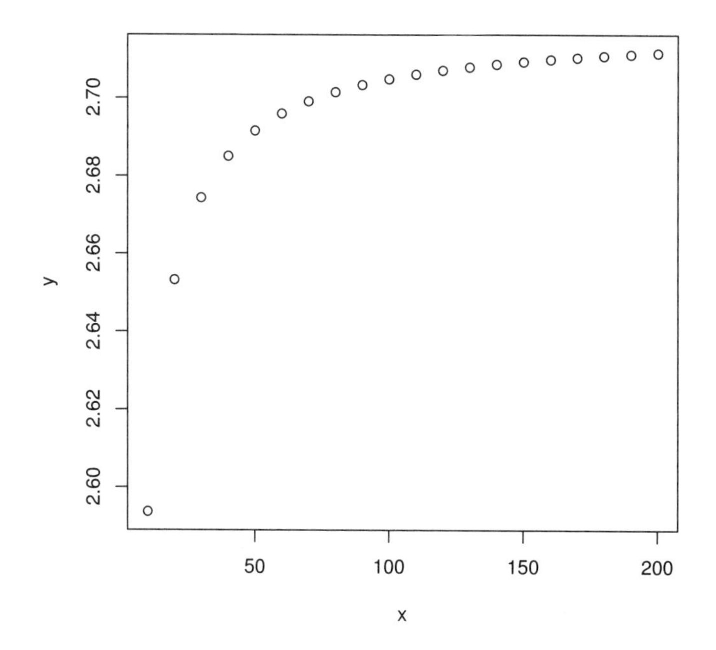

Figure 2.1 $y = (1 + 1/x)^x$; *output for Example 2.4.3.*

```
[1] FALSE  TRUE FALSE
> mean(a)                # NAs can propagate
[1] NA
> mean(a, na.rm = TRUE) # NAs can be removed
[1] 12
```

We also mention the null object, called `NULL`, which is returned by some functions and expressions. Note that `NA` and `NULL` are not equivalent. `NA` is a placeholder for something that exists but is missing. `NULL` stands for something that never existed at all.

2.6 Expressions and assignments

So far we have been using simple R commands without being very precise about what is going on. In this section we cover some useful vocabulary.

In R, the term *expression* is used to denote a phrase of code that can be executed. The following are examples of expressions.

```
> seq(10, 20, by = 3)

[1] 10 13 16 19

> 4

[1] 4

> mean(c(1, 2, 3))

[1] 2

> 1 > 2

[1] FALSE
```

If the evaluation of an expression is saved, using the `<-` operator, then the combination is called an *assignment*. The following are examples of assignments.

```
> x1 <- seq(10, 20, by = 3)
> x2 <- 4
> x3 <- mean(c(1, 2, 3))
> x4 <- 1 > 2
```

2.7 Logical expressions

A logical expression is formed using the comparison operators `<`, `>`, `<=`, `>=`, `==` (equal to), and `!=` (not equal to); and the logical operators `&` (and), `|` (or), and `!` (not). The order of operations can be controlled using parentheses `( )`. Two other comparison operators, `&&` and `||`, are introduced in Section 2.7.2.

The value of a logical expression is either `TRUE` or `FALSE`. The integers 1 and 0 can also be used to represent `TRUE` and `FALSE`, respectively (which is an example of what is called coercion).

Note that `A|B` is `TRUE` if `A` or `B` or both are `TRUE`. If you want exclusive disjunction, that is either `A` or `B` is `TRUE` but not both, then use `xor(A,B)`:

```
> c(0, 0, 1, 1) | c(0, 1, 0, 1)

[1] FALSE  TRUE  TRUE  TRUE

> xor(c(0, 0, 1, 1), c(0, 1, 0, 1))

[1] FALSE  TRUE  TRUE FALSE
```

The example above also shows that logical expressions can be applied to vectors to produce vectors of `TRUE/FALSE` values. This is particularly useful for selecting a subvector using the indexing operation, `x[subset]`.

One way of extracting a subvector is to provide an `subset` as a vector of `TRUE/FALSE` values, the same length as `x`. The result of the `x[subset]` command is that subvector of `x` for which the corresponding elements of `subset` are `TRUE`. Importantly, the argument `subset` can be generated using `x`.

For example, suppose we wished to find all those integers between 1 and 20 that are divisible by 4.

```
> x <- 1:20
> x%%4 == 0

 [1] FALSE FALSE FALSE  TRUE FALSE FALSE FALSE  TRUE FALSE FALSE FALSE
[12]  TRUE FALSE FALSE FALSE  TRUE FALSE FALSE FALSE  TRUE

> (y <- x[x%%4 == 0])

[1]  4  8 12 16 20
```

R also provides the `subset` function, for choosing a subvector of `x`. The difference between the function `subset` and using the index operator is how they handle missing values (`NA`). The `subset` function will ignore the missing index values, whereas the `x[subset]` command preserves them, for example:

```
> x <- c(1, NA, 3, 4)
> x > 2

[1] FALSE    NA  TRUE  TRUE

> x[x > 2]

[1] NA  3  4

> subset(x, subset = x > 2)

[1] 3 4
```

If you wish to know the index positions of `TRUE` elements of a logical vector `x`, then use `which(x)`.

```
> x <- c(1, 1, 2, 3, 5, 8, 13)
> which(x%%2 == 0)

[1] 3 6
```

2.7.1 Example: rounding error

Only integers and fractions whose denominator is a power of 2 can be represented exactly with the floating point representation used for storing numbers in digital computers (see Section 9.1 for more detail). All other numbers are subject to rounding error. This necessary limitation has caused many heartaches.

```
> 2 * 2 == 4
[1] TRUE
> sqrt(2) * sqrt(2) == 2
[1] FALSE
```

The problem here is that `sqrt(2)` has rounding error, which is magnified when we square it. The solution is to use the function `all.equal(x, y)`, which returns `TRUE` if the difference between `x` and `y` is smaller than some set tolerance, based on R's operational level of accuracy.

```
> all.equal(sqrt(2) * sqrt(2), 2)
[1] TRUE
```

We return to the issue of accuracy in Section 9.2.

2.7.2 *Sequential* `&&` *and* `||`

The logical operators `&&` and `||` are sequentially evaluated versions of `&` and `|`, respectively.

Suppose that `x` and `y` are logical expressions. To evaluate `x & y`, R first evaluates `x` and `y`, then returns `TRUE` if `x` and `y` are both `TRUE`, `FALSE` otherwise. To evaluate `x && y`, R first evaluates `x`. If `x` is `FALSE` then R returns `FALSE` without evaluating `y`. If `x` is `TRUE` then R evaluates `y` and returns `TRUE` if `y` is `TRUE`, `FALSE` otherwise.

Similarly, to evaluate `x || y`, R only evaluates `y` if it has to, that is, if `x` is `FALSE`.

Sequential evaluation of `x` and `y` is useful when `y` is not always well defined. For example, suppose we wish to know if $x \sin(1/x) = 0$.

```
> x <- 0
> x * sin(1/x) == 0
[1] NA
Warning message:
In sin(1/x) : NaNs produced
> (x == 0) | (sin(1/x) == 0)
[1] TRUE
Warning message:
In sin(1/x) : NaNs produced
> (x == 0) || (sin(1/x) == 0)
[1] TRUE
```

Note that `&&` and `||` only work on scalars, whereas `&` and `|` work on vectors on an element-by-element basis.

2.8 Matrices

A matrix is created from a vector using the function `matrix`, which has the form

```
matrix(data, nrow = 1, ncol = 1, byrow = FALSE).
```

Here `data` is a vector of length at most `nrow*ncol`, `nrow` and `ncol` are the number of rows and columns respectively (with default values of 1), and `byrow` can be either `TRUE` or `FALSE` (defaults to `FALSE`) and indicates whether you would like to fill the matrix up row-by-row or column-by-column, using the elements of `data`. If `length(data)` is less than `nrow*ncol` (for example, the length is 1), then `data` is reused as many times as is needed. This provides a compact way of making a matrix of zeros or ones.

To create a diagonal matrix we use `diag(x)`. To join matrices with rows of the same length (stacking vertically) use `rbind(...)`. To join matrices with columns of the same length (stacking horizontally) use `cbind(...)`.

We refer to the elements of a matrix using two indices.

```
> (A <- matrix(1:6, nrow = 2, ncol = 3, byrow = TRUE))

     [,1] [,2] [,3]
[1,]    1    2    3
[2,]    4    5    6

> A[1, 3] <- 0
> A[, 2:3]

     [,1] [,2]
[1,]    2    0
[2,]    5    6

> (B <- diag(c(1, 2, 3)))

     [,1] [,2] [,3]
[1,]    1    0    0
[2,]    0    2    0
[3,]    0    0    3
```

The usual algebraic operations, including `*`, act elementwise on matrices. To perform matrix multiplication we use the operator `%*%`. We also have a number of functions for using with matrices, for example `nrow(x)`, `ncol(x)`, `det(x)` (the determinant), `t(x)` (the transpose), and `solve(A, B)`, which returns `x` such that `A %*% x == B`. If `A` is invertible then `solve(A)` returns the matrix inverse of `A`.

```
> (A <- matrix(c(3, 5, 2, 3), nrow = 2, ncol = 2))
```

```
     [,1] [,2]
[1,]    3    2
[2,]    5    3

> (B <- matrix(c(1, 1, 0, 1), nrow = 2, ncol = 2))

     [,1] [,2]
[1,]    1    0
[2,]    1    1

> A %*% B

     [,1] [,2]
[1,]    5    2
[2,]    8    3

> A * B

     [,1] [,2]
[1,]    3    0
[2,]    5    3

> (A.inv <- solve(A))

     [,1] [,2]
[1,]   -3    2
[2,]    5   -3

> A %*% A.inv

     [,1]          [,2]
[1,]    1 -8.881784e-16
[2,]    0  1.000000e+00

> A^(-1)

          [,1]      [,2]
[1,] 0.3333333 0.5000000
[2,] 0.2000000 0.3333333
```

Observe the small error in `A %*% A.inv`. Numerical errors like this are the result of having to store real numbers in a binary format, with a finite number of bits, and are often called *rounding errors* (see Chapter 9).

Note that, in R, a matrix is stored as a vector with an added dimension attribute, which gives the number of rows and columns. The matrix elements are stored columnwise in the vector. Therefore it is possible to access the matrix elements using a single index, as follows.

```
> A[2]

[1] 5
```

If you wish to find out if an object is a matrix or vector, then you use `is.matrix(x)` and `is.vector(x)`. Of course mathematically speaking, a vector is equivalent to a matrix with one row or column, but they are treated as different types of object in R. To create a matrix `A` with one column from a vector `x`, we use `A <- as.matrix(x)`. Note that this does not change `x`.

To create a vector from the columns of a matrix `A` we use `as.vector(A)`; this just strips the dimension attribute from `A` and leaves the elements as they are (stored columnwise). This process of changing the object type is called *coercion*. In many instances R will implicitly coerce the type of an object in order to apply the operations or functions you ask it to.

Occasionally it is convenient to arrange objects in arrays of more than two dimensions. In R this is done with the `array(data, dim)` command, where `data` is a vector containing the elements of the array and `dim` is a vector whose length is the number of dimensions and whose elements give the size of the array along each dimensional axis. To fill the array you need `length(data)` equal to `prod(dim)`; see the online help for details of how the elements of `data` are indexed within the array.

2.9 The workspace

The objects that you create using R remain in existence until you explicitly delete them. To list all currently defined objects, use `ls()` or `objects()`. To remove object `x`, use `rm(x)`. To remove all currently defined objects, use `rm(list = ls())`.

To save all of your existing objects to a file called `fname` in the current working directory, use `save.image(file = "fname")`. To save specific objects (say `x` and `y`) use `save(x, y, file = "fname")`. To load a set of saved objects use `load(file = "fname")`. When you quit R you will be asked if you wish to save your workspace image, which will save your existing objects to the file `.RData` in the current working directory.

R keeps a record of all the commands you type. To save this history to the file `fname` use `savehistory(file = "fname")` and to load the history file `fname` use `loadhistory(file = "fname")`. If you save your workspace image when quitting, then your current history will be saved in `.Rhistory` in the current working directory.

2.10 Exercises

1. Give R assignment statements that set the variable z to

   (a). x^{a^b}
   (b). $(x^a)^b$

(c). $3x^3 + 2x^2 + 6x + 1$ (try to minimise the number of operations required)

(d). the digit in the second decimal place of x (hint: use `floor(x)` and/or `%%`)

(e). $z + 1$

2. Give R expressions that return the following matrices and vectors

(a). $(1, 2, 3, 4, 5, 6, 7, 8, 7, 6, 5, 4, 3, 2, 1)$

(b). $(1, 2, 2, 3, 3, 3, 4, 4, 4, 4, 5, 5, 5, 5, 5)$

(c). $\begin{pmatrix} 0 & 1 & 1 \\ 1 & 0 & 1 \\ 1 & 1 & 0 \end{pmatrix}$

(d). $\begin{pmatrix} 0 & 2 & 3 \\ 0 & 5 & 0 \\ 7 & 0 & 0 \end{pmatrix}$

3. Suppose `vec` is a vector of length 2. Interpreting `vec` as the coordinates of a point in $\mathbb{R}^2$, use R to express it in polar coordinates. You will need (at least one of) the inverse trigonometric functions: `acos(x)`, `asin(x)`, and `atan(x)`.

4. Use R to produce a vector containing all integers from 1 to 100 that are not divisible by 2, 3, or 7.

5. Suppose that `queue <- c("Steve", "Russell", "Alison", "Liam")` and that `queue` represents a supermarket queue with Steve first in line. Using R expressions update the supermarket queue as successively:

(a). Barry arrives;

(b). Steve is served;

(c). Pam talks her way to the front with one item;

(d). Barry gets impatient and leaves;

(e). Alison gets impatient and leaves.

For the last case you should not assume that you know where in the queue Alison is standing.

Finally, using the function `which(x)`, find the position of Russell in the queue.

Note that when assigning a text string to a variable, it needs to be in quotes. We formally introduce text in Section 4.1.

6. Which of the following assignments will be successful? What will the vectors `x`, `y`, and `z` look like at each stage?

```
rm(list = ls())
x <- 1
x[3] <- 3
y <- c()
y[2] <- 2
```

```
y[3] <- y[1]
y[2] <- y[4]
z[1] <- 0
```

CHAPTER 3

Basic programming

3.1 Introduction

This chapter introduces a set of basic programming constructs, which are the building blocks of most programs. Some of these tools are used by practically all programming languages, for example, conditional execution by `if` statements, and looped execution by `for` and `while` statements. Other tools, such as vector-based programming, are more specialised, but are just as important for efficient R coding. An implication is that code that seems to be efficient in another language may not be efficient in R.

A program or script is just a list of commands, which are executed one after the other. Typically a program has three parts: input, computations, output. Some would add a fourth part: documentation. When writing a program we generally do not enter each command one at a time using the R command line; instead we write the list of commands in a separate file, which we can save. However, while developing a program, you may find it very useful to type individual commands into the console to test their effect immediately.

Suppose we have a program saved as `prog.r` in the working directory. There are two main ways to run or execute the program: either we use the command `source("prog.r")` or we can just copy and paste the whole program into R. (A third way to execute a program in R is by typing the command `R CMD BATCH prog.r` into a shell.)

Commands involving input from the keyboard or output to the screen behave more predictably when we use `source`. This is because when you use `source`, R does not have to decide whether you are typing a command or typing input to a program. Also, `source` will stop processing if an error occurs, whereas pasted code will continue to run. Continuing to run may be harmless, or it may waste time, or compromise existing objects. Directory information can be prefixed to the file name if necessary. For example `source("../scripts/prog.r")` will go up one level, down into the `scripts` directory, then look for the file `prog.r`. An absolute (as opposed to relative) address will work whatever the current working directory is, for example `source("C:/Documents and Settings/odj/My Documents/spuRs/resources/scripts/prog.r")`.

There are three reasons that we choose to save our programs in a file and execute them this way: first, it allows us to easily modify the program code, to extend or correct it; second, it allows us to re-run the program with different inputs; and third, it makes sharing code straightforward.

Because R programs are run in an environment where there may already be user-defined variables, it is good programming practice to clear the workspace before running a program, to ensure the same starting point each time. Accordingly we will (try to) begin every program with the command `rm(list=ls())`, which removes all objects in the workspace.

On a notational note, in keeping with usual practice, from here on we will refer to simple variables, vectors, matrices, and arrays generically as variables: something whose name is fixed but whose value(s) varies. More generally, the term object includes variables and also user-defined functions, which we meet later.

3.1.1 Example: roots of a quadratic 1 `quad1.r`

Here is a simple example of a program for calculating the real roots of a quadratic equation. Note the use of `#` for commenting the code. Also note that when using the `source` command, the shorthand `x` for `show(x)` no longer works.

```
# program: spuRs/resources/scripts/quad1.r
# find the zeros of a2*x^2 + a1*x + a0 = 0

# clear the workspace
rm(list=ls())

# input
a2 <- 1
a1 <- 4
a0 <- 2

# calculation
root1 <- (-a1 + sqrt(a1^2 - 4*a2*a0))/(2*a2)
root2 <- (-a1 - sqrt(a1^2 - 4*a2*a0))/(2*a2)

# output
show(c(root1, root2))
```

Executing this code (running the program) produces the following output

```
> source("../scripts/quad1.r")
[1] -0.5857864 -3.4142136
```

In order to write programs that implement mathematical algorithms, we need to be able to make choices and repeat operations. These tasks are achieved using the `if` command and the `for` and `while` commands.

3.2 Branching with if

It is often useful to force the execution of some or other part of a program to depend on a condition. The `if` function has the form

```
if (logical_expression) {
    expression_1
    ...
}
```

A natural extension of the `if` command includes an `else` part:

```
if (logical_expression) {
    expression_1
    ...
} else {
    expression_2
    ...
}
```

Braces `{ }` are used to group together one or more expressions. If there is only one expression then the braces are optional.

When an `if` expression is evaluated, if `logical_expression` is `TRUE` then the first group of expressions is executed and the second group of expressions is not executed. Conversely if `logical_expression` is `FALSE` then only the second group of expressions is executed. `if` statements can be nested to create elaborate pathways through a program.

Warning: because the `else` part of an `if` statement is optional, if you type

```
if (logical_expression) {
    expression_1
    ...}
else {
    expression_2
    ...}
```

then you get an error. This is because R believes the `if` statement is finished before it sees the `else` part, which appears on a separate line. That is, R treats the `else` as the start of a new command, but there is no command that starts with an `else`, so R generates an error.

Other useful functions for conditional execution are `ifelse`, which we cover in Section 3.5, and `switch`, which allows for multiple branches.

3.2.1 Example: roots of a quadratic 2 quad2.r

Here is an improved version of our program for finding the roots of a quadratic. Try it with some different values of `a2`, `a1`, and `a0`.

```
# program spuRs/resources/scripts/quad2.r
# find the zeros of a2*x^2 + a1*x + a0 = 0

# clear the workspace
rm(list=ls())

# input
a2 <- 1
a1 <- 4
a0 <- 5

# calculate the discriminant
discrim <- a1^2 - 4*a2*a0
# calculate the roots depending on the value of the discriminant
if (discrim > 0) {
    roots <- c( (-a1 + sqrt(a1^2 - 4*a2*a0))/(2*a2),
                (-a1 - sqrt(a1^2 - 4*a2*a0))/(2*a2) )
} else {
    if (discrim == 0) {
        roots <- -a1/(2*a2)
    } else {
        roots <- c()
    }
}

# output
show(roots)
```

As an exercise the reader should try using additional `if` statements to rewrite program `quad2.r` so that it can deal with the case $a_2 = 0$ (Exercise 8).

Expressions that are grouped using braces `{ }` are viewed by R as a single expression. Similarly an `if` command is viewed as a single expression. Thus the code

```
if (logical_expression_1) {
    expression_1
    ...
} else {
    if (logical_expression_2) {
        expression_2
        ...
    } else {
        expression_3
        ...
    }
}
```

can be written equivalently (and more clearly) as

```
if (logical_expression_1) {
    expression_1
    ...
} else if (logical_expression_2) {
    expression_2
    ...
} else {
    expression_3
    ...
}
```

3.3 Looping with `for`

The `for` command has the following form, where `x` is a simple variable and `vector` is a vector.

```
for (x in vector) {
    expression_1
    ...
}
```

When executed, the `for` command executes the group of expressions within the braces `{ }`, once for each element of `vector`. The grouped expressions can use `x`, which takes on each of the values of the elements of `vector` as the loop is repeated.

3.3.1 Example: summing a vector

The following example uses a loop to sum the elements of a vector. Note that we use the function `cat` (for concatenate) to display the values of certain variables. The advantage of `cat` over `show` is that it allows us to combine text and variables together. The combination of characters `\n` (backslash-n) is used to 'print' a new line.

Also note that to sum the elements of a vector, it is more accurate and much easier (but less instructive) to use the built-in function `sum`.

```
> (x_list <- seq(1, 9, by = 2))

[1] 1 3 5 7 9

> sum_x <- 0
> for (x in x_list) {
+     sum_x <- sum_x + x
+     cat("The current loop element is", x, "\n")
+     cat("The cumulative total is", sum_x, "\n")
+ }
```

```
The current loop element is 1
The cumulative total is 1
The current loop element is 3
The cumulative total is 4
The current loop element is 5
The cumulative total is 9
The current loop element is 7
The cumulative total is 16
The current loop element is 9
The cumulative total is 25
```

```
> sum(x_list)
[1] 25
```

3.3.2 Example: n factorial 1 `nfact1.r`

The following program calculates $n!$.

```
# program: spuRs/resources/scripts/nfact1.r
# Calculate n factorial

# clear the workspace
rm(list=ls())

# Input
n <- 6

# Calculation
n_factorial <- 1
for (i in 1:n) {
    n_factorial <- n_factorial * i
}

# Output
show(n_factorial)
```

Here is the output

```
> source("../scripts/nfact1.r")
[1] 720
```

Note that we can also compute the factorial easily using `prod(1:n)`.

3.3.3 Example: pension value `pension.r`

Here is an example for calculating the value of a pension fund under compounding interest. It uses the function `floor(x)`, whose value is the largest integer smaller than `x`.

```
# program: spuRs/resources/scripts/pension.r
# Forecast pension growth under compound interest

# clear the workspace
rm(list=ls())

# Inputs
r <- 0.11               # Annual interest rate
term <- 10              # Forecast duration (in years)
period <- 1/12          # Time between payments (in years)
payments <- 100         # Amount deposited each period

# Calculations
n <- floor(term/period)  # Number of payments
pension <- 0
for (i in 1:n) {
    pension[i+1] <- pension[i]*(1 + r*period) + payments
}
time <- (0:n)*period

# Output
plot(time, pension)
```

Executing the command `source("pension.r")` produces the output given in Figure 3.1.

The next example highlights an inefficiency in `pension.r`.

3.3.4 Example: redimensioning an array

Here is an observation that you may be able to use to make some of your programs run faster. The following two programs produce the same result, but the first is faster.

Program 1

```
n <- 1000000
x <- rep(0, n)
for (i in 1:n) {
    x[i] <- i
}
```

Program 2

```
n <- 1000000
x <- 1
for (i in 2:n) {
    x[i] <- i
}
```

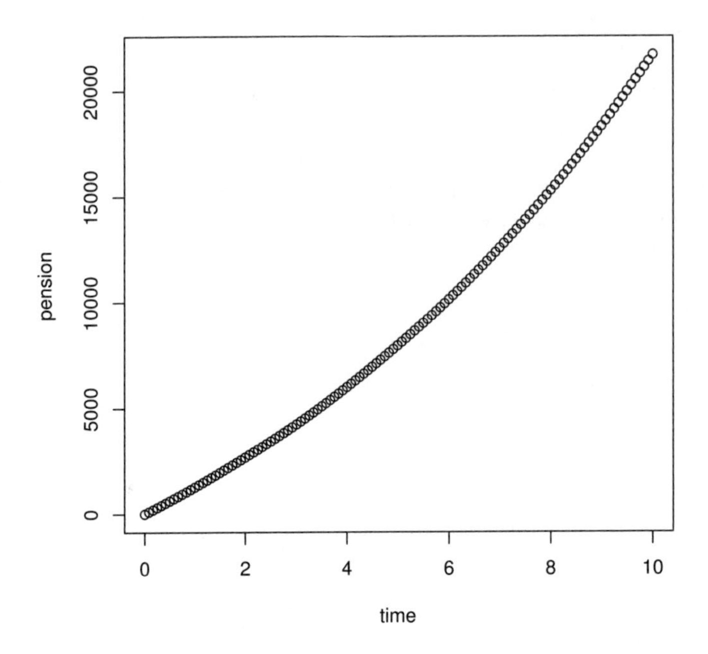

Figure 3.1 *Value of a pension fund: output from Exercise 3.3.3.*

The reason for the difference is a technical one, namely changing the size of a vector takes just about as long as creating a new vector does. In the second program, each statement `x[i] <- i` changes the length of `x` from `i - 1` to `i`, and this is what makes it slower than the first program.

The process of changing the size of a vector is known as *redimensioning* an array, while creating it 'fully-grown' is called *preallocation*. See Section 9.3 for more detail.

3.4 Looping with `while`

Often we do not know beforehand how many times we need to go around a loop. That is, each time we go around the loop, we check some condition to see if we are done yet. In this situation we use a `while` loop, which has the form

```
while (logical_expression) {
    expression_1
    ...
}
```

When a `while` command is executed, `logical_expression` is evaluated first. If it is `TRUE` then the group of expressions in braces `{ }` is executed. Control is then passed back to the start of the command: if `logical_expression` is still `TRUE` then the grouped expressions are executed again, and so on. Clearly, for the loop to stop eventually, `logical_expression` must eventually be `FALSE`. To achieve this `logical_expression` usually depends on a variable that is altered within the grouped expressions.

The `while` loop is more fundamental than the `for` loop, as we can always rewrite a `for` loop as a `while` loop.

3.4.1 Example: Fibonacci numbers `fibonacci.r`

Consider the Fibonacci numbers $F_1, F_2, \ldots$, which are defined inductively using the rules $F_1 = 1$, $F_2 = 1$, and $F_n = F_{n-1} + F_{n-2}$ for $n \geq 2$. Suppose that you wished to know the first Fibonacci number larger than 100. We can find this using a while loop as follows:

```
# program: spuRs/resources/scripts/fibonacci.r
# calculate the first Fibonacci number greater than 100

# clear the workspace
rm(list=ls())

# initialise variables
F <- c(1, 1) # list of Fibonacci numbers
n <- 2       # length of F

# iteratively calculate new Fibonacci numbers
while (F[n] <= 100) {
    # cat("n =", n, " F[n] =", F[n], "\n")
    n <- n + 1
    F[n] <- F[n-1] + F[n-2]
}

# output
cat("The first Fibonacci number > 100 is F(", n, ") =", F[n], "\n")
```

```
> source("../scripts/fibonacci.r")

The first Fibonacci number > 100 is F( 12 ) = 144

> F

 [1]   1   1   2   3   5   8  13  21  34  55  89 144
```

3.4.2 Example: compound interest `compound.r`

In this example we use a while loop to work out how long it will take to pay off a loan.

```
# program: spuRs/resources/scripts/compound.r
# Duration of a loan under compound interest

# clear the workspace
rm(list=ls())

# Inputs
r <- 0.11                # Annual interest rate
period <- 1/12           # Time between repayments (in years)
debt_initial <- 1000     # Amount borrowed
repayments <- 12         # Amount repaid each period

# Calculations
time <- 0
debt <- debt_initial
while (debt > 0) {
    time <- time + period
    debt <- debt*(1 + r*period) - repayments
}

# Output
cat('Loan will be repaid in', time, 'years\n')
```

```
> source("../scripts/compound.r")
Loan will be repaid in 13.25 years
```

3.5 Vector-based programming

It is often necessary to perform an operation upon each of the elements of a vector. R is set up so that such programming tasks can be accomplished using vector operations rather than looping. Using vector operations is more efficient computationally, as well as more concise literally.

For example, we could find the sum of the first n squares using a loop as follows:

```
> n <- 100
> S <- 0
> for (i in 1:n) {
+     S <- S + i^2
+ }
> S
```

```
[1] 338350
```

Alternatively, using vector operations we have:

```
> sum((1:n)^2)

[1] 338350
```

Of course, for the above example we can also use the formula $n(n+1)(2n+1)/6$, assuming we remember it.

The `ifelse` function performs elementwise conditional evaluation upon a vector. `ifelse(test, A, B)` takes three vector arguments: a logical expression `test`, and two expressions `A` and `B`. The function returns a vector that is a combination of the evaluated expressions `A` and `B`: the elements of `A` that correspond to the elements of `test` that are `TRUE`, and the elements of `B` that correspond to the elements of `test` that are `FALSE`. As before, if the vectors have differing lengths then R will repeat the shorter vector(s) to match the longer, if possible. An example follows.

```
> x <- c(-2, -1, 1, 2)
> ifelse(x > 0, "Positive", "Negative")

[1] "Negative" "Negative" "Positive" "Positive"
```

There are some niceties about the mode of the outcome. See `?ifelse` for more details.

Two other useful functions are `pmin` and `pmax`, which provide vectorised versions of the minimum and maximum. For example,

```
> pmin(c(1, 2, 3), c(3, 2, 1), c(2, 2, 2))

[1] 1 2 1
```

3.6 Program flow

The term *flow* is used to describe how control of a program moves from one line to another, and is determined by `if` statements, `for` loops and `while` loops (and functions, as we will see later). Given a program, we can chart its flow by numbering each line, making sure we have a single command per line, then systematically working out the order in which each line is visited. To do this we need to keep a list of all the variables in use and their values, as they can affect the flow.

Consider the following example; line numbers are given on the left

```
   # program: spuRs/resources/scripts/threexplus1.r
1  x <- 3
2  for (i in 1:3) {
3    show(x)
4    if (x %% 2 == 0) {
5      x <- x/2
6    } else {
7      x <- 3*x + 1
8    }
9  }
10 show(x)
```

Charting the flow through this program, we get the output presented in Table 3.1.

Table 3.1 *Charting the flow for program* `threexplus1.r`

line	x	i	comments
1	3		i not defined yet
2	3	1	i is set to 1
3	3	1	3 written to screen
4	3	1	`(x %% 2 == 0)` is `FALSE` so go to line 7
7	10	1	x is set to 10
8	10	1	end of else part
9	10	1	end of for loop, not finished so back to line 2
2	10	2	i is set to 2
3	10	2	10 written to screen
4	10	2	`(x %% 2 == 0)` is `TRUE` so go to line 5
5	5	2	x is set to 5
6	5	2	end of if part, go to line 9
9	5	2	end of for loop, not finished so back to line 2
2	5	3	i is set to 3
3	5	3	5 written to screen
4	5	3	`(x %% 2 == 0)` is `FALSE` so go to line 7
7	16	3	x is set to 16
8	16	3	end of else part
9	16	3	end of for loop, finished so continue to line 10
10	16	3	16 written to screen

This is exactly what the computer does when it executes a program: it keeps track of its current position in the program and maintains a list of variables and their values. *Whatever line you are currently at, if you know all the variables then you always know which line to go to next.*

3.6.1 Pseudo-code

Pseudo-code is used to describe shorthand and/or informally written programs. Pseudo-code does not conform to the strict syntax (grammatical rules) of any particular programming language, but it does use variables, arrays, if statements and loops. That is, it contains enough information to work out how control will flow through the program.

As you learn other high-level programming languages, you will see that the fundamental programming structures—such as variables, arrays, if statements, and loops—are common to all of them. Pseudo-code pays attention to these fundamentals but ignores the details. It is a useful way of describing algorithms without worrying about all the bookkeeping required of a full program. Case study 21.3 (on inventory) gives a lengthy example of a program, explained at different levels of detail using pseudo-code.

3.7 Basic debugging

You will spend a lot of time correcting errors in your programs. To find an error or bug, you need to be able to see how your variables change as you move through the branches and loops of your code. An effective and simple way of doing this is to include statements like `cat("var =", var, "\n")` throughout the program, to display the values of variables such as `var` as the program executes. Once you have the program working you can delete these or just comment them so they are not executed.

For example, if we wanted to see how the variable i changed in the program above, we could add a line as follows:

```
# program: spuRs/resources/scripts/threexplus1.r
x <- 3
for (i in 1:3) {
  show(x)
  cat("i = ", i, "\n")
  if (x %% 2 == 0) {
    x <- x/2
  } else {
    x <- 3*x + 1
  }
}
show(x)
```

Running the program gives the following output

```
> source("../scripts/threexplus1.r")

[1] 3
i =  1
```

```
[1] 10
i =  2
[1] 5
i =  3
[1] 16
```

It is good programming style to solve the simplest possible version of the problem at hand, and then add complexity only as it becomes necessary. Although such an organic approach seems slow at first blush, it provides considerable protection against the complexities that inevitably accrue as the full exercise takes shape.

It is also very helpful to make dry runs of your code, using simple starting conditions for which you know what the answer should be. These dry runs should ideally use short and simple versions of the final program, so that analysis of the output can be kept as simple as possible. Graphs and summary statistics of intermediate outcomes can be very revealing, and the code to create them is easily commented out for production runs.

Careful use of indentation will improve the readability of your code considerably. Indentation can be used to reinforce the overall structure of the code, for example, where do loops and conditional statements begin and end? Some text editors, for example, the Emacs family, provide syntactically aware indentation, which facilitates writing such code.

3.8 Good programming habits

Good programming is clear rather than clever. Being clever is good, but given a choice, being clear is preferable. The reason for this is that in practice much more time is spent correcting and modifying programs than is ever spent writing them, and if you are to be successful in either correcting or modifying a program, you will need it to be clear.

You will find that even programs you write yourself can be very difficult to understand after only a few weeks have passed.

We find the following to be useful guidelines: start each program with some comments giving the name of the program, the author, the date it was written, and what the program does. A description of what a program does should explain what all the inputs and outputs are.

Variable names should be descriptive, that is, they should give a clue as to what the value of the variable represents. Avoid using reserved names or function names as variable names (in particular `t`, `c`, and `q` are all function names in R). You can find out whether or not your preferred name for an object is already in use by the `exists` function.

Use blank lines to separate sections of code into related parts, and use indenting to distinguish the inside part of an if statement or a for or while loop.

Document the programs that you use in detail, ideally with citations for specific algorithms. There is no worse feeling than returning to undocumented code that had been written several years earlier to try to find and then explain an anomaly.

3.9 Exercises

1. Consider the function $y = f(x)$ defined by

x	≤ 0	$\in (0,1]$	> 1
$f(x)$	$-x^3$	x^2	$\sqrt{x}$

   Supposing that you are given x, write an R expression for y using if statements.

   Add your expression for y to the following program, then run it to plot the function f.

```
# input
x.values <- seq(-2, 2, by = 0.1)

# for each x calculate y
n <- length(x.values)
y.values <- rep(0, n)
for (i in 1:n) {
    x <- x.values[i]
    # your expression for y goes here
    y.values[i] <- y
}

# output
plot(x.values, y.values, type = "l")
```

   Your plot should look like Figure 3.2. Do you think f has a derivative at 1? What about at 0?

   We remark that it is possible to vectorise the program above, using the ifelse function.

2. Let $h(x,n) = 1 + x + x^2 + \cdots + x^n = \sum_{i=0}^{n} x^i$. Write an R program to calculate $h(x,n)$ using a for loop.

3. The function $h(x,n)$ from Exercise 2 is the finite sum of a geometric sequence. It has the following explicit formula, for $x \neq 1$,

$$h(x,n) = \frac{1 - x^{n+1}}{1 - x}.$$

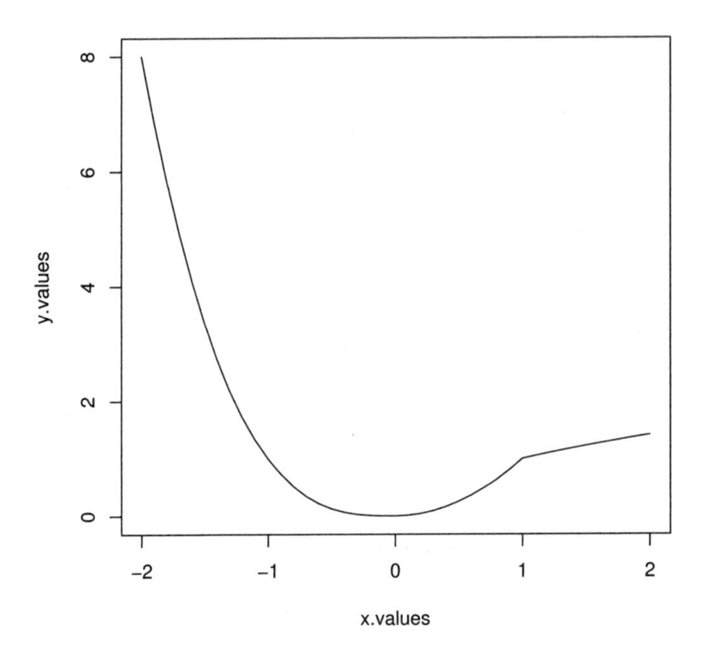

Figure 3.2 *The graph produced by Exercise 1.*

Test your program from Exercise 2 against this formula using the following values

x	n	$h(x, n)$
0.3	55	1.428571
6.6	8	4243335.538178

You should use the computer to calculate the formula rather than doing it yourself.

4. First write a program that achieves the same result as in Exercise 2 but using a `while` loop. Then write a program that does this using vector operations (and no loops).

   If it doesn't already, make sure your program works for the case $x = 1$.

5. To rotate a vector $(x, y)^T$ anticlockwise by θ radians, you premultiply it by the matrix

$$\begin{pmatrix} \cos(\theta) & -\sin(\theta) \\ \sin(\theta) & \cos(\theta) \end{pmatrix}$$

   Write a program in R that does this for you.

6. Given a vector `x`, calculate its geometric mean using both a for loop and vector operations. (The geometric mean of $x_1, \ldots, x_n$ is $\left(\prod_{i=1}^n x_i\right)^{1/n}$.)

You might also like to have a go at calculating the harmonic mean, $\left(\sum_{i=1}^{n} 1/x_i\right)^{-1}$, and then check that if the x_i are all positive, the harmonic mean is always less than or equal to the geometric mean, which is always less than or equal to the arithmetic mean.

7. How would you find the sum of every third element of a vector `x`?
8. How does program `quad2.r` (Exercise 3.2.1) behave if `a2` is 0 and/or `a1` is 0? Using `if` statements, modify `quad2.r` so that it gives sensible answers for all possible (numerical) inputs.
9. Chart the flow through the following two programs.

(a). The first program is a modification of the example from Section 3.6, where x is now an array. You will need to keep track of the value of each element of x, namely $x[1]$, $x[2]$, etc.

```
# threexplus1array.r
x <- 3
for (i in 1:3) {
  show(x)
  if (x[i] %% 2 == 0) {
    x[i+1] <- x[i]/2
  } else {
    x[i+1] <- 3*x[i] + 1
  }
}
show(x)
```

(b). The second program implements the Lotka-Volterra model for a 'predator-prey' system. We suppose that $x(t)$ is the number of prey animals at the start of a year t (rabbits) and $y(t)$ is the number of predators (foxes), then the Lotka-Volterra model is:

$$\begin{aligned} x(t+1) &= x(t) + b_r \cdot x(t) - d_r \cdot x(t) \cdot y(t); \\ y(t+1) &= y(t) + b_f \cdot d_r \cdot x(t) \cdot y(t) - d_f \cdot y(t); \end{aligned}$$

where the parameters are defined by:

b_r is the natural birth rate of rabbits in the absence of predation;
d_r is the death rate per encounter of rabbits due to predation;
d_f is the natural death rate of foxes in the absence of food (rabbits);
b_f is the efficiency of turning predated rabbits into foxes.

```
# program spuRs/resources/scripts/predprey.r
# Lotka-Volterra predator-prey equations
br <- 0.04   # growth rate of rabbits
dr <- 0.0005 # death rate of rabbits due to predation
df <- 0.2    # death rate of foxes
bf <- 0.1    # efficiency of turning predated rabbits into foxes
x <- 4000
y <- 100
while (x > 3900) {
```

```
    # cat("x =", x, " y =", y, "\n")
    x.new <- (1+br)*x - dr*x*y
    y.new <- (1-df)*y + bf*dr*x*y
    x <- x.new
    y <- y.new
}
```

Note that you do not actually need to know anything about the program to be able to chart its flow.

10. Write a program that uses a loop to find the minimum of a vector `x`, without using any predefined functions like `min(...)` or `sort(...)`.

    You will need to define a variable, `x.min` say, in which to keep the smallest value you have yet seen. Start by assigning `x.min <- x[1]` then use a `for` loop to compare `x.min` with `x[2]`, `x[3]`, etc. If/when you find `x[i] < x.min`, update the value of `x.min` accordingly.

11. Write a program to merge two sorted vectors into a single sorted vector.

    Do not use the `sort(x)` function, and try to make your program as efficient as possible. That is, try to minimise the number of operations required to merge the vectors.

12. The game of craps is played as follows. First, you roll two six-sided dice; let x be the sum of the dice on the first roll. If $x = 7$ or 11 you win, otherwise you keep rolling until either you get x again, in which case you also win, or until you get a 7 or 11, in which case you lose.

    Write a program to simulate a game of craps. You can use the following snippet of code to simulate the roll of two (fair) dice:

    ```
    x <- sum(ceiling(6*runif(2)))
    ```

13. Suppose that $(x(t), y(t))$ has polar coordinates $(\sqrt{t}, 2\pi t)$. Plot $(x(t), y(t))$ for $t \in [0, 10]$. Your plot should look like Figure 3.3.

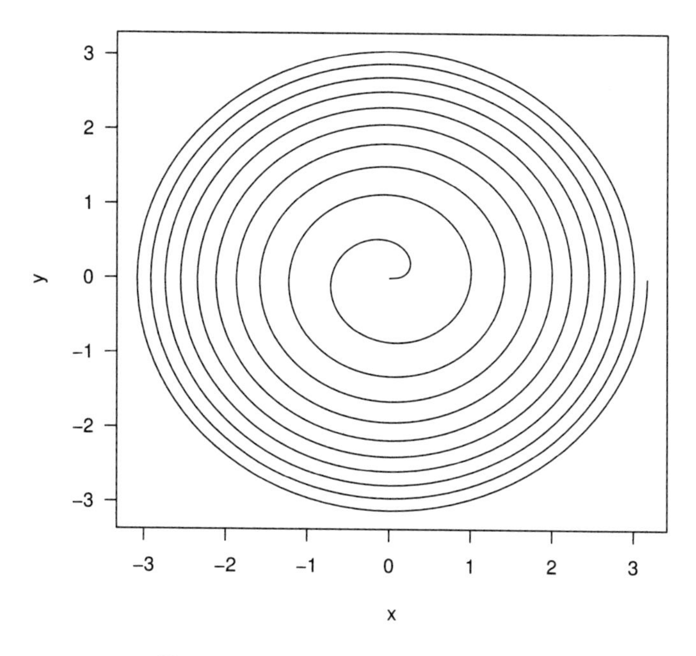

Figure 3.3 *The output from Exercise 13.*

CHAPTER 4

I/O: Input and Output

This chapter describes some of the infrastructure that R provides for importing data for subsequent analysis, and for saving and displaying the results of that analysis. A further discussion of data input appears in Chapter 6, in the context of dataframes, and in Chapter 7 we give more details on the construction of graphical output.

Computer programs excel at processing large amounts of data. To facilitate data processing we need to be able to read input directly from a file. It is sometimes useful to be able to write output to a file too. This chapter covers writing to and reading from plain text files, and creating graphics.

Another important aspect of input and output (I/O) is dealing with alphanumeric characters, so that we can read and write text as well as numbers. Accordingly we will distinguish between different *modes* of objects, such as character, numeric, and logical. We can determine the mode of an object by using the `mode` function.

4.1 Text

So far we have seen objects of `numeric` mode and `logical` mode (`TRUE/FALSE`). A *string* of characters is said to be of mode `character`.

Character strings are denoted using either double quotes `" "` or single quotes `' '`. Strings can be arranged into vectors and matrices just like numbers. We can also paste strings together using `paste(..., sep)`. Here `sep` is an optional input (with default `" "`) that determines which padding character is to be placed between the strings (which are input where `...` appears).

```
> x <- "Citroen SM"
> y <- "Jaguar XK150"
> z <- "Ford Falcon GT-HO"
> (wish.list <- paste(x, y, z, sep = ", "))
[1] "Citroen SM, Jaguar XK150, Ford Falcon GT-HO"
```

Special characters can be included in strings using the escape character `\`. Use `\"` for `"`; `\n` for a newline; `\t` for a tab; `\b` for a backspace; and `\\` for `\`.

If a character string can be understood as a number, then `as.numeric(x)` coerces it to be that number. Use `as.character(x)` to coerce a number into a character string, though note that R will often do this for you as required. A generally more useful method of converting a number to a character string is to use the function `format(x, digits, nsmall, width)`. `digits`, `nsmall`, and `width` are all optional: `nsmall` suggests how many decimal places to use; `digits` suggests how many significant digits to include; and `width` suggests how long the total character string should be. Note that R will quite happily override your suggested values for `digits`, `nsmall`, and `width`. This can be avoided by using the function `round(x, k)` to round `x` to `k` digits before you use `format`.

As we have already seen, the command `cat` displays concatenated character strings.

The following example shows how to use formatted output to print a table of numbers. The program writes out the first `n` powers of the number `x`.

```
# program spuRs/resources/scripts/powers.r
# display powers 1 to n of x

# input
x <- 7
n <- 5

# display powers
cat("Powers of", x, "\n")
cat("exponent    result\n\n")

result <- 1
for (i in 1:n) {
  result <- result * x
  cat(format(i, width = 8),
      format(result, width = 10),
      "\n", sep = "")
}
```

It produces the following output:

```
> source("../scripts/powers.r")

Powers of 7
exponent    result

       1         7
       2        49
       3       343
       4      2401
       5     16807
```

Functions `format` and `paste` also take vector input. Thus the program above could be vectorised as follows:

```
> cat(paste(format(1:n, width = 8), format(x^(1:n), width = 10),
+     "\n"), sep = "")
       1         7
       2        49
       3       343
       4      2401
       5     16807
```

Greater control is available for expressing numbers as character strings, using the `sprintf` and `formatC` functions. See the built-in help for details.

4.2 Input from a file

R provides a number of ways to read data from a file, the most flexible of which is the `scan` function. We use `scan` to read a *vector* of values from a file. `scan` has a large number of options, of which we only need a few at this point. It has the form

```
scan(file = "", what = 0, n = -1, sep = "", skip = 0, quiet = FALSE)
```

`scan` returns a vector. All the parameters are optional; the defaults are indicated above.

`file` gives the file to read from. The default `""` indicates read from the keyboard (see Section 4.3).

`what` gives an example of the mode of data to be read, with a default of `0` for numeric data. Use `" "` for character data.

`n` gives the number of elements to read. If `n = -1` then `scan` keeps reading until the end of the file.

`sep` allows you to specify the character that is used to separate values, such as `","`. The default `""` has the special meaning of allowing any amount of white space (including tabs) to separate values. Note that a newline/return always separates values.

`skip` is the number of lines to skip before you start reading, default of 0. This is useful if your file includes some lines of description before the data starts.

`quiet` controls whether or not `scan` reports how many values it has read, default `FALSE`.

If you try to read more items than are left in the file, by specifying `n`, then `scan` returns a vector of reduced length, possibly of length 0.

To find out what files are in directory `dir.name`, use `dir(path = "dir.name")`, or equivalently `list.files(path = "dir.name")`. The directory address can be relative to the current working directory or an absolute address. `path` has the default value `"."`, denoting the current working directory.

It is common for data to be arranged in tables, with columns corresponding to variables and rows corresponding to separate observations. Tabular data is conveniently stored in a text file, with each line corresponding to a row, and values separated by a specific character, such as a comma. R provides specific functions to conveniently read such files, in particular `read.table`. We will discuss these in Section 6.2.

4.2.1 Example: file input `quartiles1.r`

The following program reads a vector of numbers from a file then calculates their median, 1st quartile and 3rd quartile. The $100p$-th percentage point of a sample is defined to be the smallest sample point x such that at least a fraction p of the sample is less than or equal to x. The first quartile is the 25% point of a sample, the third quartile the 75% point, and the median is the 50% point. (Note that some definitions of the quartiles and median vary slightly from these.)

For this example the file `data1.txt` was created beforehand using a text editor, and is stored in the directory `../data`, which is a sibling to the working directory (that is, it has the same parent directory as the working directory).

```
# program: spuRs/resources/scripts/quartiles1.r
# Calculate median and quartiles.

# Clear the workspace
rm(list=ls())

# Input
# We assume that the file file_name consists of numeric values
# separated by spaces and/or newlines
file_name = "../data/data1.txt"

# Read from file
data <- scan(file = file_name)

# Calculations
n <- length(data)
data.sort <- sort(data)
data.1qrt <- data.sort[ceiling(n/4)]
data.med <- data.sort[ceiling(n/2)]
data.3qrt <- data.sort[ceiling(3*n/4)]

# Output
cat("1st Quartile:", data.1qrt, "\n")
```

```
cat("Median:      ", data.med, "\n")
cat("3rd Quartile:", data.3qrt, "\n")
```

Suppose that the file `data1.txt` has the following single line

```
8 9 3 1 2 0 7 4 5 6
```

Running the program then produces the following output:

```
> source("../scripts/quartiles1.r")

1st Quartile: 2
Median:       4
3rd Quartile: 7
```

As for many statistical operations, R has a built-in function for calculating quartiles, though using a different definition to the one above. Here is a solution to the problem above using the built-in function `quantile`. We leave it to the reader to find a definition of the input arguments, using the help function.

```
> quantile(scan("../data/data1.txt"), (0:4)/4)

  0%  25%  50%  75% 100%
0.00 2.25 4.50 6.75 9.00
```

4.3 Input from the keyboard

`scan` can be used to read from the keyboard if the input `file` is given the value `""` (the default). Use an empty line to denote the end of the input. Keyboard input only works if `scan` is invoked interactively, or executed using `source` (or within a function: see Chapter 5). If you copy and paste commands containing `scan(file = "")`, then R will interpret the lines following `scan(file = "")` as input rather than as commands.

To read a single line of text from the keyboard R provides the command `readline(prompt)`, which takes the optional character input `prompt` (default `""`). Like `scan`, `readline` also only works properly if executed using `source` (or within a function).

4.3.1 Example: roots of a quadratic 2b `quad2b.r`

Here is yet another version of our program for finding the roots of a quadratic, where we now take the input from the keyboard, using `readline`.

```
# program spuRs/resources/scripts/quad2b.r
# find the zeros of a2*x^2 + a1*x + a0 = 0
```

```
# clear the workspace
rm(list=ls())

# input
cat("find the zeros of a2*x^2 + a1*x + a0 = 0\n")
a2 <- as.numeric(readline("a2 = "))
a1 <- as.numeric(readline("a1 = "))
a0 <- as.numeric(readline("a0 = "))

# calculate the discriminant
discrim <- a1^2 - 4*a2*a0
# calculate the roots depending on the value of the discriminant
if (discrim > 0) {
    roots <- (-a1 + c(1,-1) * sqrt(a1^2 - 4*a2*a0))/(2*a2)
} else {
    if (discrim == 0) {
        roots <- -a1/(2*a2)
    } else {
        roots <- c()
    }
}

# output
if (length(roots) == 0) {
    cat("no roots\n")
} else if (length(roots) == 1) {
    cat("single root at", roots, "\n")
} else {
    cat("roots at", roots[1], "and", roots[2], "\n")
}
```

Here it is in action

```
> source("quad2b.r")

find the zeros of a2*x^2 + a1*x + a0 = 0
a2 = 2
a1 = 2
a0 = 0
roots at 0 and -1

> source("quad2b.r")

find the zeros of a2*x^2 + a1*x + a0 = 0
a2 = 2
a1 = 0
a0 = 2
no roots
```

4.4 Output to a file

R provides a number of commands for writing output to a file. We will generally use `write` or `write.table` for writing numeric values and `cat` for writing text, or a combination of numeric and character values.

The command `write` has the form

```
write(x, file = "data", ncolumns = if(is.character(x)) 1 else 5,
      append = FALSE)
```

Here `x` is the vector to be written. If `x` is a matrix or array then it is converted to a vector (column by column) before being written. The other parameters are optional.

`file` gives the file to write or append to, as a character string. The default `"data"` writes to a file called `data` in the current working directory. To write to the screen use `file = ""`.

`ncolumns` gives the number of columns in which to write the vector `x`. The default is 5 for numbers and 1 for characters. Note that the vector is written row by row.

`append` indicates whether to append to or overwrite the file. The default is `FALSE`.

Because `write` converts matrices to vectors before writing them, using it to write a matrix to a file can cause unexpected results. Since R stores its matrices by column, you should pass the transpose of the matrix to `write` if you want the output to reflect the matrix structure.

```
> (x <- matrix(1:24, nrow = 4, ncol = 6))

     [,1] [,2] [,3] [,4] [,5] [,6]
[1,]    1    5    9   13   17   21
[2,]    2    6   10   14   18   22
[3,]    3    7   11   15   19   23
[4,]    4    8   12   16   20   24

> write(t(x), file = "../results/out.txt", ncolumns = 6)
```

Here is what the file `out.txt` looks like:

```
1 5 9 13 17 21
2 6 10 14 18 22
3 7 11 15 19 23
4 8 12 16 20 24
```

A more flexible command for writing to a file is `cat`, which has the form

```
cat(..., file = "", sep = " ", append = FALSE)
```

... is a list of expressions (separated by commas) that are coerced into character strings, concatenated, and then written.

file gives the file to write or append to, as a character string. The default "" writes to the screen.

sep is a character string that is inserted between the written objects, with default value " ".

append indicates whether to append to or overwrite the file, with default FALSE.

Note that cat does not automatically write a newline after the expressions .... If you want a newline you must explicitly include the string "\n".

R also provides functions to write objects in specific formats, for example write.table for writing data in a table (see Section 6.2 for details). There is also the very useful dump, which creates a text representation of almost any R object that can subsequently be read by source. For example

```
> x <- matrix(rep(1:5, 1:5), nrow = 3, ncol = 5)
> dump("x", file = "../results/x.txt")
> rm(x)
> source("../results/x.txt")
> x

     [,1] [,2] [,3] [,4] [,5]
[1,]    1    3    4    4    5
[2,]    2    3    4    5    5
[3,]    2    3    4    5    5
```

4.5 Plotting

We have already seen plot(x, y, type) used to plot one vector against another, with the x values on the x-axis and the y values on the y-axis. In fact the input y is optional, and if omitted then x is plotted against 1:length(x) (so you get the x values on the y-axis and 1:length(x) on the x-axis). Some other useful optional parameters are xlab, ylab, and main, which all take character strings and are used to label the x-axis, y-axis and the whole plot respectively.

To add points (x[1], y[1]), (x[2], y[2]), ... to the current plot, use points(x, y). To add lines instead use lines(x, y). Vertical or horizontal lines can be drawn using abline(v = xpos) and abline(h = ypos). Both points and lines take the optional input col, which determines the colour ("red", "blue", etc.). The complete list of available colours can be obtained by the colours function (or colors). To add the text labels[i] at the point (x[i], y[i]), use text(x, y, labels). The optional input pos is used to indicate where to position the labels in relation to the points. (Use help(text) to see the possible values of `pos`.) If the current plot does not have a title, then `title(main)` will provide one (here `main` is a character string).

As an example we plot part of the parabola $y^2 = 4x$, as well as its focus and directrix. We make use of the surprisingly useful input `type = "n"`, which results in the graph dimensions being established, and the axes being drawn, but nothing else.

```
> x <- seq(0, 5, by = 0.01)
> y.upper <- 2 * sqrt(x)
> y.lower <- -2 * sqrt(x)
> y.max <- max(y.upper)
> y.min <- min(y.lower)
> plot(c(-2, 5), c(y.min, y.max), type = "n", xlab = "x",
+     ylab = "y")
> lines(x, y.upper)
> lines(x, y.lower)
> abline(v = -1)
> points(1, 0)
> text(1, 0, "focus (1, 0)", pos = 4)
> text(-1, y.min, "directrix x = -1", pos = 4)
> title("The parabola y^2 = 4*x")
```

The output is given in Figure 4.1

One way of having more than one plot visible is to open additional graphics devices. In a Windows environment this is done by using the command `windows()` before each additional plot. In Unix use the command `X11()`, and for MacOS `quartz()`, instead. See `?dev.new` and `?dev.control` for more information.

Alternatively you can create a grid of plots in a single graphics window using the commands `par(mfrow = c(nr, nc))` or `par(mfcol = c(nr, nc))`. The command `par` is used to set many different parameters that control how graphics are produced. Setting `mfrow = c(nr, nc)` creates a grid of plots with `nr` rows and `nc` columns, which is filled row by row. `mfcol` is similar but fills the plots column by column.

The following example illustrates `mfrow` and the function `curve`, which is used to plot the function $x \sin(x)$ over different ranges.

```
> par(mfrow = c(2, 2))
> curve(x * sin(x), from = 0, to = 100, n = 1001)
> curve(x * sin(x), from = 0, to = 10, n = 1001)
> curve(x * sin(x), from = 0, to = 1, n = 1001)
> curve(x * sin(x), from = 0, to = 0.1, n = 1001)
> par(mfrow = c(1, 1))
```

The output is given in Figure 4.2.

We return to the subject of plotting in Chapter 7.

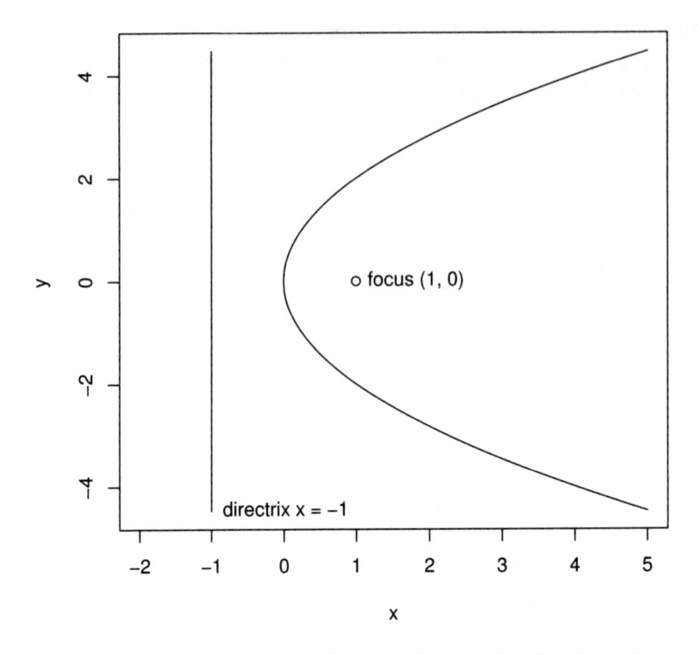

Figure 4.1 *A plot built up in stages. Refer to Section 4.5 for the code to produce this diagram.*

4.6 Exercises

1. Here are the first few lines of the files `age.txt` and `teeth.txt`, taken from the database of a statistically minded dentist:

```
ID      Age
1       18
2       19
3       17
.       .
.       .
.       .

ID      Num Teeth
1       28
2       27
3       32
.       .
.       .
.       .
```

   Write a program in R to read each file, and then write an amalgamated list to the file `age_teeth.txt`, of the following form:

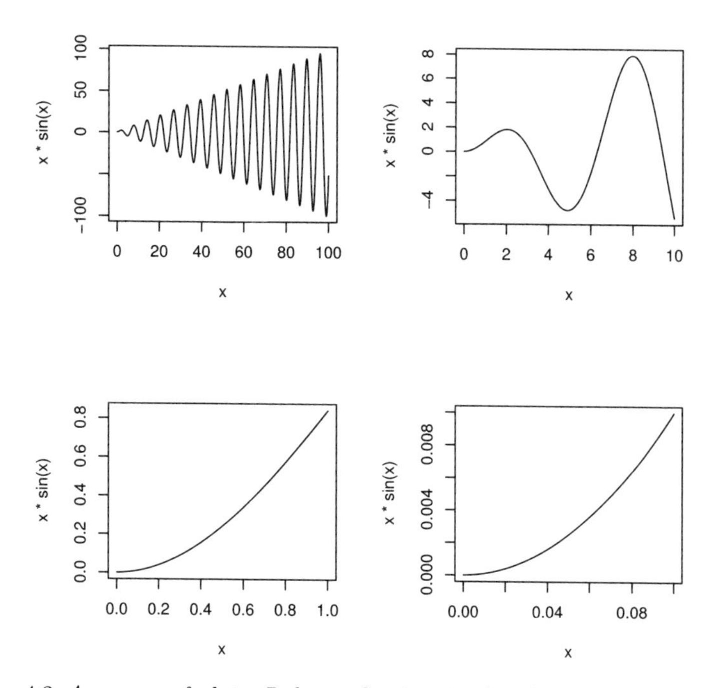

Figure 4.2 *An array of plots. Refer to Section 4.5 for the code to produce this diagram.*

```
ID      Age     Num Teeth
1       18      28
2       19      27
3       17      32
.       .       .
.       .       .
.       .       .
```

2. The function `order(x)` returns a permutation of `1:length(x)` giving the order of the elements of `x`. For example

```
> x <- c(1.1, 0.7, 0.8, 1.4)
> (y <- order(x))
[1] 2 3 1 4
> x[y]
[1] 0.7 0.8 1.1 1.4
```

Using `order` or otherwise, modify your program from Exercise 1 so that the output file is ordered by its second column.

3. Devise a program that outputs a table of squares and cubes of the numbers `1` to `n`. For `n <- 7` the output should be as follows:

```
> source("../scripts/square_cube.r")

 number  square     cube

       1       1       1
       2       4       8
       3       9      27
       4      16      64
       5      25     125
       6      36     216
       7      49     343
```

4. Write an R program that prints out the standard multiplication table:

```
> source("../scripts/mult_table.r")
      [,1] [,2] [,3] [,4] [,5] [,6] [,7] [,8] [,9]
 [1,]    1    2    3    4    5    6    7    8    9
 [2,]    2    4    6    8   10   12   14   16   18
 [3,]    3    6    9   12   15   18   21   24   27
 [4,]    4    8   12   16   20   24   28   32   36
 [5,]    5   10   15   20   25   30   35   40   45
 [6,]    6   12   18   24   30   36   42   48   54
 [7,]    7   14   21   28   35   42   49   56   63
 [8,]    8   16   24   32   40   48   56   64   72
 [9,]    9   18   27   36   45   54   63   72   81
```

   Hint: generate a matrix `mtable` that contains the table, then use `show(mtable)`.

5. Use R to plot the hyperbola $x^2 - y^2/3 = 1$, as in Figure 4.3.

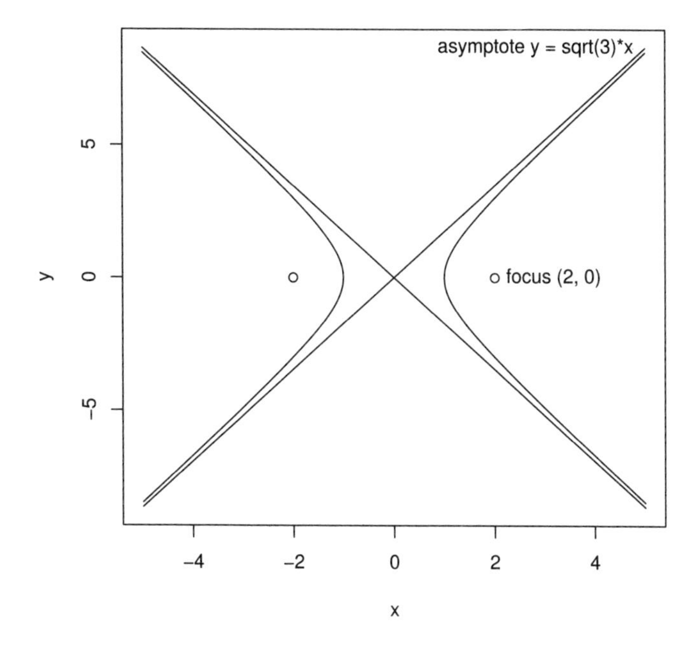

Figure 4.3 *The hyperbola $x^2 - y^2/3 = 1$; see Exercise 5.*

CHAPTER 5

Programming with functions

In this chapter we cover the creation of functions, the rules that they must follow, and how they relate to and communicate with the environments from which they are called. We also present some tips on the construction of efficient functions, with especial reference to how functions are treated in R.

Functions are one of the main building blocks for large programs: they are an essential tool for structuring complex algorithms. In some other programming languages *procedures* and *subroutines* play the same role as functions in R.

5.1 Functions

A function has the form

```
name <- function(argument_1, argument_2, ...) {
    expression_1
    expression_2
    ...
    return(output)
}
```

Here `argument_1`, `argument_2`, etc., are the names of variables and `expression_1`, `expression_2`, and `output` are all regular R expressions. `name` is the name of the function. Note that some functions have no arguments, and that the braces are only necessary if the function comprises more than one expression.

To call or run the function we type

```
name(x1, x2, ...)
```

The value of this expression is the value of the expression `output`. To calculate the value of `output` the function first copies the value of `x1` to `argument_1`, `x2` to `argument_2`, and so on.[1] The arguments then act as variables within the function. We say that the arguments have been *passed* to the function. Next

[1] If fact, to save time R only makes a new copy of an argument if its value is changed within the function. However, to understand how a function works it suffices to think that all the arguments are copied when the function is called.

the function evaluates the grouped expressions contained in the braces { }; the value of the expression output is returned as the value of the function.

A function may have more than one return statement, in which case it stops after executing the first one it reaches. If there is no statement return(output) then the value returned by the function is the value of the last expression in the braces (as long as it is not assigned to a variable).

A function *always* returns a value. For some functions the value returned is unimportant, for example if the function has written its output to a file then there may be no need to return a value as well. In such cases one usually omits the return statement, or returns NULL.

If, when called, the value returned by a function (or any expression) is not assigned to a variable, then it is printed. The expression invisible(x) has the same value as x, but its value is not printed.

5.1.1 Example: roots of a quadratic 3 quad3.r

As an example we write our program for finding the roots of a quadratic as a function. The command rm(list=ls()) has no effect on the main workspace if executed inside a function, so we have moved it outside. (The reason behind this should become clear in Section 5.2.)

Note that the name of the function does not have to match the name of the program file, but when a program consists of a single function this is conventional.

```
# program spuRs/resources/scripts/quad3.r

quad3 <- function(a0, a1, a2) {
  # find the zeros of a2*x^2 + a1*x + a0 = 0
  if (a2 == 0 && a1 == 0 && a0 == 0) {
    roots <- NA
  } else if (a2 == 0 && a1 == 0) {
    roots <- NULL
  } else if (a2 == 0) {
    roots <- -a0/a1
  } else {
    # calculate the discriminant
    discrim <- a1^2 - 4*a2*a0
    # calculate the roots depending on the value of the discriminant
    if (discrim > 0) {
        roots <- (-a1 + c(1,-1) * sqrt(a1^2 - 4*a2*a0))/(2*a2)
    } else if (discrim == 0) {
        roots <- -a1/(2*a2)
    } else {
        roots <- NULL
```

```
    }
  }
  return(roots)
}
```

To use the function we first load it (using `source` or by copying and pasting into R), then call it, supplying suitable arguments.

```
> rm(list = ls())
> source("../scripts/quad3.r")
> quad3(1, 0, -1)

[1] -1  1

> quad3(1, -2, 1)

[1] 1

> quad3(1, 1, 1)

NULL
```

The most important advantage of using a function is that once it is loaded, it can be used again and again without having to reload it. User-defined functions can be used in the same way as predefined functions are used in R. In particular they can be used within other functions.

The second most important use of functions is to break down a programming task into smaller logical units. Large programs are typically made up of a number of smaller functions, each of which does a simple well-defined task.

5.1.2 Example: n choose r `n_choose_r.r`

The number of ways that you can choose r things from a set of n, ignoring the order in which you choose them, is n choose r, which we write as $\binom{n}{r}$.

As is well known, $\binom{n}{r} = \frac{n!}{r!(n-r)!}$. One way to write a function for calculating $\binom{n}{r}$, is to first write a function to calculate $n!$, and then use it within our function for $\binom{n}{r}$.

```
# program spuRs/resources/scripts/n_choose_r.r

n_factorial <- function(n) {
    # Calculate n factorial
    n_fact <- prod(1:n)
    return(n_fact)
}

n_choose_r <- function(n, r) {
```

```
    # Calculate n choose r
    n_ch_r <- n_factorial(n)/n_factorial(r)/n_factorial(n-r)
    return(n_ch_r)
}
```

Here it is in action.

```
> rm(list = ls())
> source("../scripts/n_choose_r.r")
> n_choose_r(4, 2)

[1] 6

> n_choose_r(6, 4)

[1] 15
```

As an aside we note that $\binom{n}{r}$ can be defined for any real value of n and non-negative integer value of r, using the definition $\binom{n}{r} = n(n-1)\cdots(n-r+1)/r!$. This generalisation is useful for defining certain probability distributions (amongst other things).

Finally, note that more efficient functions that achieve the same goal are available in R; specifically, `choose` and `factorial`.

5.1.3 Example: Winsorised mean `wmean.r`

Let $\mathbf{x} = \{x_1, x_2, \ldots, x_n\}$ be a sample of real numbers and let $x_{(1)} \leq x_{(2)} \leq \cdots \leq x_{(n)}$ be the ordered sample. The k-th trimmed mean of $\mathbf{x}$ is defined as

$$\bar{x}_k = \frac{x_{(k+1)} + \cdots + x_{(n-k)}}{n - 2k}.$$

That is, we discard the k smallest and k largest values then take the average. The trimmed mean is less susceptible to outliers than the untrimmed mean.

The k-th Winsorised mean is defined as

$$w_k = \frac{(k+1)x_{(k+1)} + x_{(k+2)} + \cdots + x_{(n-k-1)} + (k+1)x_{(n-k)}}{n}.$$

That is, instead of discarding the k-th largest and k-th smallest values, we replace them by $x_{(n-k)}$ and $x_{(k+1)}$, respectively. The Winsorised mean can be used when you think that your sample may contain occasional extraordinary values, either because of errors or because you are not measuring what you think you are measuring (this would be a conceptual rather than a measurement error).

Here is a function for calculating the k-th Winsorised mean.

```
# program spuRs/resources/scripts/wmean.r

wmean <- function(x, k) {
    # calculate the k-th Windsorised mean of the vector x
    x <- sort(x)
    n <- length(x)
    x[1:k] <- x[k+1]
    x[(n-k+1):n] <- x[n-k]
    return(mean(x))
}
```

Here it is in practice.

```
> source("../scripts/wmean.r")
> x <- c( 8.244, 51.421, 39.020, 90.574, 44.697,
+        83.600, 73.760, 81.106, 38.811, 68.517)
> mean(x)

[1] 57.975

> wmean(x, 2)

[1] 59.8773

> x.err <- x
> x.err[1] <- 1000
> mean(x.err)

[1] 157.1506

> wmean(x.err, 2)

[1] 65.9695
```

5.1.4 Program flow using functions

When a function is executed the computer sets aside space for the function variables, makes a copy of the function code, then transfers control to the function. When the function is finished the output of the function is passed back to the main program, then the copy of the function and all its variables are deleted. We illustrate this using the program below. Note that we have numbered the lines of the function `swap` separately from the main program. The flow of the program is charted in Table 5.1.

```
    # swap.r

f1  swap <- function(x) {
      # swap values of x[1] and x[2]
f2    y <- x[2]
f3    x[2] <- x[1]
```

Table 5.1 *Control flow for the program* `swap.r`

	main program	function swap (1st)		function swap (2nd)		
line	x	x	y	x	y	comments
p1	(7, 8, 9)					
f1	(7, 8, 9)	(7, 8)				control transferred to swap from p2
f2	(7, 8, 9)	(7, 8)	8			
f3	(7, 8, 9)	(7, 7)	8			
f4	(7, 8, 9)	(8, 7)	8			
f5	(7, 8, 9)	(8, 7)	8			swap returns (8, 7); control returned to line p2, function variables deleted
p2	(8, 7, 9)					
f1	(8, 7, 9)			(7, 9)		control transferred to swap from p3
f2	(8, 7, 9)			(7, 9)	9	
f3	(8, 7, 9)			(7, 7)	9	
f4	(8, 7, 9)			(9, 7)	9	
f5	(8, 7, 9)			(9, 7)	8	swap returns (9, 7); control returned to line p3, function variables deleted
p3	(8, 9, 7)					

```
f4     x[1] <- y
f5     return(x)
f6  }

p1  x <- c(7, 8, 9)
p2  x[1:2] <- swap(x[1:2])
p3  x[2:3] <- swap(x[2:3])
```

5.2 Scope and its consequences

Arguments and variables defined within a function exist *only* within that function. That is, if you define and use a variable `x` inside a function, it does not exist outside the function. If variables with the same name exist inside and outside a function, then they are separate and do not interact at all. You can think of a function as a separate environment that communicates with the outside world only through the values of its arguments and its output expression.[2] For example if you execute the command `rm(list=ls())` inside a function (which is only rarely a good idea), you only delete those objects that are defined inside the function.

```
> test <- function(x) {
+     y <- x + 1
+     return(y)
+ }
> test(1)

[1] 2

> x

Error: Object "x" not found

> y

Error: Object "y" not found

> y <- 10
> test(1)

[1] 2

> y

[1] 10
```

That part of a program in which a variable is defined is called its *scope*. Restricting the scope of variables within a function provides an assurance that calling the function will not modify variables outside the function, except by assigning the returned value.

Beware however, the scope of a variable is not symmetric. That is, variables defined inside a function cannot be seen outside, but variables defined outside the function *can* be seen inside the function (provided there is not a variable with the same name defined inside). This arrangement allows for elegant programming in certain situations (in particular when programming recursively, see Section 5.5), but it also makes it possible to write a function whose behaviour depends on the context within which it is run. Consider the following example:

```
> test2 <- function(x) {
+     y <- x + z
+     return(y)
+ }
> z <- 1
> test2(1)
```

[2] This statement is not entirely accurate, but provides a useful model.

```
[1] 2

> z <- 2
> test2(1)

[1] 3
```

The moral of this example is that it is advisable to ensure that the variables you use in a function either are arguments, or have been defined in the function. Exercise 4 gives a subtle example of what can go wrong. Conversely, an example where we deliberately make use of this aspect of scoping, to simplify our coding, is given in Section 12.4.1.

5.3 Optional arguments and default values

To give the argument `argument_1` the default value `x1` we use `argument_1 = x1` within the function definition. If an argument has a default value then it may be omitted when calling the function, in which case the default is used.

If you omit an argument then there is possible ambiguity regarding which arguments are assigned to which variables. To avoid this R assigns arguments to variables from the left, unless an argument is named.

```
> test3 <- function(x = 1, y = 1, z = 1) {
+     return(x * 100 + y * 10 + z)
+ }
> test3(2, 2)

[1] 221

> test3(y = 2, z = 2)

[1] 122
```

5.4 Vector-based programming using functions

We have mentioned that many R functions are vectorised, meaning that given vector input the function acts on each element separately, and a vector output is returned. This is a very powerful aspect of R that allows for compact, efficient, and readable code.

To further facilitate vector-based programming, R provides a family of powerful and flexible functions that enable the vectorisation of user-defined functions: `apply`, `sapply`, `lapply`, `tapply`, and `mapply`.

The effect of `sapply(X, FUN)` is to apply function `FUN` to every element of vector `X`. That is, `sapply(X, FUN)` returns a vector whose `i`-th element is the value of the expression `FUN(X[i])`. If `FUN` has arguments other than `X[i]`, then they can be included using `sapply(X, FUN, ...)`, which returns `FUN(X[i], ...)` as the `i`-th element. That is, the arguments `...` are passed directly from `sapply` to `FUN`, thus allowing you to use a function with more than one argument, though note that the values of the arguments `...` are the same each time. To vectorise over more than one argument, use `mapply`.

If you wish to apply a function that takes a vector argument to each of the rows (or columns) of a matrix, then use the function `apply`, which is a more flexible but more complex version of `sapply`.

We cover `tapply` in Section 6.4.1, and provide more detail about `sapply` and `lapply` in Section 6.4.2. See also `help(apply)`.

5.4.1 Example: density of primes `primedensity.r`

Here we give an example of `sapply` in action. The idea is to write a function `prime` that tests if a given integer is prime. We then use `sapply` to apply `prime` to the vector `2:n`, so that we know all the primes less than or equal to `n`.

Let $\rho(n)$ be the number of primes less than or equal to n. Both Legendre and Gauss famously asserted that

$$\lim_{n\to\infty} \frac{\rho(n)\log(n)}{n} \to 1.$$

The result was eventually proved some time later by Hadamard and de la Vallée Poussin in 1896. The proof is hard, but we can easily check the result numerically. Our program uses the function `cumsum(x)`, which returns the cumulative sums of `x` as a vector. We apply it to a logical vector of TRUE/FALSE values, which R coerces into a 1/0 vector before computing the cumulative sum.

```
# spuRs/resources/scripts/primedensity.r
# estimate the density of primes (using a very inefficient algorithm)

# clear the workspace
rm(list=ls())

prime <- function(n) {
    # returns TRUE if n is prime
    # assumes n is a positive integer
    if (n == 1) {
        is.prime <- FALSE
    } else if (n == 2) {
        is.prime <- TRUE
    } else {
```

```
        is.prime <- TRUE
        for (m in 2:(n/2)) {
            if (n %% m == 0) is.prime <- FALSE
        }
    }
    return(is.prime)
}

# input
# we consider primes <= n
n <- 1000

# calculate the number of primes <= m for m in 2:n
# num.primes[i] == number of primes <= i+1
m.vec <- 2:n
primes <- sapply(m.vec, prime)
num.primes <- cumsum(primes)

# output
# plot the actual prime density against the theoretical limit
par(mfrow = c(1, 2))
plot(m.vec, num.primes/m.vec, type = "l",
    main = "prime density", xlab = "n", ylab = "")
lines(m.vec, 1/log(m.vec), col = "red")

plot(m.vec, num.primes/m.vec*log(m.vec), type = "l",
    main = "prime density * log(n)", xlab = "n", ylab = "")
par(mfrow = c(1, 1))
```

Executing the command `source("primedensity.r")` gives the output of Figure 5.1.

We see that at the point $n = 1000$ the prime density $\rho(n)/n$ is not particularly close to $1/\log(n)$, though the rate of decay looks correct. To see better convergence you will need to take much larger n, however this will take a long time as it takes longer and longer to check each number to see if it is prime.

It does not help that the algorithm we used is inefficient. The function `prime` can be made more efficient in two ways. First, we need only check for factors up to $\sqrt{n}$, since if $n = ab$ then at least one of a and b is less than or equal to $\sqrt{n}$. Second, once we find one factor we don't need to keep checking. Incorporating these two refinements we get the following:

```
# program spuRs/resources/scripts/prime.r

prime <- function(n) {
    # returns TRUE if n is prime
    # assumes n is a positive integer
    if (n == 1) {
        is.prime <- FALSE
```

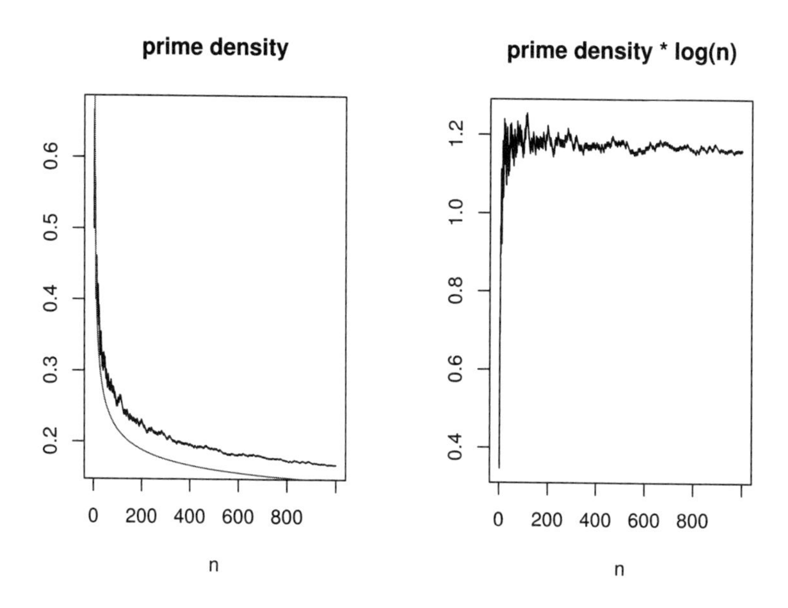

Figure 5.1 *The density of primes. Output from Example 5.4.1.*

```
    } else if (n == 2) {
        is.prime <- TRUE
    } else {
        is.prime <- TRUE
        m <- 2
        m.max <- sqrt(n)  # only want to calculate this once
        while (is.prime && m <= m.max) {
            if (n %% m == 0) is.prime <- FALSE
            m <- m + 1
        }
    }
    return(is.prime)
}
```

However, if what you really want to do is not just check that n is prime, but rather find all the primes less than or equal to n, then a much more efficient algorithm is the 'Sieve of Eratosthenes' (ca. 240 BC). An implementation of Eratosthenes' algorithm is given in Section 5.5.

5.5 Recursive programming

Recursive programming is a powerful programming technique, made possible by functions. A recursive program is simply one that calls itself. This is useful because many algorithms are recursive in nature.

5.5.1 *Example: n factorial 2* `nfact2.r`

We can write $n!$ as $n*((n-1)!)$. We implement this recursive definition below. Note that the program uses `cat` statements to provide some feedback, and we have numbered the lines for the purpose of charting the program flow.

```
   # function nfact2.r

1  nfact2 <- function(n) {
       # calculate n factorial
2      if (n == 1) {
3          cat("called nfact2(1)\n")
4          return(1)
5      } else {
6          cat("called nfact2(", n, ")\n", sep = "")
7          return(n*nfact2(n-1))
8      }
9  }
```

```
> source("../scripts/nfact2.r")
> nfact2(6)
```

```
called nfact2(6)
called nfact2(5)
called nfact2(4)
called nfact2(3)
called nfact2(2)
called nfact2(1)
[1] 720
```

When you chart the flow through a recursive function, it is important to remember that when a function is called, a *new* copy of the function is created with a *new* set of function variables. For example, calling `nfact2(3)` gives the program flow shown in Table 5.2. We write $i.j$ to indicate line j within the i-th nested function call.

5.5.2 *Example: Sieve of Eratosthenes* `primesieve.r`

The Sieve of Eratosthenes is an algorithm for finding all of the primes less than or equal to a given number n. It works as follows:

Table 5.2 *Control flow through the function* `nfact2`

	nfactorial (1st call)	nfactorial (2nd call)	nfactorial (3rd call)	
line	n	n	n	comments
1.1	3			
1.2	3			$n \neq 1$ so go to line 6
1.6	3			print 'called nfactorial(3)'
2.1	3	2		nfactorial(2) called on line 1.7
2.2	3	2		$n \neq 1$ so go to line 6
2.6	3	2		print 'called nfactorial(2)'
3.1	3	2	1	nfactorial(1) called on line 2.7
3.2	3	2	1	$n = 1$ so go to line 3
3.3	3	2	1	print 'called nfactorial(1)'
3.4	3	2	1	return 1, delete variables, return control to line 2.7
2.7	3	2		return 2, delete variables, return control to line 1.7
1.7	3			return 6, delete variables, return control to calling line

1. Start with the list $2, 3, \ldots, n$ and largest known prime $p = 2$.
2. Remove from the list all elements that are multiples of p (but keep p itself).
3. Increase p to the smallest element of the remaining list that is larger than the current p.
4. If p is larger than $\sqrt{n}$ then stop, otherwise go back to step 2.

Here is a recursive implementation of the algorithm. You may find that it takes you some time to understand how it works.

```
# program spuRs/resources/scripts/primesieve.r
# loadable spuRs function

primesieve <- function(sieved, unsieved) {
  # finds primes using the Sieve of Eratosthenes
  # sieved: sorted vector of sieved numbers
  # unsieved: sorted vector of unsieved numbers

  # cat("sieved", sieved, "\n")
  # cat("unsieved", unsieved, "\n")
  p <- unsieved[1]
```

```
  n <- unsieved[length(unsieved)]
  if (p^2 > n) {
      return(c(sieved, unsieved))
  } else {
      unsieved <- unsieved[unsieved %% p != 0]
      sieved <- c(sieved, p)
      return(primesieve(sieved, unsieved))
  }
}
```

Here it is in action:

```
> rm(list = ls())
> source("../scripts/primesieve.r")
> primesieve(c(), 2:200)
```

```
 [1]   2   3   5   7  11  13  17  19  23  29  31  37  41  43  47  53
[17]  59  61  67  71  73  79  83  89  97 101 103 107 109 113 127 131
[33] 137 139 149 151 157 163 167 173 179 181 191 193 197 199
```

It can be shown that the Sieve of Eratosthenes uses $O(n(\log n)(\log\log n))$ operations to find all the primes less than or equal to n. (The notation $g(x) = O(f(x))$ means there exists a constant c such that $\lim_{x\to\infty} g(x)/f(x) \le c$. In other words $g(x)$ grows no faster than a constant times $f(x)$.) You should try to calculate how many operations are used by the algorithm given in Example 5.4.1. You will see that it is much less efficient.

5.6 Debugging functions

Often code will be used in circumstances under which you cannot control the type of input (numeric, character, logical, etc.). Unexpected input can lead to undesirable consequences, for example, the function could fail to work and the user may not know why. Worse still, the function could seem to work but return plausible nonsense, and the user may be none the wiser. It can be worth performing simple checks on the input to be sure that it conforms to your expectations. (Useful considerations here are: what will your function do if the input is the wrong type, or the right type but incomplete?) The `stop` function is useful in these circumstances: `stop("Your message here.")` will cease processing and print the message to the user.

The `browser` function is very useful to invoke inside your own functions. The command `browser()` will temporarily stop the program, and allow you to inspect its objects. You can also step through the code, executing one expression at a time.

When in the browser environment, R commands can be entered and evaluated as normally, but some commands have specific new interpretations. The important ones are:

n enters the step-through debugger. In step-through mode,

- n evaluates the current step and prints the next step to be evaluated. The return key has the same effect.
- c continues evaluation from the next expression to the end of the current set of expressions, whether that be the end of the current loop or the end of the function (cont has the same effect).
- Q stops evaluation and exits the browser, returning the user to the top-level prompt.

c stops the browser and continues evaluation, starting at the next statement (the return key and cont both have the same effect).

A commented example of its application follows. my_fun attempts to multiply its input by the (undefined) variable z

```
> my_fun <- function(x) {
+     browser()
+     y <- x * z
+     return(y)
+ }

> my_fun(c(1,2,3))

Called from: my_fun(c(1,2,3))
Browse[1]>
```

browser catches the execution and presents us with a prompt. Using n, we will step through the function one line at a time. At each point, R shows us the next line to be evaluated. We signify our input using curly braces; thus: { Enter }.

```
Browse[1]> n

debug: y <- x * z
Browse[1]>

Browse[1]> { Enter }

Error in my_fun(c(1, 2, 3)) : object "z" not found
```

The result makes it clear to us that the problem in our function is in the line y <- x * z. Here, the problem is obvious: the code calls for an object z, which does not exist. In any case, we can run the function again, return to that point in the proceedings, and take a look around.

```
> my_fun(c(1,2,3))

Called from: my_fun(c(1,2,3))
Browse[1]>
```

```
Browse[1]> n

debug: y <- x * z
Browse[1]>
```

We know that there is a problem here. We identify and examine the objects to locate the problem.

```
Browse[1]> ls()

[1] "x"
Browse[1]>

Browse[1]> Q

>
```

It is clear that something is missing in the environment.

See ?browser and ?debug for more information, and note that we provide more advice on debugging in Section 8.3.

5.7 Exercises

1. The (Euclidean) length of a vector $v = (a_0, \ldots, a_k)$ is the square root of the sum of squares of its coordinates, that is $\sqrt{a_0^2 + \cdots + a_k^2}$. Write a function that returns the length of a vector.
2. In Exercise 3.9.2 you wrote a program to calculate $h(x, n)$, the sum of a finite geometric series. Turn this program into a *function* that takes two arguments, x and n, and returns $h(x, n)$.

   Make sure you deal with the case $x = 1$.
3. In this question we simulate the rolling of a die. To do this we use the function `runif(1)`, which returns a 'random' number in the range (0,1). To get a random integer in the range $\{1, 2, 3, 4, 5, 6\}$, we use `ceiling(6*runif(1))`, or if you prefer, `sample(1:6,size=1)` will do the same job.

   (a). Suppose that you are playing the gambling game of the Chevalier de Méré. That is, you are betting that you get at least one six in 4 throws of a die. Write a program that simulates one round of this game and prints out whether you win or lose.

   Check that your program can produce a different result each time you run it.

   (b). Turn the program that you wrote in part (a) into a function `sixes`, which returns `TRUE` if you obtain at least one six in n rolls of a fair die, and returns `FALSE` otherwise. That is, the argument is the number of rolls n, and the value returned is `TRUE` if you get at least one six and `FALSE` otherwise.

   How would you give n the default value of 4?

(c). Now write a program that uses your function `sixes` from part (b), to simulate N plays of the game (each time you bet that you get at least 1 six in n rolls of a fair die). Your program should then determine the proportion of times you win the bet. This proportion is an estimate of the *probability* of getting at least one 6 in n rolls of a fair die.

Run the program for $n = 4$ and $N = 100$, 1000, and 10000, conducting several runs for each N value. How does the *variability* of your results depend on N?

The probability of getting no 6's in n rolls of a fair die is $(5/6)^n$, so the probability of getting at least one is $1 - (5/6)^n$. Modify your program so that it calculates the theoretical probability as well as the simulation estimate and prints the difference between them. How does the *accuracy* of your results depend on N?

You may find the `replicate` function useful here.

(d). In part (c), instead of processing the simulated runs as we go, suppose we first store the results of every game in a file, then later postprocess the results.

Write a program to write the result of all N runs to a textfile `sixes_sim.txt`, with the result of each run on a separate line. For example, the first few lines of the textfile could look like

```
TRUE
FALSE
FALSE
TRUE
FALSE
.
.
```

Now write another program to read the textfile `sixes_sim.txt` and again determine the proportion of bets won.

This method of saving simulation results to a file is particularly important when each simulation takes a very long time (hours or days), in which case it is good to have a record of your results in case of a system crash.

4. Consider the following program and its output

```
# Program spuRs/resources/scripts/err.r

# clear the workspace
rm(list=ls())

random.sum <- function(n) {
    # sum of n random numbers
    x[1:n] <- ceiling(10*runif(n))
    cat("x:", x[1:n], "\n")
    return(sum(x))
```

```
}

x <- rep(100, 10)
show(random.sum(10))
show(random.sum(5))

> source("../scripts/err.r")

x: 8 5 4 2 10 6 8 9 3 2
[1] 57
x: 2 2 3 5 9
[1] 521
```

Explain what is going wrong and how you would fix it.

5. For $r \in [0, 4]$, the *logistic map* of $[0, 1]$ into $[0, 1]$ is defined as $f(x) = rx(1-x)$.

   Given a point $x_1 \in [0, 1]$ the sequence $\{x_n\}_{n=1}^{\infty}$ given by $x_{n+1} = f(x_n)$ is called the *discrete dynamical system* defined by f.

   Write a function that takes as parameters x_1, r, and n, generates the first n terms of the discrete dynamical system above, and then plots them.

   The logistic map is a simple model for population growth subject to resource constraints: if x_n is the population size at year n, then x_{n+1} is the size at year $n+1$. Type up your code, then see how the system evolves for different starting values x_1 and different values of r.

   Figure 5.2 gives some typical output.

6. The Game of Life is a cellular automaton and was devised by the mathematician J.H. Conway in 1970. It is played on a grid of cells, each of which is either alive or dead. The grid of cells evolves in time and each cell interacts with its eight neighbours, which are the cells directly adjacent horizontally, vertically, and diagonally.

   At each time step cells change as follows:

   - A live cell with fewer than two neighbours dies of loneliness.
   - A live cell with more than three neighbours dies of overcrowding.
   - A live cell with two or three neighbours lives on to the next generation.
   - A dead cell with exactly three neighbours comes to life.

   The initial pattern constitutes the first generation of the system. The second generation is created by applying the above rules simultaneously to every cell in the first generation: births and deaths all happen simultaneously. The rules continue to be applied repeatedly to create further generations.

   Theoretically the Game of Life is played on an infinite grid, but in practice we use a finite grid arranged as a torus. That is, if you are in the left-most column of the grid then your left-hand neighbours are in the right-most column, and if you are in the top row then your neighbours above are in the bottom row.

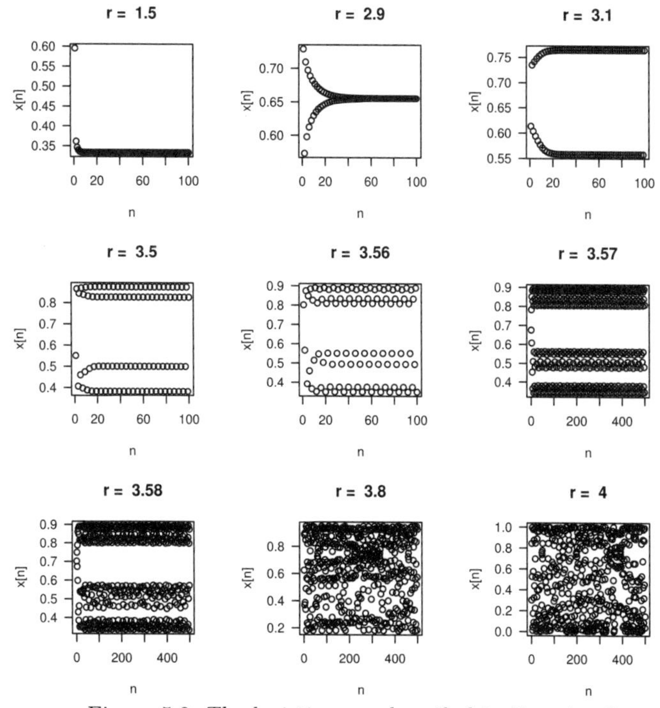

Figure 5.2 *The logistic map described in Exercise 5.*

Here is an implementation of the Game of Life in R. The grid of cells is stored in a matrix `A`, where `A[i,j]` is 1 if cell (i, j) is alive and 0 otherwise.

```
# program spuRs/resources/scripts/life.r

neighbours <- function(A, i, j, n) {
    # A is an n*n 0-1 matrix
    # calculate number of neighbours of A[i,j]
    .
    .
    .
}

# grid size
n <- 50

# initialise lattice
A <- matrix(round(runif(n^2)), n, n)

finished <- FALSE
while (!finished) {
    # plot
```

```
        plot(c(1,n), c(1,n), type = "n", xlab = "", ylab = "")
        for (i in 1:n) {
            for (j in 1:n) {
                if (A[i,j] == 1) {
                    points(i, j)
                }
            }
        }

        # update
        B <- A
        for (i in 1:n) {
            for (j in 1:n) {
                nbrs <- neighbours(A, i, j, n)
                if (A[i,j] == 1) {
                    if ((nbrs == 2) | (nbrs == 3)) {
                        B[i,j] <- 1
                    } else {
                        B[i,j] <- 0
                    }
                } else {
                    if (nbrs == 3) {
                        B[i,j] <- 1
                    } else {
                        B[i,j] <- 0
                    }
                }
            }
        }
        A <- B

        ## continue?
        #input <- readline("stop? ")
        #if (input == "y") finished <- TRUE
    }
```

Note that this program contains an infinite loop! To stop it you will need to use the escape or stop button (Windows or Mac) or control-C (Unix). Alternatively, uncomment the last two lines. To get the program to run you will need to complete the function `neighbours(A, i, j, n)`, which calculates the number of neighbours of cell (i, j). (The program `forest_fire.r` in Section 21.2.3 uses a similar function of the same name, which you may find helpful.)

Once you get the program running, you might like to initialise it using the glider gun, shown in Figure 5.3 (see `glidergun.r` in the `spuRs` package). Many other interesting patterns have been discovered in the Game of Life.[3]

7. The number of ways you can choose r things from a set of n, ignoring the

[3] M. Gardner, *Wheels, Life, and Other Mathematical Amusements.* Freeman, 1985.

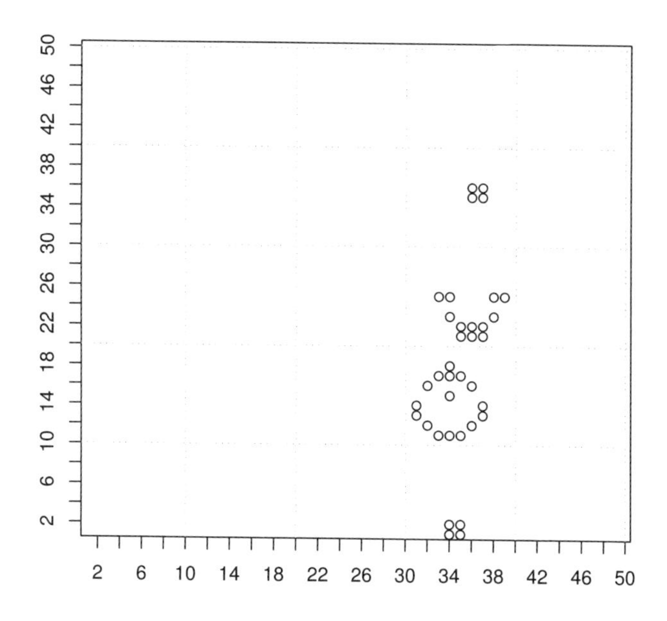

Figure 5.3 *The glider gun, from Exercise 6.*

order in which they are chosen, is $\binom{n}{r} = n!/(r!(n-r)!)$. Let x be the first element of the set of n things. We can partition the collection of possible size r subsets into those that contain x and those that don't: there must be $\binom{n-1}{r-1}$ subsets of the first type and $\binom{n-1}{r}$ subsets of the second type. Thus

$$\binom{n}{r} = \binom{n-1}{r-1} + \binom{n-1}{r}.$$

Using this and the fact that $\binom{n}{n} = \binom{n}{0} = 1$, write a recursive function to calculate $\binom{n}{r}$.

8. A classic puzzle called the *Towers of Hanoi* uses a stack of rings of different sizes, stacked on one of 3 poles, from the largest on the bottom to the smallest on top (so that no larger ring is on top of a smaller ring). The object is to move the stack of rings from one pole to another by moving one ring at a time so that larger rings are never on top of smaller rings.

   Here is a recursive algorithm to accomplish this task. If there is only one ring, simply move it. To move n rings from the pole `frompole` to the pole `topole`, first move the top $n-1$ rings from `frompole` to the remaining `sparepole`, then move the last and largest from `frompole` to the empty `topole`, then move the $n-1$ rings on `sparepole` to `topole` (on top of the largest).

The following program implements this algorithm. For example, if there are initially 8 rings, we then move them from pole 1 to pole 3 by calling `moverings(8,1,3)`.

```
# Program spuRs/resources/scripts/moverings.r

# Tower of Hanoi

moverings <- function(numrings, frompole, topole) {
  if (numrings == 1) {
    cat("move ring 1 from pole", frompole,
        "to pole", topole, "\n")
  } else {
    sparepole <- 6 - frompole - topole # clever
    moverings(numrings - 1, frompole, sparepole)
    cat("move ring", numrings, "from pole", frompole,
        "to pole", topole, "\n")
    moverings(numrings - 1, sparepole, topole)
  }
  return(invisible(NULL))
}
```

Check that the algorithm works for the cases `moverings(3, 1, 3)` and `moverings(4, 1, 3)`, then satisfy yourself that you understand why it works.

Use mathematical induction to show that, using this algorithm, moving a stack of n rings will require exactly $2^n - 1$ individual movements.

CHAPTER 6

Sophisticated data structures

As a programming language that has its roots in statistical analysis, it is natural that R will have provided sophisticated structures for the storage and manipulation of data. In Chapter 2, we presented some primitive object types that R uses to represent data. In this chapter we study R's more sophisticated data structures—lists and dataframes—that simplify data representation, manipulation, and analysis. The dataframe is like a matrix but extended to allow for different object modes in different columns, and the list is a general data storage object that can house pretty much any other kind of R object. We also introduce the factor, which is a special kind of variable that is used to represent categorical objects.

6.1 Factors

Statisticians typically recognise three basic types of variable: numeric, ordinal, and categorical. Both ordinal and categorical variables take values from some finite set, but the set is ordered for ordinal variables. For example in an experiment one might grade the level of physical activity as low, medium, or high, giving an ordinal measurement. An example of a categorical variable is hair colour. In R the data type for ordinal and categorical vectors is *factor*. The possible values of a factor are referred to as its *levels*.

There are two reasons for using factors. The first is that the behaviour of many statistical models depend on the type of input and output variables, so we need some way of distinguishing numeric, ordinal, and categorical variables. The second is that factors can be stored very efficiently.

In practice, a factor is not terribly different from a character vector, except that the elements of a factor can take only a limited number of values (which R keeps a record of), and in statistical routines R is able to treat a factor differently than a character vector. To create a factor we apply the function `factor` to some vector `x`. By default the distinct values of `x` become the levels, or we can specify them using the optional `levels` argument. The latter allows us to have more levels than just those in `x`, which is useful if we wish to change some of the values later. We check whether or not an object `x` is a factor using `is.factor(x)`, and list its levels using `levels(x)`.

```
> hair <- c("blond", "black", "brown", "brown", "black", "gray",
+           "none")
> is.character(hair)

[1] TRUE

> is.factor(hair)

[1] FALSE

> hair <- factor(hair)
> levels(hair)

[1] "black" "blond" "brown" "gray"  "none"

> hair <- factor(hair, levels = c("black", "gray", "brown",
+                                 "blond", "white", "none"))
> table(hair)

hair
black  gray brown blond white  none
    2     1     2     1     0     1
```

Note the use of the function `table` to calculate the number of times each level of the factor appears. `table` can be applied to other modes of vectors as well as factors. The output of the `table` function is a one-dimensional array (as opposed to a vector). If more than one vector is passed to `table`, then it produces a multidimensional array.

By default R arranges the levels of a factor alphabetically. If you specify the levels yourself, then R uses the ordering that you provide. Beware that, alphabetically speaking, the string `10` is less than the string `2`, which can lead to unexpected results if your levels start with numbers.

To create an ordered factor we just include the option `ordered = TRUE` in the factor command. In this case it is usual to specify the levels of the factor yourself, as that determines the ordering.

```
> phys.act <- c("L", "H", "H", "L", "M", "M")
> phys.act <- factor(phys.act, levels = c("L", "M", "H"),
+     ordered = TRUE)
> is.ordered(phys.act)

[1] TRUE

> phys.act[2] > phys.act[1]

[1] TRUE
```

Often abbreviations or numerical codes are used to represent the levels of a factor. You can change the names of the levels using the `labels` argument. If you do this then it is good practice to specify the levels too, so you know which label goes with which level.

```
> phys.act <- factor(phys.act, levels = c("L", "M", "H"),
+                labels = c("Low", "Medium", "High"), ordered = TRUE)
> table(phys.act)

phys.act
   Low Medium   High
     2      2      2

> which(phys.act == "High")

[1] 2 3
```

Even though R usefully reports the results of operations upon factors by the levels that we assign to them, R represents factors internally as integers. This can cause considerable heartache if you treat factors carelessly, since R can coerce the factor into a numeric vector without telling you.

```
> hair

[1] blond black brown brown black gray  none
Levels: black gray brown blond white none

> as.vector(hair)

[1] "blond" "black" "brown" "brown" "black" "gray"  "none"

> as.numeric(hair)

[1] 4 1 3 3 1 2 6

> c(hair, 5)

[1] 4 1 3 3 1 2 6 5

> x <- factor(c(0.8, 1.1, 0.7, 1.4, 1.4, 0.9))
> as.numeric(x)               # does not recover x

[1] 2 4 1 5 5 3

> as.numeric(levels(x))[x]    # does recover x

[1] 0.8 1.1 0.7 1.4 1.4 0.9

> as.numeric(as.character(x)) # does recover x

[1] 0.8 1.1 0.7 1.4 1.4 0.9
```

A final point to be aware of is that if you take a subset of a factor, you may end up with missing levels, which can cause problems with some statistical procedures. One solution is to define the factor again using the `factor` function, to force the recalculation of the levels. Alternatively you can pass the `drop = TRUE` argument to the subscripting operator.

```
> table(hair[hair == "gray" | hair == "none"])
```

```
black  gray brown blond white  none
    0     1     0     0     0     1

> table(hair[hair == "gray" | hair == "none", drop = TRUE])

gray none
   1    1
```

6.2 Dataframes

We have already seen how to work in R with numbers, strings, and logical values. We have also worked with homogeneous collections of such objects, grouped into numeric, character, or logical vectors. The defining characteristic of the vector data structure in R is that all components must be of the same mode. Obviously to work with datasets from real experiments we need a way to group data of differing modes. Imagine for example a forestry experiment in which we randomly selected a number of plots and then from each plot selected a number of trees. For each tree we measured its height and diameter (which are numeric), and also the species of tree (which is a character string).

Plot	Tree	Species	Diameter (cm)	Height (m)
2	1	DF	39	20.5
2	2	WL	48	33.0
3	2	GF	52	30.0
3	5	WC	36	20.7
3	8	WC	38	22.5
⋮	⋮	⋮	⋮	⋮

As experimental data collated in a table looks like an array, you may be tempted to represent it in R as a matrix. But in R matrices cannot contain heterogeneous data (data of different modes). Lists and dataframes are able to store much more complicated data structures than matrices.

A dataframe is a list that is tailored to meet the practical needs of representing multivariate datasets. It is a list of vectors restricted to be of equal length. Each vector—or column—corresponds to a variable in an experiment, and each row corresponds to a single observation or experimental unit. Each vector can be of any of the basic modes of object.

Large dataframes are usually read into R from a file, using the function `read.table`, which has the form:

```
read.table(file, header = FALSE, sep = "")
```

`read.table` returns a dataframe. There are many more optional arguments, we have given the two most important. As always, use the built-in help for more details: `?read.table`.

`file` is the name of the file to be read from. The name can be relative to the current working directory or absolute.

It is assumed that each row of the file corresponds to the observations of a single trial. Thus there must be the same number of values in each row. They may be of different modes, but the pattern of modes must be the same in each row.

`header` indicates whether or not the first line of the file is a line of text giving the variable names.

`sep` gives the character used to separate values in each row. The default `""` has the special interpretation that a variable amount of white space (spaces, tabs, or returns) can separate values.

There are two commonly used variants of `read.table`. `read.csv(file)` is for comma-separated data and is equivalent to `read.table(file, header = TRUE, sep = ",")`. `read.delim(file)` is for tab-delimitated data and is equivalent to `read.table(file, header = TRUE, sep = "\t")`.

If a header is present, it is used to name the columns of the dataframe. You can assign your own column names after reading the dataframe (using the `names` function, see below) or when you read it in, using the `col.names` argument, which should be assigned a character vector the same length as the number of columns. If there is no header and no `col.names` argument, then R uses the names `"V1"`, `"V2"`, etc.

The experiment described at the start of this section was conducted at Upper Flat Creek, part of the University of Idaho Experimental Forest. The results are given in the file `ufc.csv`, the first few lines of which are given below. Note that dbh stands for diameter at breast height.

```
"plot","tree","species","dbh.cm","height.m"
2,1,"DF",39,20.5
2,2,"WL",48,33
3,2,"GF",52,30
3,5,"WC",36,20.7
3,8,"WC",38,22.5
```

We note that the values are comma-separated and there is a header line. Thus we read in the data using

```
> ufc <- read.csv("../data/ufc.csv")
```

We use the `head` and `tail` functions to examine the object

```
> head(ufc)

  plot tree species dbh.cm height.m
1    2    1      DF     39     20.5
2    2    2      WL     48     33.0
3    3    2      GF     52     30.0
```

```
4    3    5      WC     36     20.7
5    3    8      WC     38     22.5
6    4    1      WC     46     18.0
```

```
> tail(ufc)
```

```
    plot tree species dbh.cm height.m
331  143    1      GF   28.0     21.0
332  143    2      GF   33.0     20.5
333  143    7      WC   47.8     20.5
334  144    1      GF   10.2     16.0
335  144    2      DF   31.5     22.0
336  144    4      WL   26.5     25.0
```

Each column, or variable, in a dataframe has a unique name. We can extract that variable by means of the dataframe name, the column name, and a dollar sign, viz:

```
> x <- ufc$height.m
> x[1:5]
```

```
[1] 20.5 33.0 30.0 20.7 22.5
```

We can also use the notation `[[?]]` to extract columns. For example `ufc$height.m`, `ufc[[5]]`, and `ufc[["height.m"]]` are all equivalent.

You can extract the elements of a dataframe directly using matrix indexing:

```
> ufc[1:5, 5]
```

```
[1] 20.5 33.0 30.0 20.7 22.5
```

To select more than one of the variables in a dataframe, in other words to subset the dataframe, we use the notation `[?]`. We can also use names in this situation: `ufc[4:5]` is equivalent to `ufc[c("dbh.cm", "height.m")]`.

```
> diam.height <- ufc[4:5]
> diam.height[1:5, ]
```

```
  dbh.cm height.m
1     39     20.5
2     48     33.0
3     52     30.0
4     36     20.7
5     38     22.5
```

```
> is.data.frame(diam.height)
```

```
[1] TRUE
```

The result of selecting columns using `[?]` is another dataframe. This can sometimes cause confusion when you select only one variable.

```
> x <- ufc[5]
> x[1:5]
```

```
Error in `[.data.frame`(x, 1:5) : undefined columns selected
```

When extracting variables using `[[?]]`, we can only do so *one at a time.* Selecting a column using `[[?]]` preserves the mode of the object that is being extracted, whereas `[?]` keeps the mode of the object from which the extraction is being made.

```
> mode(ufc)
```

```
[1] "list"
```

```
> mode(ufc[5])
```

```
[1] "list"
```

```
> mode(ufc[[5]])
```

```
[1] "numeric"
```

As well as reading in a dataframe from a file, we can construct one from a collection of vectors and/or existing dataframes using the function `data.frame`, which has the form:

```
data.frame(col1 = x1, col2 = x2, ..., df1, df2, ...)
```

Here `col1`, `col2`, etc., are the column names (given as character strings without quotes) and `x1`, `x2`, etc., are vectors of equal length. `df1`, `df2`, etc., are dataframes, whose columns must be the same length as the vectors `x1`, `x2`, etc. Column names may be omitted, in which case R will choose a name for you.

We can also create a new variable within a dataframe, by naming it and assigning it a value. For example, for many tree species, the shape of a mature trunk can be modelled as an elliptic paraboloid, which gives a volume of height times cross-sectional area at breast height divided by two. That is, exactly half the volume of a cylinder of the same height and diameter. We can calculate this and add it to the `ufc` dataframe as follows:

```
> ufc$volume.m3 <- pi * (ufc$dbh.cm/200)^2 * ufc$height/2
> mean(ufc$volume.m3)
```

```
[1] 1.93294
```

Equivalently one could assign to `ufc[6]` or `ufc["volume.m3"]` or `ufc[[6]]` or `ufc[["volume.m3"]]`.

The command `names(df)` will return the names of the dataframe `df` as a vector of character strings. To change the names of `df` you pass a vector of character strings to `names(df)`. For example:

```
> (ufc.names <- names(ufc))

[1] "plot"      "tree"      "species"   "dbh.cm"    "height.m"
[6] "volume.m3"

> names(ufc) <- c("P", "T", "S", "D", "H", "V")
> names(ufc)

[1] "P" "T" "S" "D" "H" "V"

> names(ufc) <- ufc.names
```

Note that if `df` is a dataframe then `names(df)` is *not* a variable, even though we can assign a value to it. Technically speaking `names(df)` is called an *attribute*. In general the values an attribute can take are determined by the mode of object it is attached to. In our case we must have exactly one name for each column of the dataframe, and they must all be different.

Another example of an attribute is the `dim` (dimension) of a matrix. Provided the total number of elements remains the same, we can change the shape of a matrix just by changing the `dim` attribute; R will reassign values from the old matrix to the new one column by column. However you should beware that even though `dim(df)` will return the number of rows and columns of a dataframe, `dim(df) <- c(x, y)` will just generate errors. The reason is that `dim` is not an attribute of a dataframe; the function `dim` has been extended to dataframes purely for convenience. We say more about attributes in general in Section 8.4.

In addition to column or variable names, a dataframe also has row names. By default the rows are named `"1"`, `"2"`, `"3"`, etc., when the dataframe is created, however both `read.table` and `data.frame` take the optional argument `row.names`, which you can use to specify the row names. The command `row.names(df)` will return the row names of dataframe `df` as a character vector. Like `names`, `row.names` is an attribute of a dataframe, so you can change the row names of `df` by making an assignment to `row.names(df)`.

As with column names, if you delete a row then the names of the remaining rows are unchanged.

The function `subset` is a convenient tool for selecting the rows of a dataframe, especially when combined with the operator `%in%`. For example, suppose we are only interested in the height of trees of species DF (Douglas Fir) or GF (Grand Fir):

```
> fir.height <- subset(ufc, subset = species %in% c("DF", "GF"),
+                      select = c(plot, tree, height.m))
> head(fir.height)

  plot tree height.m
1    2    1     20.5
```

```
3    3  2    30.0
7    4  2    17.0
8    5  2    29.3
9    5  4    29.0
10   6  1    26.0
```

For vectors `x` and `y` (of the same mode), the expression `x %in% y` returns a logical vector the same length as `x`, whose i-th element is `TRUE` if and only if `x[i]` is an element of `y`. We say that the `%in%` operator is performing many-to-many matching. The `subset` argument takes a logical vector and determines which rows are selected. The `select` argument takes a vector of columns, which are those selected. Note that the vector is of columns, not column names. Also note that the expressions assigning values to `subset` and `select` can directly use the columns of the target dataframe, which is given as the first argument.

To write a dataframe to a file we use

```
write.table(x, file = "", append = FALSE, sep = " ",
            row.names = TRUE, col.names = TRUE)
```

Details of the arguments follow. For a complete list type `?write.table`.

`x` is the dataframe to be written.

`file` is the (name and) address of the file to write to. It will be created if it does not exist. The default is to write to the screen.

`append` indicates whether or not to append to or overwrite the file.

`sep` is the character used to separate values within a row. Rows are separated by new lines.

`row.names` either a logical value indicating whether or not to include the existing row names as the first column, or a character vector of row names.

`col.names` either a logical value indicating whether or not to include the existing column names as the first row, or a character vector of column names.

We can identify the *complete* rows from a two-dimensional object such as a dataframe (that is, rows that have no missing values) via the `complete.cases` command. We can easily remove rows with missing values using the `na.omit` function.

6.2.1 Attaching

For your convenience, R allows you to `attach` a dataframe to the workspace. When attached, the variables in the dataframe can be referred to without being prefixed by the name of the dataframe.

```
> attach(ufc)
> max(height.m[species == "GF"])

[1] 47
```

To detach the dataframe `df` use the command `detach(df)`. When you attach a dataframe R actually makes a copy of each variable, which is deleted when the dataframe is detached. Thus, if you change an attached variable you *do not* change the dataframe.

```
> height.m <- 0 #vandalism
> max(height.m)

[1] 0

> max(ufc$height.m)

[1] 47

> detach(ufc)
> max(ufc$height.m)

[1] 47
```

6.3 Lists

We have seen that a vector is an indexed set of objects. All the elements of a vector have to be of the same type—numeric, character, or logical—which is called the *mode* of the vector. Like a vector, a list is an indexed set of objects (and so has a length), but unlike a vector the elements of a list can be of different types, including other lists! The mode of a list is `list`.

A list is just a generic container for other objects and the power and utility of lists comes from this generality. A list might contain an individual measurement, a vector of observations on a single response variable, a dataframe, or even a list of dataframes containing the results of several experiments. In R lists are often used for collecting and storing complicated function output. Lists become invaluable devices as we become more comfortable with R, and start to think of different ways to solve our problems. Dataframes are special kinds of lists.

A list is created using the `list(...)` command, with comma-separated arguments. Single square brackets are used to select a sublist; double square brackets are used to extract a single element.

```
> my.list <- list("one", TRUE, 3, c("f", "o", "u", "r"))
> my.list[[2]]

[1] TRUE
```

```
> mode(my.list[[2]])

[1] "logical"

> my.list[2]

[[1]]
[1] TRUE

> mode(my.list[2])

[1] "list"

> my.list[[4]][1]

[1] "f"

> my.list[4][1]

[[1]]
[1] "f" "o" "u" "r"
```

When displaying a list, R uses double square brackets `[[1]]`, `[[2]]`, etc., to indicate list elements, then single square brackets `[1]`, `[2]`, etc., to indicate vector elements.

The elements of a list can be named when the list is created, using arguments of the form `name1 = x1`, `name2 = x2`, etc., or they can be named later by assigning a value to the `names` attribute. Unlike a dataframe, the elements of a list do not have to be named. Names can be used (within quotes) when indexing with single or double square brackets, or they can be used (with or without quotes) after a dollar sign to extract a list element.

```
> my.list <- list(first = "one", second = TRUE, third = 3,
+     fourth = c("f", "o", "u", "r"))
> names(my.list)

[1] "first"  "second" "third"  "fourth"

> my.list$second

[1] TRUE

> names(my.list) <- c("First element", "Second element",
+     "Third element", "Fourth element")
> my.list$"Second element"

[1] TRUE

> x <- "Second element"
> my.list[[x]]

[1] TRUE
```

Note the deployment of double quotes to extract the nominated element of the list, even though the name includes spaces. Single quotes and backticks will serve the same purpose.

To *flatten* a list `x`, that is convert it to a vector, we use `unlist(x)`.

```
> x <- list(1, c(2, 3), c(4, 5, 6))
> unlist(x)

[1] 1 2 3 4 5 6
```

If the list object itself comprises lists, then these lists are also flattened, unless the argument `recursive = FALSE` is set.

Many functions produce list objects as their output. For example, when we fit a least squares regression, the regression object itself is a list, and can be manipulated using list operations. Least squares regression fits a straight line $y = ax + b$ to a set of observations $\{(x_i, y_i)\}_{i=1}^n$. In R this can be achieved using the `lm` function,

```
> lm.xy <- lm(y ~ x, data = data.frame(x = 1:5, y = 1:5))
> mode(lm.xy)

[1] "list"

> names(lm.xy)

 [1] "coefficients" "residuals"     "effects"       "rank"
 [5] "fitted.values" "assign"       "qr"            "df.residual"
 [9] "xlevels"       "call"         "terms"         "model"
```

At this point we are not interested in how the straight line is fitted, but we observe that `lm` returns a list: the first element (called `coefficients`) is a vector giving a and b; the second element (called `residuals`) is a vector giving $y_i - ax_i - b$ for all i; the third element (called `fitted.values`) is a vector giving $ax_i + b$; and so on.

6.3.1 Example: Australian rules football

The Victorian Football League (VFL) was founded in 1897, then in 1990 became the Australian Football League (AFL). Teams that have played in the VFL and AFL, and the years in which they won the premiership, are presented in Table 6.1.

We can store this data as a list, where each element is a vector of dates, named according to the name of the team.

Table 6.1 *VFL/AFL teams and the years in which they have won the premiership*

Adelaide	1997, 1998
Carlton	1906, 1907, 1908, 1914, 1915, 1938, 1945, 1947, 1968, 1970, 1972, 1979, 1981, 1982, 1987, 1995
Collingwood	1902, 1903, 1910, 1917, 1919, 1927, 1928, 1929, 1930, 1935, 1936, 1953, 1958, 1990
Essendon	1897, 1901, 1911, 1912, 1923, 1924, 1942, 1946, 1949, 1950, 1962, 1965, 1984, 1985, 1993, 2000
Fitzroy/Brisbane Lions	1898, 1899, 1904, 1905, 1913, 1916, 1922, 1944, 2001, 2002, 2003
Footscray/Western Bulldogs	1954
Fremantle	
Geelong	1925, 1931, 1937, 1951, 1952, 1963, 2007
Hawthorn	1961, 1971, 1976, 1978, 1983, 1986, 1988, 1989, 1991, 2008
Melbourne	1900, 1926, 1939, 1940, 1941, 1948, 1955, 1956, 1957, 1959, 1960, 1964
North Melbourne/Kangaroos	1975, 1977, 1996, 1999
Port Adelaide	2004
Richmond	1920, 1921, 1932, 1934, 1943, 1967, 1969, 1973, 1974, 1980
Saint Kilda	1966
South Melbourne/Sydney	1909, 1918, 1933, 2005
West Coast	1992, 1994, 2006

```
> premierships <- list(
+   Adelaide = c(1997, 1998),
+   Carlton = c(1906, 1907, 1908, 1914, 1915, 1938, 1945, 1947,
+               1968, 1970, 1972, 1979, 1981, 1982, 1987, 1995),
+   Collingwood = c(1902, 1903, 1910, 1917, 1919, 1927, 1928, 1929,
+                   1930, 1935, 1936, 1953, 1958, 1990),
+   Essendon = c(1897, 1901, 1911, 1912, 1923, 1924, 1942, 1946,
+                1949, 1950, 1962, 1965, 1984, 1985, 1993, 2000),
+   Fitzroy_Brisbane = c(1898, 1899, 1904, 1905, 1913, 1916, 1922, 1944,
+                        2001, 2002, 2003),
+   Footscray_W.B. = c(1954),
+   Fremantle = c(),
+   Geelong = c(1925, 1931, 1937, 1951, 1952, 1963, 2007),
+   Hawthorn = c(1961, 1971, 1976, 1978, 1983, 1986, 1988, 1989, 1991, 2008),
+   Melbourne = c(1900, 1926, 1939, 1940, 1941, 1948, 1955, 1956,
+                 1957, 1959, 1960, 1964),
+   N.Melb_Kangaroos = c(1975, 1977, 1996, 1999),
+   PortAdelaide = c(2004),
+   Richmond = c(1920, 1921, 1932, 1934, 1943, 1967, 1969, 1973,
+                1974, 1980),
+   StKilda = c(1966),
+   S.Melb_Sydney = c(1909, 1918, 1933, 2005),
```

```
+   WestCoast = c(1992, 1994, 2006)
+ )
```

To summarise the structure of a list (or dataframe), use `str()`

```
> str(premierships)

List of 16
 $ Adelaide        : num [1:2] 1997 1998
 $ Carlton         : num [1:16] 1906 1907 1908 1914 1915 ...
 $ Collingwood     : num [1:14] 1902 1903 1910 1917 1919 ...
 $ Essendon        : num [1:16] 1897 1901 1911 1912 1923 ...
 $ Fitzroy_Brisbane: num [1:11] 1898 1899 1904 1905 1913 ...
 $ Footscray_W.B.  : num 1954
 $ Fremantle       : NULL
 $ Geelong         : num [1:7] 1925 1931 1937 1951 1952 ...
 $ Hawthorn        : num [1:10] 1961 1971 1976 1978 1983 ...
 $ Melbourne       : num [1:12] 1900 1926 1939 1940 1941 ...
 $ N.Melb_Kangaroos: num [1:4] 1975 1977 1996 1999
 $ PortAdelaide    : num 2004
 $ Richmond        : num [1:10] 1920 1921 1932 1934 1943 ...
 $ StKilda         : num 1966
 $ S.Melb_Sydney   : num [1:4] 1909 1918 1933 2005
 $ WestCoast       : num [1:3] 1992 1994 2006
```

A natural question to ask is who won the premiership on a given year.

```
> year <- 1967
> for (i in 1:length(premierships)) {
+     if (year %in% premierships[[i]]) {
+         winner <- names(premierships)[i]
+     }
+ }
> winner

[1] "Richmond"
```

In the next section we see how to vectorise this example.

6.4 The `apply` family

R provides many techniques for manipulating lists and dataframes. In particular R has several functions that allow you to easily apply a function to all or selected elements of a list or dataframe.

6.4.1 `tapply`

`tapply` is a lovely function that allows us to vectorise the application of a function to subsets of data. In conjunction with factors, this can make for some exceptionally efficient code. It has the form

```
tapply(X, INDEX, FUN, ...),
```

where the additional arguments are as follows:

`X` is the target vector to which the function will be applied;

`INDEX` is a factor, the same length as `X`, which is used to group the elements of `X` (Note that `INDEX` will be automatically coerced to a factor if it is not one already);

`FUN` is the function to be applied. It is applied to subvectors of `X` corresponding to a single level of `INDEX`.

`tapply` returns a one-dimensional array the same length as `levels(INDEX)`, whose *i*-th element is the result of applying `FUN` to `X[INDEX == levels(INDEX)[i]]` (plus any additional arguments given by `...`).

As an example, consider again the Upper Flat Creek data. Using `tapply` we obtain average height by species as follows:

```
> tapply(ufc$height.m, ufc$species, mean)

      DF       GF       WC       WL
25.30000 24.34322 23.48777 25.47273
```

We can reduce the noise as follows:

```
> round(tapply(ufc$height.m, ufc$species, mean), digits = 1)

  DF   GF   WC   WL
25.3 24.3 23.5 25.5
```

To find out how many examples we have of each species we could use `table`, or equivalently:

```
> tapply(ufc$species, ufc$species, length)

 DF  GF  WC  WL
 57 118 139  22
```

The argument `INDEX` can also be a list of factors, in which case the output is an array with dimensions given by the length of each factor, with each element given by applying `FUN` to a subset of `X` indexed by a specific factor combination. For example, we can average height by species and plot:

```
> ht.ps <- tapply(ufc$height.m, ufc[c("plot", "species")], mean)
> round(ht.ps[1:5,], digits=1)

    species
plot   DF GF   WC WL
   2 20.5 NA   NA 33
   3   NA 30 21.6 NA
   4 17.0 NA 18.0 NA
   5 29.3 29   NA NA
   6 26.0 NA 28.2 NA
```

Note from the missing values that most plots contain only a couple of different species.

6.4.2 *Applying functions to lists* `lapply` *and* `sapply`

We have used the `sapply` and `apply` commands to apply a function to a vector or an array, for example to calculate the row and column totals for a matrix. To apply a function to a list we use either `sapply` or `lapply`.

The `lapply(X, FUN, ...)` function applies the function `FUN` to each element of the list `X` and returns a list. The `sapply(X, FUN, ...)` function applies the function `FUN` to each element of `X`, which can be a list or a vector, and by default will try to return the results in a vector or a matrix, if this makes sense, otherwise in a list. Extra parameters can be passed to `FUN` by way of the `...`.

For example, to obtain the mean diameter, height, and volume of trees in the Upper Flat Creek dataset:

```
> lapply(ufc[4:6], mean)

$dbh.cm
[1] 37.41369

$height.m
[1] 24.22560

$volume.m3
[1] 1.93294

> sapply(ufc[4:6], mean)

   dbh.cm  height.m volume.m3
 37.41369  24.22560   1.93294
```

Note that the output of the command `sapply(ufc[4:6], mean)` is a vector with a names attribute.

Using the VFL/AFL premiership data, here is a vectorized way to find who won in 1967.

```
> in.1967 <- function(x) return(1967 %in% x)
> names(premierships)[sapply(premierships, in.1967)]
```

```
[1] "Richmond"
```

Again using `sapply` we can easily calculate the number of premierships won by each team

```
> sort(sapply(premierships, length))
```

```
      Fremantle   Footscray_W.B.      PortAdelaide           StKilda
              0                1                 1                 1
       Adelaide        WestCoast N.Melb_Kangaroos      S.Melb_Sydney
              2                3                 4                 4
        Geelong          Hawthorn         Richmond Fitzroy_Brisbane
              7               10                10                11
      Melbourne      Collingwood           Carlton          Essendon
             12               14                16                16
```

To restrict the list of premierships to the post-1990 AFL era, we can use `lapply`

```
> AFL <- function(x) x[x >= 1990]
> premierships.AFL <- lapply(premierships, AFL)
> str(premierships.AFL)
```

```
List of 16
 $ Adelaide        : num [1:2] 1997 1998
 $ Carlton         : num 1995
 $ Collingwood     : num 1990
 $ Essendon        : num [1:2] 1993 2000
 $ Fitzroy_Brisbane: num [1:3] 2001 2002 2003
 $ Footscray_W.B.  : num(0)
 $ Fremantle       : NULL
 $ Geelong         : num 2007
 $ Hawthorn        : num [1:2] 1991 2008
 $ Melbourne       : num(0)
 $ N.Melb_Kangaroos: num [1:2] 1996 1999
 $ PortAdelaide    : num 2004
 $ Richmond        : num(0)
 $ StKilda         : num(0)
 $ S.Melb_Sydney   : num 2005
 $ WestCoast       : num [1:3] 1992 1994 2006
```

To restrict the list to premierships between the years 1970 and 1979 we can do the following:

```
> between.years <- function(x, a, b) x[a <= x & x <= b]
> premierships.1970s <- lapply(premierships, between.years,
+     1970, 1979)
```

6.4.3 Example: tree growth

A sample of 66 Grand Fir trees (*Abies grandis*) was selected from national forests around northern and central Idaho. The trees were selected to be dominant in their environment, with no visible evidence of crown damage, forks, broken tops, etc. For each tree the habitat type and the national forest from which it came were recorded. We have data from nine national forests and six different habitat types.[1]

For each tree the height, diameter, and age were measured (age was measured using tree rings), then the tree was split lengthways to determine the height and diameter of the tree at *any* age. In this instance height and diameter were recorded for the age the tree was felled and then at ten-year periods going back in time. The diameter of the tree was measured at a height of 1.37 m (4′6″), which is called *breast height* in forestry. The height refers to the height of the main trunk only.

The data are provided in the comma-separated file `treegrowth.csv`, with each row giving diameter at breast height (dbh) in inches and height in feet, for a single tree at a given age. This dataset is provided in the package that accompanies this book.

For example, here are the rows relevant to the first two trees:

```
> treeg <- read.csv("../data/treegrowth.csv")
> treeg[1:15, ]

   tree.ID forest habitat dbh.in height.ft age
1        1      4       5   14.6      71.4  55
2        1      4       5   12.4      61.4  45
3        1      4       5    8.8      40.1  35
4        1      4       5    7.0      28.6  25
5        1      4       5    4.0      19.6  15
6        2      4       5   20.0     103.4 107
7        2      4       5   18.8      92.2  97
8        2      4       5   17.0      80.8  87
9        2      4       5   15.9      76.2  77
10       2      4       5   14.0      70.7  67
11       2      4       5   11.7      56.6  57
12       2      4       5   10.6      43.0  47
13       2      4       5    8.0      35.6  37
14       2      4       5    6.2      29.3  27
15       2      4       5    3.4      16.2  17
```

An alternative way of structuring the data is to collect the measurements for each tree together in a single variable. We will use a list whose elements are

[1] A.R. Stage, 1963. A mathematical approach to polymorphic site index curves for grand fir. *Forest Science* **9**, 167–180.

the tree ID number; forest code; habitat code; and three vectors giving age, dbh, and height measurements. Each tree record will then be a single element of a larger list called `trees`.

```
> trees <- list() #list of trees
> n <- 0 #number of trees in the list of trees
> #start collecting information on current tree
> current.ID <- treeg$tree.ID[1]
> current.age <- treeg$age[1]
> current.dbh <- treeg$dbh.in[1]
> current.height <- treeg$height.ft[1]
> for (i in 2:dim(treeg)[1]) {
+   if (treeg$tree.ID[i] == current.ID) {
+     #continue collecting information on current tree
+     current.age <- c(treeg$age[i], current.age)
+     current.dbh <- c(treeg$dbh.in[i], current.dbh)
+     current.height <- c(treeg$height.ft[i], current.height)
+   } else {
+     #add previous tree to list of trees
+     n <- n + 1
+     trees[[n]] <- list(tree.ID = current.ID,
+                        forest = treeg$forest[i-1],
+                        habitat = treeg$habitat[i-1],
+                        age = current.age,
+                        dbh.in = current.dbh,
+                        height.ft = current.height)
+     #start collecting information on current tree
+     current.ID <- treeg$tree.ID[i]
+     current.age <- treeg$age[i]
+     current.dbh <- treeg$dbh.in[i]
+     current.height <- treeg$height.ft[i]
+   }
+ }
> #add final tree to list of trees
> n <- n + 1
> trees[[n]] <- list(tree.ID = current.ID,
+                    forest = treeg$forest[i],
+                    habitat = treeg$habitat[i],
+                    age = current.age,
+                    dbh.in = current.dbh,
+                    height.ft = current.height)
```

Let's see how the data on the first two trees is now structured.

```
> str(trees[1:2])
```

```
List of 2
 $ :List of 6
  ..$ tree.ID  : int 1
  ..$ forest   : int 4
```

```
  ..$ habitat  : int 5
  ..$ age      : int [1:5] 15 25 35 45 55
  ..$ dbh.in   : num [1:5] 4 7 8.8 12.4 14.6
  ..$ height.ft: num [1:5] 19.6 28.6 40.1 61.4 71.4
 $ :List of 6
  ..$ tree.ID  : int 2
  ..$ forest   : int 4
  ..$ habitat  : int 5
  ..$ age      : int [1:10] 17 27 37 47 57 67 77 87 97 107
  ..$ dbh.in   : num [1:10] 3.4 6.2 8 10.6 11.7 14 15.9 17 18.8 20
  ..$ height.ft: num [1:10] 16.2 29.3 35.6 43 56.6 ...
```

Here we used loops to split the data up. Phil Spector suggested a more compact solution that we provide below, with an interesting twist.

```
> getit <- function(name, x) {
+     if (all(x[[name]] == x[[name]][1])) {
+         x[[name]][1]
+     }
+     else {
+         x[[name]]
+     }
+ }
> repts <- function(x) {
+     res <- lapply(names(x), getit, x)
+     names(res) <- names(x)
+     res
+ }
> trees.ps <- lapply(split(treeg, treeg$tree.ID), repts)
> str(trees.ps[1:2])
```

```
List of 2
 $ 1:List of 6
  ..$ tree.ID  : int 1
  ..$ forest   : int 4
  ..$ habitat  : int 5
  ..$ dbh.in   : num [1:5] 14.6 12.4 8.8 7 4
  ..$ height.ft: num [1:5] 71.4 61.4 40.1 28.6 19.6
  ..$ age      : int [1:5] 55 45 35 25 15
 $ 2:List of 6
  ..$ tree.ID  : int 2
  ..$ forest   : int 4
  ..$ habitat  : int 5
  ..$ dbh.in   : num [1:10] 20 18.8 17 15.9 14 11.7 10.6 8 6.2 3.4
  ..$ height.ft: num [1:10] 103.4 92.2 80.8 76.2 70.7 ...
  ..$ age      : int [1:10] 107 97 87 77 67 57 47 37 27 17
```

Suppose now that we would like to plot a curve of height versus age for each tree. First we need to know the maximum age and height so that we can set up the plot region.

```
> max.age <- 0
> max.height <- 0
> for (i in 1:length(trees)) {
+     if (max(trees[[i]]$age) > max.age)
+         max.age <- max(trees[[i]]$age)
+     if (max(trees[[i]]$height.ft) > max.height)
+         max.height <- max(trees[[i]]$height.ft)
+ }
```

Alternatively, here is a more concise way of calculating `max.age` and `max.height`, using `sapply`.

```
> my.max <- function(x, i) max(x[[i]]) #max of element i of list x
> max.age <- max(sapply(trees, my.max, "age"))
> max.height <- max(sapply(trees, my.max, "height.ft"))
```

The plotting is now straightforward. See Figure 6.1 for the output.

```
> plot(c(0, max.age), c(0, max.height), type = "n", xlab = "age (years)",
+     ylab = "height (feet)")
> for (i in 1:length(trees)) lines(trees[[i]]$age, trees[[i]]$height.ft)
```

In the next chapter we will present functions that can create graphics like that presented in Figure 6.1 directly from the dataframe.

6.5 Exercises

1. From the `spuRs` package you can obtain the dataset `ufc.csv`, with forest inventory observations from the University of Idaho Experimental Forest. Try to answer the following questions:

   (a). What are the species of the three tallest trees? Of the five fattest trees? (Use the `order` command.)

   (b). What are the mean diameters by species?

   (c). What are the two species that have the largest third quartile diameters?

   (d). What are the two species with the largest median slenderness (height/diameter) ratios? How about the two species with the smallest median slenderness ratios?

   (e). What is the identity of the tallest tree of the species that was the fattest on average?

2. Create a list in R containing the following information:

   - your full name,
   - gender,
   - age,
   - a list of your 3 favourite movies,

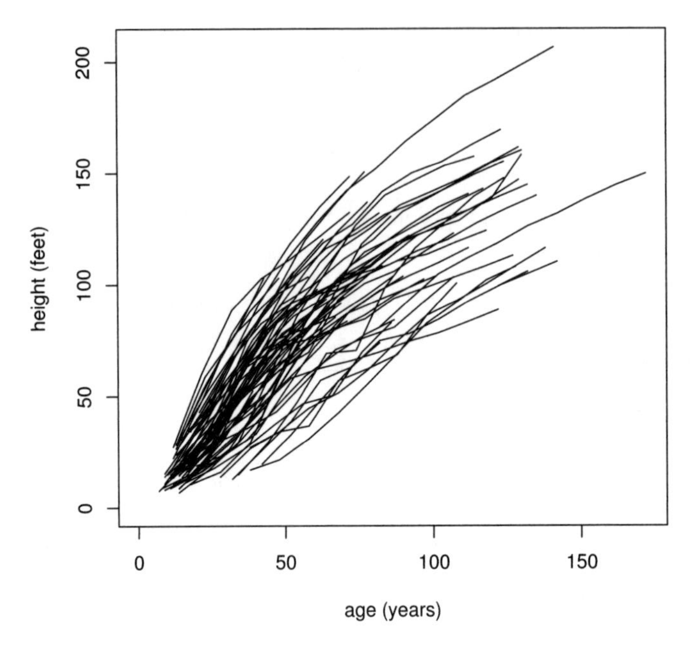

Figure 6.1 *Height against age for all 66 trees in the tree growth dataset. See Example 6.4.3.*

- the answer to the question 'Do you support the United Nations?', and
- a list of the birth day and month of your immediate family members including you (identified by first name).

Do the same for three close friends, then write a program to check if there are any shared birthdays or names in the four lists.

Produce a table of birthdays by birth month and a table of the mean number of immediate family members by gender.

3. Using the tree growth data (Section 6.4.3, available from the `spuRs` package), plot tree age versus height for each tree, broken down by habitat type. That is, create a grid of 5 plots, each showing the trees from a single habitat.

   For the curious, the habitats corresponding to codes 1 through 5 are: Ts/Pach, Ts/Op, Th/Pach, AG/Pach, and PA/Pach. These codes refer respectively to the climax tree species, which is the most shade-tolerant species that can grow on the site, and the dominant understorey plant. Ts refers to *Thuja plicata* and *Tsuga heterophylla*, Th refers to just *Thuja plicata*, AG is *Abies grandis*, PA is *Picea engelmanii* and *Abies lasiocarpa*, Pach is *Pachistima myrsinites*, and Op is the nasty *Oplopanaz horridurn*.

Abies grandis is considered a major climax species for AG/Pach, a major seral species for Th/Pach and PA/Pach, and a minor seral species for Ts/Pach and Ts/Op. Loosely speaking, a community is *seral* if there is evidence that at least some of the species are temporary, and *climax* if the community is self-regenerating.[2]

4. Pascal's triangle.

   Suppose we represent Pascal's triangle as a list, where item n is row n of the triangle. For example, Pascal's triangle to depth four would be given by

```
list(c(1), c(1, 1), c(1, 2, 1), c(1, 3, 3, 1))
```

   The n-th row can be obtained from row $n-1$ by adding all adjacent pairs of numbers, then prefixing and suffixing a 1.

   Write a function that, given Pascal's triangle to depth n, returns Pascal's triangle to depth $n+1$. Verify that the eleventh row gives the binomial coefficients $\binom{10}{i}$ for $i = 0, 1, \ldots, 10$.

5. Horse racing.

   The following is an excerpt from the file `racing.txt` (available in the `spuRs` archive), which has details of nine horse races run in the U.K. in July 1998.

```
1 0 54044 4.5   53481 4     53526 4     53526 3.5   53635 3    53792
1 1 54044 1.375 53481 1.5   53635 1.5   53635 1.375 53928 1.25 54026
1 0 54044 1.75  53481 1.625 53792 1.625 53792 1.75  53936
1 0 54044 14    53481 20    53635 20    53635 16    53868 20   54026
1 0 54044 20    53481 25    53635 25    53635
1 0 54044 33    53481 50    53635 50    53635 66    53929
1 0 54044 20    53481 25    53635 25    53635 33    53792 50   54045
2 1 55854 6     55709 7     56157 7     56157
2 0 55854 6     55138 6.5   55397 6.5   55397 7     55825 7    56157
...
```

   In each row, the first number gives the race number. There is one line for each horse in each race. The next number is 0 or 1 depending on whether the horse lost or won the race. Numbers then come in pairs (t_i, p_i), $i = 1, 2, \ldots$, where t_i is a time and p_i a price. That is, the odds on the horse at time t_i were $p_i : 1$.

   Import this data into an object with the following structure:

   - A list with one element per race.
   - Each race is a list with one element per horse.
   - Each horse is a list with three elements: a logical variable indicating win/loss, a vector of times, and a vector of prices.

[2] R. Daubenmire, 1952. Forest Vegetation of Northern Idaho and Adjacent Washington, and Its Bearing on Concepts of Vegetation Classification, *Ecological Monographs* **22**, 301–330.

Write a function that, given a single race, plots log price against time for each horse, on the same graph. Highlight the winning horse in a different colour.

6. Here is a recursive program that prints all the possible ways that an amount x (in cents) can be made up using Australian coins (which come in 5, 10, 20, 50, 100, and 200 cent denominations). To avoid repetition, each possible decomposition is ordered.

```
# Program spuRs/resources/scripts/change.r

change <- function(x, y.vec = c()) {
  # finds possible ways of making up amount x using Australian coins
  # x is given in cents and we assume it is divisible by 5
  # y.vec are coins already used (so total amount is x + sum(y.vec))
  if (x == 0) {
    cat(y.vec, "\n")
  } else {
    coins <- c(200, 100, 50, 20, 10, 5)
    new.x <- x - coins
    new.x <- new.x[new.x >= 0]
    for (z in new.x) {
      y.tmp <- c(y.vec, x - z)
      if (identical(y.tmp, sort(y.tmp))) {
        change(z, y.tmp)
      }
    }
  }
  return(invisible(NULL))
}
```

Rewrite this program so that instead of writing its output to the screen it returns it as a list, where each element is a vector giving a possible decomposition of x.

CHAPTER 7

Better graphics

7.1 Introduction

One major selling point for R is that it has better graphics capabilities than many of the commercial alternatives. Whether you want quick graphs that help you understand the structure of your data, or publication-quality graphics that accurately communicate your message to your readers, R will suffice. You can get an initial overview of R's graphic capabilities by typing `demo(graphics)`.

This chapter provides a deeper exposition of the graphical capabilities of R, building on the modest offering in Chapter 4. We cover the construction of simple graphics, in terms of the individual pieces that make up the default plot. We discuss the graphics parameters that are used to fine-tune individual graphs and the relationships between multiple graphics on a page. We show how to save graphical objects in various formats (pdf, postscript, etc.). Finally, we present some specific R graphical tools for the presentation of multivariate data (lattice graphs, which simplify the construction of conditioning plots) and some 3D-graphic construction tools.

This chapter will demonstrate some of R's graphical capacity using a forest inventory dataset, taken from the Upper Flat Creek stand of the University of Idaho Experimental Forest. We read the data, then print a summary of its structure using `str`:

```
> ufc <- read.csv("../data/ufc.csv")
> str(ufc)

'data.frame':        336 obs. of  5 variables:
 $ plot    : int  2 2 3 3 3 4 4 5 5 6 ...
 $ tree    : int  1 2 2 5 8 1 2 2 4 1 ...
 $ species : Factor w/ 4 levels "DF","GF","WC",..: 1 4 2 3 3 3 1 1 2 1 ...
 $ dbh.cm  : num  39 48 52 36 38 46 25 54.9 51.8 40.9 ...
 $ height.m: num  20.5 33 30 20.7 22.5 18 17 29.3 29 26 ...
```

The variables `height.m` and `dbh.cm` are tree height in metres, and the tree bole diameter in centimetres, measured at 1.37 metres from the ground, respectively. The latter is called 'diameter at breast height' in forestry in the USA, hence the acronym *dbh*.[1]

The graphics start at a very simple level, for example

```
> plot(ufc$dbh.cm, ufc$height.m)
```

will open a graphical window and draw a scatterplot of dbh against height for the Upper Flat Creek data, labelling the axes appropriately. A modest addition will provide more informative axis labels (Figure 7.1).

```
> plot(ufc$dbh.cm, ufc$height.m, xlab = "Diameter (cm)",
+      ylab = "Height (m)")
```

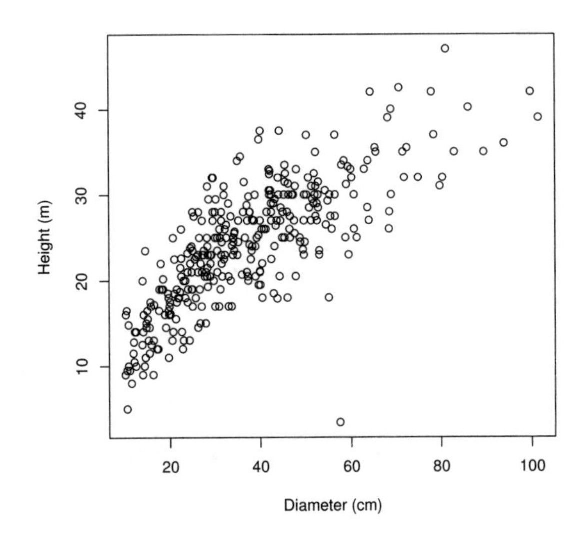

Figure 7.1 *Diameter/Height plot for all species of Upper Flat Creek inventory data. Each point represents a tree.*

The `plot` command offers a wide variety of options for customising the graphic. Each of the following arguments can be used within the `plot` statement, singly or together, separated by commas.

`type = "?"` determines the type of plot, with options:

- `"p"` for points (the default);
- `"l"` for lines;
- `"b"` for both, with gaps in the lines for the points;

[1] 'Breast height' for forestry in most other countries is 1.3 metres. Presumably, US foresters are taller.

- "c" for the lines part alone of "b", which is useful if you want to combine lines with other kinds of symbols;
- "o" for both lines and points 'overplotted', that is, without gaps in the lines;
- "h" for vertical lines, giving a 'histogram' like plot;
- "s" for a step function, going across then up;
- "S" for a step function, going up then across;
- "n" for no plotting.

`xlim = c(a,b)` will set the lower and upper limits of the x-axis to be a and b, respectively. Note that we have to know a and b to make this work!

`ylim = c(a,b)` will set the lower and upper limits of the y-axis to be a and b, respectively.

`xlab = "X axis label goes in here"` provides the label for the x-axis.

`ylab = "Y axis label goes in here"` provides the label for the y-axis.

`main = "Plot title goes in here"` provides the plot title.

`pch = k` determines the shape of points, with `k` taking a value from 1 to 25.

`lwd = ?` line width, default 1.

`col = "?"` colour for lines and points. R knows about many colours, such as `"tomato"`, `"deepskyblue"`, and `"slategray"`; type `colours()` or `colors()` for a list. When overlaying plots it is useful to be able to use different colours (and shapes).

7.2 Graphics parameters: par

In order to describe the effects of changing different graphics parameters, we need to distinguish between graphics devices and plots. We can think of a graphics device as being a platform upon which the plot is created. If we create a plot, then a default graphics device is automatically opened for the plot to appear upon. To create a graphics device without a plot, we call the function that is specific to our operating system (that is, `windows` for Windows, `quartz` for Mac, and `X11` for Unix).

R keeps a list of *graphics parameters*, which control how graphics devices appear. To get a complete list with their current values, type `par()`. `pch`, `lwd` and `col` are all examples of graphics parameters. To get the value of a specific parameter, for example `pch`, type `par("pch")`. Some graphics parameters can apply to one or more plots, and others only make sense when applied to graphics devices. For example, to change the symbol for a single plot, we could include the argument `pch = 2` in the call to the plot function. However, we could also make this change for the device.

To change a graphics parameter for the graphics device, we use the `par` command. Here are some useful examples.

`par(mfrow = c(a,b))` where a and b are integers, will create a matrix of plots on one page, with a rows and b columns. These will be filled by rows; use `mfcol` if you wish to fill them by columns.

`par(mar = c(bottom, left, top, right))` will create space around each plot, in which to write axis labels and titles. Measurements are in units of character widths.

`par(oma = c(bottom, left, top, right))` will create space around the matrix of plots (an outer margin). Measurements are in units of character widths.

`par(las = 1)` rotates labels on the y-axis to be horizontal rather than vertical.

`par(pty = "s")` forces the plot shape to be square. The alternative is that the plot shape is mutable, which is the default, and corresponds to `pty = "m"`.

`par(new = TRUE)` when inserted between plots will plot the next one in the same place as the previous, effecting an overlay. It will not match the axes unless forced to do so. Increasing the `mar` parameters for the second plot sufficiently will force it to be printed *inside* the first plot.

`par(cex = x)` magnifies all plotted symbols and text by a factor x. Also, finer-grained control is available from `cex.axis` for the axis annotations, `cex.lab` for the x and y labels, `cex.main` for the title, and `cex.sub` for the subtitle text.

`par(bty = "?")` determines the type of box that is drawn about plots. Options are `"o"`, `"l"`, `"7"`, `"c"`, `"u"`, `"]"`, or `"n"` for nothing.

Note that the `par` function can accept multiple arguments in a call. For example, to arrange plots in a 3 by 2 grid, with a 4-character margin at the left and bottom of each plot, and a 1-character margin to the top and right of each plot, and with horizontal labels on the y-axis, we would use

```
> par(mfrow = c(3,2), mar = c(4,4,1,1), las = 1)
```

When used to change the value of a graphics parameter, the `par` command returns a list of the *old* values invisibly, that is, without printing them. This allows us to customise the graphics parameters, create a graph, then restore the original state, by means of the following simple commands:

```
> opar <- par( {comma separated par instructions go here} )
> plot( {plot instructions go here} )
> par(opar)
```

For example,

```
> opar <- par(mfrow = c(3,2), mar = c(4,4,1,1), las = 1)
> plot( {plot instructions go here} )
```

The content of `opar` looks like this:

```
> opar

$mfrow
[1] 1 1

$mar
[1] 5.1 4.1 4.1 2.1

$las
[1] 0
```

We then return the graphics parameters to their original state via:

```
> par(opar)
```

7.3 Graphical augmentation

A traditional plot can be augmented using any of a number of different tools after its creation.

The infrastructure of the plot can be altered. For example, axes may be omitted in the initial plot call, using the `axes = FALSE` argument, and added afterwards using the `axis` function, which provides greater control and flexibility over the locations and format of the tickmarks, and locations, format, and content of the axis labels. The plot frame can be added using the `box` function. Text can be located in the margins of the plot, to label certain areas or to augment tick labels, using the `mtext` function. Text can be placed in the plot using the `text` function. A legend can be added using the `legend` function, which includes a very useful legend location argument, as shown below. Additions can also be made to the content of the plot, using the `points`, `lines`, and `abline` functions, among others. A number of these different steps are detailed below, and the development is shown in Figure 7.2.

1. Start by creating the plot object, which sets up the dimensions of the space, but omit any plot objects for the moment.

   ```
   > opar1 <- par(las = 1, mar = c(4, 4, 3, 2))
   > plot(ufc$dbh.cm, ufc$height.m, axes = FALSE, xlab = "",
   +      ylab = "", type = "n")
   ```

2. Next, we add the points. Here we use different colours and symbols for different heights of trees: those that are realistic, and those that are not, which may reflect measurement errors. We use the vectorised `ifelse` function.

   ```
   > points(ufc$dbh.cm, ufc$height.m,
   +         col = ifelse(ufc$height.m > 4.9, "darkseagreen4", "red"),
   +         pch = ifelse(ufc$height.m > 4.9, 1, 3))
   ```

3. Then we add axes. The following are the simplest possible calls, we have much greater flexibility than shown here. We can also control the locations of the tickmarks, and their labels, we can overlay different axes, change colour, and so on. As usual, `?axis` provides the details.

```
> axis(1)
> axis(2)
```

4. We can next add axis labels using margin text (switching back to vertical direction for the y-axis text).

```
> opar2 <- par(las = 0)
> mtext("Diameter (cm)", side = 1, line = 3)
> mtext("Height (m)", side = 2, line = 3)
```

5. Wrap the plot in the traditional frame. As before, we can opt to use different line types and different colours.

```
> box()
```

6. Finally, we add a legend.

```
> legend(x = 60, y = 15,
+        c("Normal trees", "A weird tree"),
+        col=c("darkseagreen3", "red"),
+        pch=c(1, 3),
+        bty="n")
```

Note the first two arguments: the location of the legend can also be expressed relative to the graph components, for example, by `"bottomright"`. `legend` has other useful options, see the help file for more details.

7. If we wish, we can return the graphics environment to its previous state.

```
> par(opar1)
```

Finally, we mention the `playwith` package, which provides interaction with graphical objects at a level unattainable in base R.

7.4 Mathematical typesetting

It is often useful to provide more sophisticated axis labels and plot titles. R permits the use of mathematical typesetting anywhere that you can add text to a graph, through a straightforward interface. In a call to plot, the arguments for `main`, `sub`, and `xlab` and `ylab` can be character strings or expressions (or names, or calls, see `?title` for more details). When expressions are used for these arguments, they are interpreted as mathematical expressions, and the output is formatted according to some specific rules. See `?plotmath` for the syntax, rules, and examples of the mathematical markup language

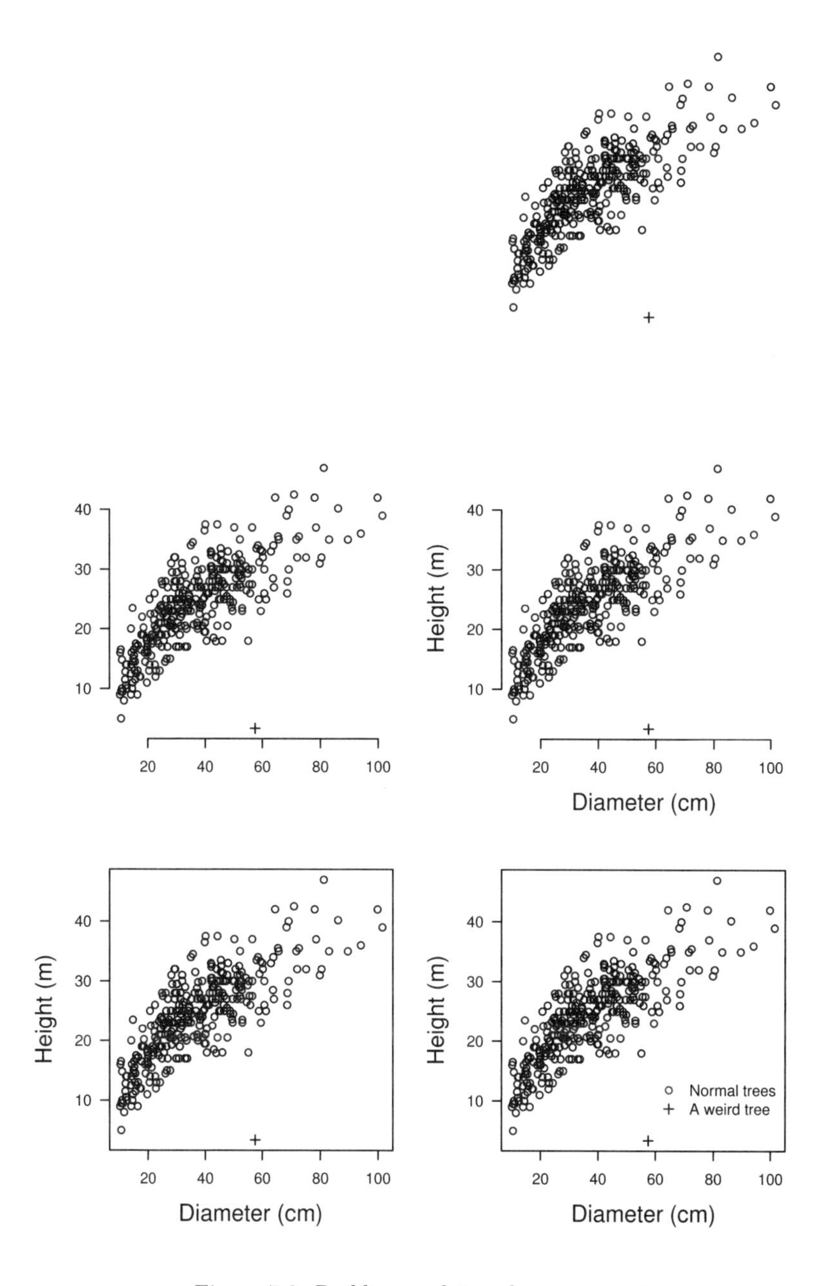

Figure 7.2 *Building a plot up by components.*

(MML). Also, run `demo(plotmath)` and examine the code and graphical output demonstrated therein.

Figure 7.3 shows some examples, including Greek lettering, mathematical typesetting, and printing the values of variables. In this example we also make use of the `curve` function, for plotting the graph of a function, and the `par(usr)` command, to change the co-ordinates of the existing plot. We change co-ordinates to simplify the placement of text within the plot.

Constructing the labels is complicated by the fact that they comprise up to three different types of objects: strings that we wish R to preserve and print, expressions that we wish R to interpret as a mathematical markup language (MML), and expressions that we wish R to evaluate at the time of execution, and then print the outcome. Various combinations of these objects can be constructed via `paste`. However, we need a way to distinguish these three different modes in the label; we will use the functions `expression` and `bquote`. It is important to note that the `expression` function is playing a different role when constructing labels than when it is used outside the context of plot functions. Here it is being used solely to alert R that the enclosed text should be interpreted as an MML expression.

```
> curve(100*(x^3-x^2)+15, from=0, to=1,
+       xlab = expression(alpha),
+       ylab = expression(100 %*% (alpha^3 - alpha^2) + 15),
+       main = expression(paste("Function : ",
+           f(alpha) == 100 %*% (alpha^3 - alpha^2) + 15)))
> myMu <- 0.5
> mySigma <- 0.25
> par(usr = c(0, 1, 0, 1)) # Change coordinates within plot
> text(0.1, 0.1, bquote(sigma[alpha] == .(mySigma)), cex=1.25)
> text(0.6, 0.6, paste("(The mean is ", myMu, ")", sep=""), cex=1.25)
> text(0.5, 0.9,
+      bquote(paste("sigma^2 = ", sigma^2 == .(format(mySigma^2, 2)))))
```

Thus,

```
xlab = expression(alpha)
```

tells R to interpret `alpha` in the context of the MML, producing an α as the label for the x-axis. The MML will interpret more complicated strings of instructions, for example:

```
ylab = expression(100 %*% (alpha^3 - alpha^2) + 15)
```

produces the label on the y-axis.

When necessary, we mix mathematical expressions and character strings into a single expression by using the `paste` function. Thus,

```
main = expression(paste("Function : ",
          f(alpha) == 100 %*% (alpha^3 - alpha^2) + 15)))
```

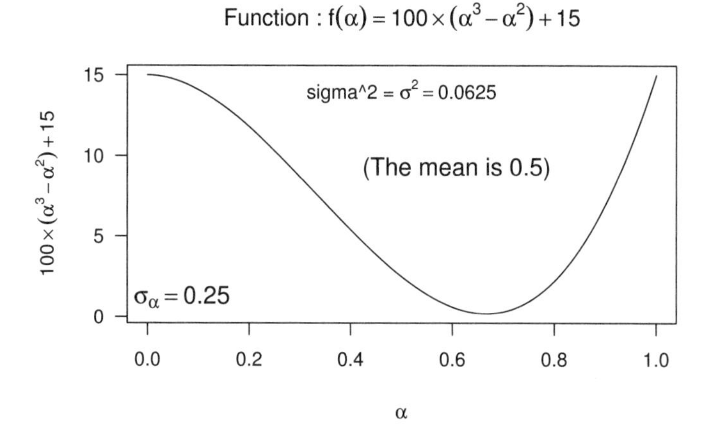

Figure 7.3 *An example of mathematical typesetting.*

will interleave the `"Function : "` string with the `f(alpha) == 100 %*% (alpha^3 - alpha^2) + 15` expression using `paste`, and then interpret the whole as an expression.

A further complication is that sometimes we wish to first evaluate some portions of the expression; for example, we may wish to annotate a graph using an expression that contains a variable whose value is only known at the time of execution.

In order to allow for evaluation of only parts of the expression, we use `bquote`. The `bquote` function can be called with a single argument, which should be a language object, for example an expression. `bquote` finds all of the terms in the argument that are wrapped in `.()`, that is, parentheses preceded by a period, and evaluates those terms. It then returns the argument with those wrapped terms replaced by their output. For example, in Figure 7.3, we ask for

```
bquote(sigma[alpha] == .(mySigma))
```

which performs the following steps.

1. Search "`sigma[alpha] == .(mySigma)`" for `.()`, finding `mySigma`. Evaluate `mySigma` and replace it with its output.
2. Return `sigma[alpha] == 0.25`, which is of mode call. The `text` function accepts objects of mode call.

A *call* is a specific kind of unevaluated function. See `?call` for more details.

Our example is trivial, but of course `bquote` can evaluate any (evaluable) expression. A slightly more complicated example blends all three kinds of elements together: strings, MML expressions, and expressions for evaluation. We blend these elements together using bquote and paste as follows:

```
bquote(paste("sigma^2 = ", sigma^2 == .(format(mySigma^2, 2))))
```

See ?bquote for more information, specifically referring to evalauting the expression in other environments, about which we say more in Section 8.2.

7.5 Permanence

Producing more permanent graphics is very simple. We merely need to wrap the plotting commands in code that opens and closes the relevant graphics device, specifying the name and address of the file to be created. For example, to create a graphic as a pdf file, which can be imported into various documents and is well accepted on the Internet, we do the following:

```
> pdf(file = "graphic.pdf", width = 4, height = 3)
> plot(ufc$dbh.cm, ufc$height.m, main = "UFC trees",
+       xlab = "Dbh (cm)", ylab = "Height (m)")
> dev.off()
```

Note that to close a graphics device, we call the dev.off function. This call is essential in order to be able to use the constructed graphic in other applications. If the dev.off call is omitted, then the operating system will not allow any interaction with the pdf.

All plotting between the pdf command and the dev.off command will appear in the pdf file graphic.pdf in the current working directory. The height and width arguments are in units of inches. Multiple plots will appear by default as separate pages in the saved pdf document. That is, if we include more than one plot statement between the pdf and dev.off, then the resulting pdf will have more than one page. This facility simplifies the task of storing graphical output from an operation that is embedded in a loop. If unique files are required for each plot, then we supply onefile = FALSE.

Using the commands postscript, jpeg, png, or bmp, we can also produce graphics in the formats that correspond to these names. The jpeg, png, and bmp graphics are all raster-style graphics, which may translate poorly when included in a document. In contrast, the pdf, postscript, and windows metafile (win.metafile, available on Windows) formats allow for vector-style graphics, which are scaleable, and better suited to integration in documents.

On Windows, we can use the win.metafile function to create Windows metafile graphics. You can also copy directly from the plot window to the clipboard as either a metafile or a bmp (bitmap) image, by right-clicking the window and selecting Copy as metafile or Copy as bitmap. Either can then be pasted directly into a Word document, for example.

7.6 Grouped graphs: lattice

Trellis graphics are a data visualisation framework developed at the Bell Labs,[2] which have been implemented in R as the lattice package.[3] Trellis graphics are a set of techniques for displaying multidimensional data. They allow great flexibility for producing *conditioning* plots; that is, plots obtained by conditioning on the value of one of the variables.

We load the lattice package by means of the library function, which is explained in greater detail in Section 8.1.

```
> library(lattice)
```

In a conditioning plot the observations are divided into collections according to the value of a conditioning variable, and each collection is plotted in its own panel in the graph. In the material that follows, we shall use *panel* to describe a portion of the plot that contains its own, possibly unique, axes, and a portion of the observations as determined by the conditioning variable. The nature of the graphics that are produced in each panel depends on which lattice function is being used to create the graphic.

In Figure 7.4 we illustrate the use of the lattice package using the ufc data. We illustrate how dbh (diameter at breast height) and height vary by species. That is, our plots are conditioned on the value of the variable species.

Top left graphic: A density plot

```
> densityplot(~ dbh.cm | species, data = ufc)
```

Top right graphic: Box and whiskers plot

```
> bwplot(~ dbh.cm | species, data = ufc)
```

Bottom left graphic: Histogram

```
> histogram(~ dbh.cm | species, data = ufc)
```

Bottom right graphic: Scatterplot

```
> xyplot(height.m ~ dbh.cm | species, data = ufc)
```

All four commands require a *model*, which is described using ~ and |. If a dataframe is passed to the function, using the argument data, then the column names of the dataframe can be used for describing the model. We interpret y ~ x | a as saying we want y as a function of x, divided up by the different levels of a. If a is not a factor then a factor will be created by coercion. If we are just interested in x we still include the ~ symbol, so that R knows that we are specifying a model. If we wish to provide within-panel conditioning on a second variable, then we use the group argument.

[2] W.S. Cleveland, *Visualizing Data*. Hobart Press, 1993.
[3] Principally authored by Deepayan Sarkar.

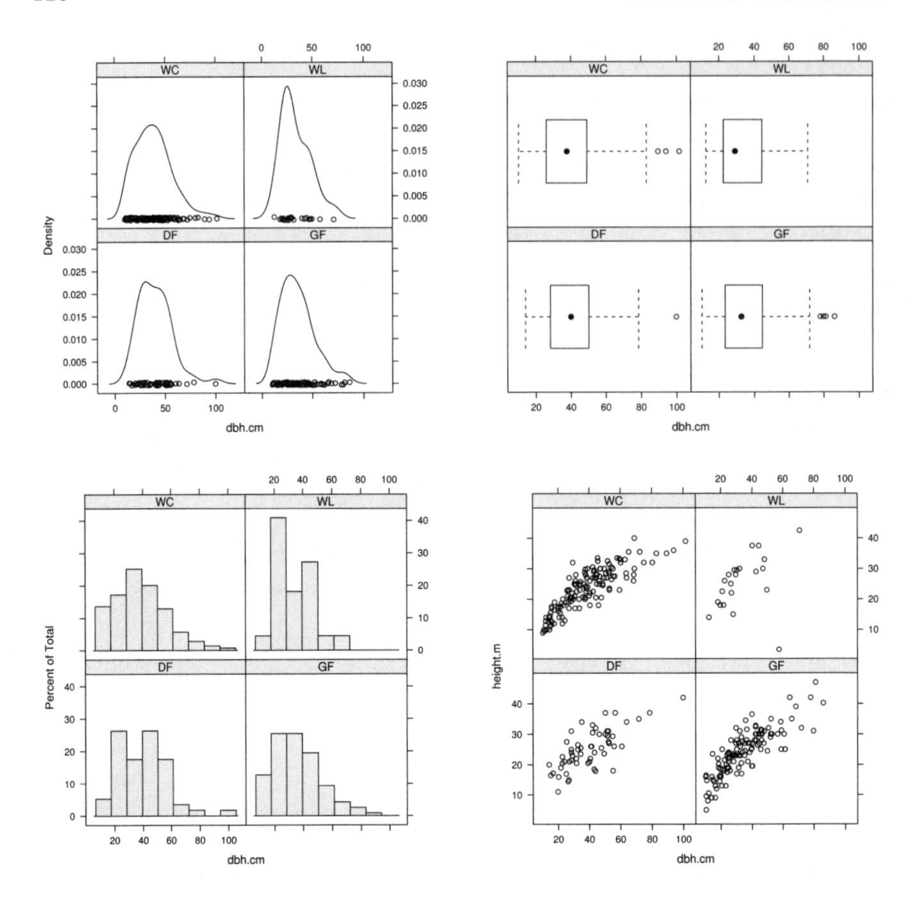

Figure 7.4 *Example lattice plots, showing diameter information conditioned on species. The top left graphic shows the empirical density curves of the diameters, the top right graphic shows box plots, the bottom left shows histograms, and the bottom right shows a scatterplot of tree height against diameter.*

The first three plots, created using `densityplot`, `bwplot` (box and whiskers plot), and `histogram`, all attempt to display the same information, namely the distribution of values taken on by the variable dbh, divided up according to the value of the variable species. Thus, all three plots use the same formula. In the fourth plot we plot height as a function of dbh. For a description of the types of plots available through the `lattice` package, type `?lattice`.

In order to display numerous lattice objects on a graphics device, we call the `print` function with the `split` and `more` arguments.

`split` takes a vector of four integers: the first pair denote the location of the lattice object, and the second pair provide the intended dimensions of the graphics device, analogously to `mfrow` in the `par` function, but with columns *first*.

`more` is logical, and tells the device whether to expect more lattice objects (`TRUE`) or not (`FALSE`).

Thus, to place a lattice object (called `my.lat`, for example) in the top-right corner of a 3-row, 2-column graphics device, and allow for more objects, we would write:

```
> print(my.lat, split = c(2,1,2,3), more = TRUE)
```

Don't forget to use `more = FALSE` in the last object for the device!

We can also fine-tune the relative location of the objects using the `position` argument. See `?print.trellis` for further details.

Graphics produced by `lattice` are highly customisable. For example, suppose we wish to plot the diameter against the height for species WC (Western Red Cedar) and GF (Grand Fir), and add a regression line to each graph, then we could do it by using a custom `panel` function, as follows (the output is given in Figure 7.5):

```
> xyplot(height.m ~ dbh.cm | species,
+        data = ufc,
+        subset = species %in% list("WC", "GF"),
+        panel = function(x, y, ...) {
+          panel.xyplot(x, y, ...)
+          panel.abline(lm(y~x), ...)
+        },
+        xlab = "Diameter (cm)",
+        ylab = "Height (m)"
+        )
```

In the lattice function `xyplot` we have used four new arguments. The application of `xlab` and `ylab` should be clear. The other two require some explanation. The `subset` argument is a logical expression or a vector of integers, and is used to restrict the data that are plotted. We can also change the order of the panels using the `index.cond` argument (see `?xyplot` for more details). The panels are plotted bottom left to top right by default, or top left to bottom right if the argument `as.table = TRUE` is supplied.

The `panel` argument accepts a function, the purpose of which is to control the appearance of the plot in each panel. The panel function should have one input argument for each variable in the model, not including the conditioning variable. The function should have `x` as the first argument and `y` as the second. Here our model uses two variables, `height.m` and `dbh.cm`, so the panel function has two inputs. To produce a plot the panel function uses particular functions, denoted by the `panel.` prefix, where the inputs for the

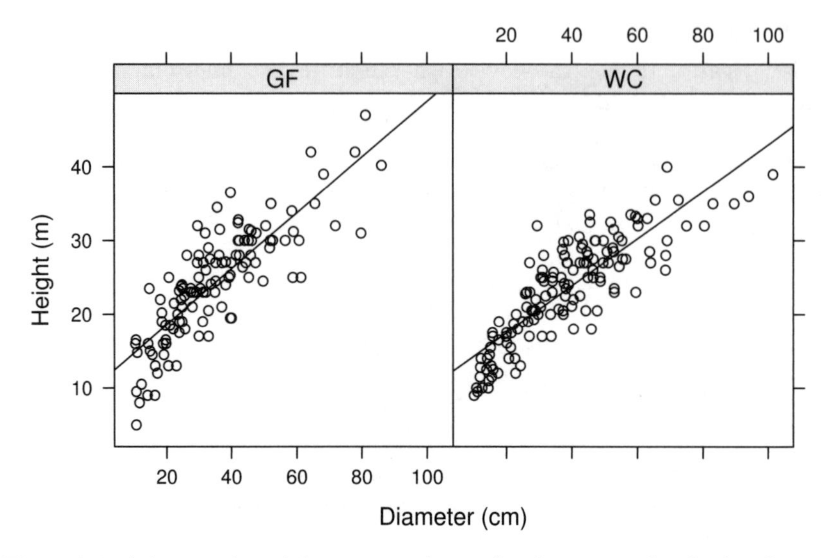

Figure 7.5 *A lattice plot of diameter at breast height against height for the species Grand Fir (GF) and Western Red Cedar (WC). Each plot includes a linear regression line.*

function `panel.ftn` are the same as the inputs for the function `ftn` (where `ftn` stands for some function or other). In addition to `panel.xyplot` and `panel.abline`, you can use `panel.points`, `panel.lines`, `panel.text`, and many others. In particular, the many and various types of plot provided by the `lattice` package all have `panel.ftn` versions. Very often you will find that the panel function is defined on the fly (as here) and uses the `panel.?` function corresponding to the lattice function we are using, as here. Greater complications ensue when the `groups` argument has been invoked to create conditioning within each panel. See `?xyplot` for more details.

To see what sort of commands we might add to a panel function, `?xyplot` is very helpful, providing very detailed information about the various options, and some attractive examples that you can adapt to your own uses.

Note the inclusion of the ... argument for the `panel` function and the calls to `panel.xyplot` and `panel.abline`. This inclusion allows the calling functions to pass any other relevant but unspecified arguments to the functions being called. It is not always necessary, but it prevents the occasional misunderstanding. We also remark that this particular graphic could also be produced by using the `type` argument to `xyplot`. In this case, we would omit the `panel` argument altogether, but include `type = c("p","r")`, representing points and regression line. Other useful options to pass to `type` are `l` for lines and `smooth` for a loess smooth curve.

A common problem that new users have is in creating a pdf with lattice graphics. We need to use the print function in order to have the lattice objects appear in the pdf. Also, note that lattice and traditional graphical objects control the graphics devices differently, and are immiscible unless care is taken. In general we would recommend using just one system for any given graphics device.

The creator of the lattice package, Deepayan Sarkar, has written a very useful book on the use of lattice for graphical representation (D. Sarkar, *Lattice: Multivariate Data Visualization with R.* Springer, 2008).

7.7 3D-plots

R provides considerable functionality for constructing 3D-graphics, using either the base graphics engine or the lattice package. We recommend using the lattice package, because the data can be supplied to the lattice functions in a familiar structure: observations in rows and variables in columns, unlike that required by the base 3D-graphics engine. The base graphics engine assumes the observations are on a grid, then requires variables x and y, which are the x and y co-ordinate of each observation, and a variable z, which is a matrix containing the values to be plotted.

We demonstrate some 3D lattice graphics using the Upper Flat Creek forest data. The tree measurements were collected on a systematic grid of measurement plots within the forest stand.[4] We have estimated the volume of the tree boles (trunks), in cubic metres per hectare, for each plot. We have assigned those volumes to the plot locations. The outcome is stored in the dataset ufc-plots.csv.

```
> ufc.plots <- read.csv("../data/ufc-plots.csv")
> str(ufc.plots)

'data.frame':        144 obs. of  6 variables:
 $ plot     : int  1 2 3 4 5 6 7 8 9 10 ...
 $ north.n  : int  12 11 10 9 8 7 6 5 4 3 ...
 $ east.n   : int  1 1 1 1 1 1 1 1 1 1 ...
 $ north    : num  1542 1408 1274 1140 1006 ...
 $ east     : num  83.8 83.8 83.8 83.8 83.8 ...
 $ vol.m3.ha: num  0 63.4 195.3 281.7 300.1 ...

> library(lattice)
```

In this dataset, the volume at each location is vol.m3.ha, and the plot location is stored in metres in the east and north variables. The plot location on the grid is reported by the east.n and north.n variables. We use the information about plot locations to provide spatial summaries of the variable of interest (Figure 7.6).

[4] For those interested, the measurement plots were variable-radius plots with basal area factor 7 m^2ha^{-1}.

```
> contourplot(vol.m3.ha ~ east * north,
+   main = expression(paste("Volume (", m^3, ha^{-1}, ")", sep = "")),
+   xlab = "East (m)", ylab = "North (m)",
+   region = TRUE,
+   aspect = "iso",
+   col.regions = gray((11:1)/11),
+   data = ufc.plots)

> wireframe(vol.m3.ha ~ east * north,
+   main = expression(paste("Volume (", m^3, ha^{-1}, ")", sep = "")),
+   xlab = "East (m)", ylab = "North (m)",
+   data = ufc.plots)
```

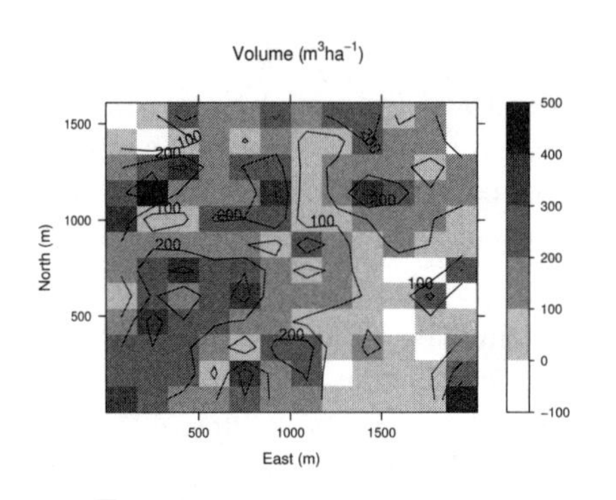

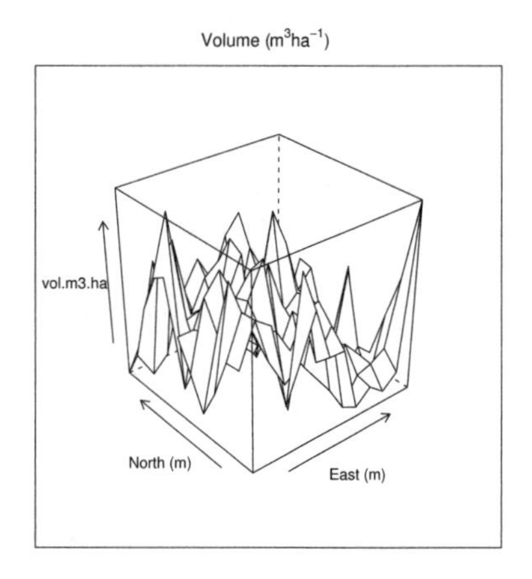

Figure 7.6 *Example 3D lattice plots, showing volume information conditioned on plot location. The left panel shows a contour-plot, the right panel shows a wireframe.*

To learn more about base-graphics 3D-plots, run the demonstrations `demo(persp)` and `demo(image)` and look at the examples presented.

7.8 Exercises

1. The *slenderness* of a tree is defined as the ratio of its height over its diameter, both in metres.[5] Slenderness is a useful metric of a tree's growing history, and indicates the susceptibility of the tree to being knocked over in high winds. Although every case must be considered on its own merits, assume that a slenderness of 100 or above indicates a high-risk tree for these data. Construct a boxplot, a histogram, and a density plot of the slenderness ratios by species for the Upper Flat Creek data. Briefly discuss the advantages and disadvantages of each graphic.

[5] For example, Y. Wang, S. Titus, and V. LeMay. 1998. Relationships between tree slenderness coefficients and tree or stand characteristics for major species in boreal mixedwood forests. *Canadian Journal of Forest Research* **28**, 1171–1183.

2. Among the data accompanying this book, there is another inventory dataset, called `ehc.csv`. This dataset is also from the University of Idaho Experimental Forest, but covers the East Hatter Creek stand. The inventory was again a systematic grid of plots. Your challenge is to produce and interpret Figure 7.4 using the East Hatter Creek data.

3. Regression to the mean.

   Consider the following very simple genetic model. A population consists of equal numbers of two sexes: male and female. At each generation men and women are paired at random, and each pair produces exactly two offspring, one male and one female. We are interested in the distribution of height from one generation to the next. Suppose that the height of both children is just the average of the height of their parents, how will the distribution of height change across generations?

   Represent the heights of the current generation as a dataframe with two variables, `m` and `f`, for the two sexes. The command `rnorm(100, 160, 20)` will generate a vector of length 100, according to the normal distribution with mean 160 and standard deviation 20 (see Section 16.5.1). We use it to randomly generate the population at generation 1:

```
pop <- data.frame(m = rnorm(100, 160, 20), f = rnorm(100, 160, 20))
```

   The command `sample(x, size = length(x))` will return a random sample of size `size` taken from the vector `x` (without replacement). (It will also sample with replacement, if the optional argument `replace` is set to `TRUE`.) The following function takes the dataframe `pop` and randomly permutes the ordering of the men. Men and women are then paired according to rows, and heights for the next generation are calculated by taking the mean of each row. The function returns a dataframe with the same structure, giving the heights of the next generation.

```
next.gen <- function(pop) {
  pop$m <- sample(pop$m)
  pop$m <- apply(pop, 1, mean)
  pop$f <- pop$m
  return(pop)
}
```

   Use the function `next.gen` to generate nine generations, then use the lattice function `histogram` to plot the distribution of male heights in each generation, as in Figure 7.7. The phenomenon you see is called *regression to the mean.*

   Hint: construct a dataframe with variables height and generation, where each row represents a single man.

4. Reproduce Figure 6.1 using `lattice` graphics.

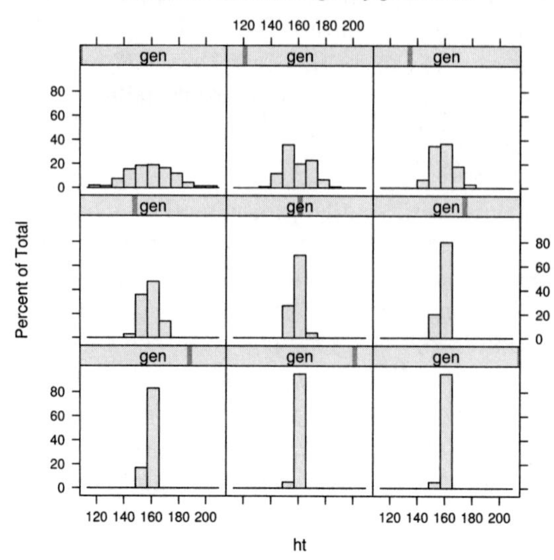

Figure 7.7 *Regression to the mean: male heights across nine generations. See Exercise 3.*

CHAPTER 8

Pointers to further programming techniques

This chapter briefly mentions some more advanced aspects of programming in R. We introduce the management of and interaction with packages. We present details about how R arranges the objects that are created within the workspace, and within functions that we are running (frames and environments). We provide further suggestions for debugging your own functions, and we present some of the infrastructure that R provides for object-oriented programming. Finally we demonstrate how to include in R, code that has been compiled using another computer language, for example C.

8.1 Packages

A package is an archive of files that conforms to a certain format and structure and that provides extra functionality, usually extending R in a particular direction. The R community has produced many high-quality R packages for performing specific tasks, usually statistical in nature. The HTML help facility called by `help.start()` gives details of the packages that are installed on your computer.

8.1.1 Package management

Any package is in one of three states: installed and loaded, installed but not loaded, or not installed. A package that is loaded is directly available to your R session, a package that is installed is available for loading but its contents are not available until it is loaded, and a package that has not been installed cannot be loaded.

Find out which packages are loaded using `sessionInfo`:

```
> sessionInfo()

R version 2.8.0 (2008-10-20)
i386-unknown-freebsd7.0
```

```
locale:
C

attached base packages:
[1] stats     graphics  grDevices utils     datasets  methods
[7] base
```

The `sessionInfo` function is useful because it also provides the version of R and the operating system for which it is compiled.

Packages are divided into three groups: base, recommended, and other. *Base* packages are installed along with R, and their objects are always available. For example, we can use the `lm` command from the `stats` package without explicitly loading anything. *Recommended* packages are installed along with R but must be loaded before they can be used. *Other* packages are not installed by default, and must be installed separately.

A non-base package can be loaded using the `library` function:

```
> library(lattice)
```

`sessionInfo` confirms that `lattice` has been loaded:

```
> sessionInfo()

R version 2.8.0 (2008-10-20)
i386-unknown-freebsd7.0

locale:
C

attached base packages:
[1] stats     graphics  grDevices utils     datasets  methods
[7] base

other attached packages:
[1] lattice_0.17-15

loaded via a namespace (and not attached):
[1] grid_2.8.0
```

Note that the argument to `library` is an object name, not a character string. It is possible to pass a vector of package names, encoded as character strings, but in order to do so we must add the `character.only = TRUE` argument, which tells R that the package name can be a string.

If a package is not installed then the `library` function produces an error. If the install status is uncertain (for example if you are writing a function that requires the package), then use the `require` function, which returns `FALSE` if the package is not installed, rather than an error.

Installing all available packages would be a waste of space and time, as you would never use most of them. Similarly, loading all installed packages every time you start R would take some time, so by default R only loads the base packages when it starts and requires the user to load any others as and when they are needed.

Mostly the `require` and `library` functions do not need any arguments other than the package name. If the package has been installed in a non-standard place, for example, if it has been installed by a user who lacks write authority to the default installation directory, then the location of the package on the user's hard drive must be passed to the function by the `lib.loc` argument.

The command to find out what packages are available for loading is `installed.packages`. The output of the function is quite verbose, but we only need the first column. For example, on the computer used to compile this book, the first five installed packages are:

```
> installed.packages()[1:5, 1]

  Brobdingnag       CarbonEL          DAAG        Design         Ecdat
"Brobdingnag"     "CarbonEL"        "DAAG"      "Design"       "Ecdat"
```

In the discussion that follows, we mention a repository. This is a software storage resource at which one or more packages are available, and it may be local or remote. The most commonly used repositories are the CRAN (Comprehensive R Archive Network) mirrors, which can be accessed from the R website. All the following commands will work for any suitable repository.

All the packages that are available at a repository, and whose requirements are matched by the currently running version of R, can be listed using the command `available.packages`.

A package that is available in the repository but has not yet been installed may be installed using the `install.packages` function. If we include the argument `dependencies = TRUE`, then the function will also install packages that are necessary to run the package or packages of interest; such packages are called dependencies.

If the user is constrained in terms of write access to the installation directory, then the packages can be downloaded to a directory nominated in the `destdir` argument and installed in a directory nominated in the `lib` argument. The latter directory must then be passed to the `require` or `library` functions using the `lib.loc` argument.

For example, if while using Microsoft Windows we wish to download the `spuRs` package to `D:/tmp`, install it into `D:/lib`, and then use it in our R session, we would use the following steps. We shall assume that the directories exist and are writeable by the user.

```
> install.packages("spuRs", destdir="D:/tmp", lib="D:/lib")
> library(spuRs, lib.loc="D:/lib")
```

If a locally installed package has a dependency that is also locally installed then the call to `library` will fail. A simple solution is to load the dependency first, using `library`. Alternatively, the location of locally installed packages can be provided to R using the `.libPaths` function.

The status of the packages that are installed can be compared with the repository using the `old.packages` function, and easily updated using the `update.packages` function.

In versions of R that support a GUI, such as `Rgui.exe` in Microsoft Windows, it is also possible to load and install packages using the GUI. In many operating systems it is also possible to install packages from the shell. In each case refer to the R manuals to learn more.

8.1.2 Package construction

The 'R Installation and Administration' manual is invaluable reading.

Constructing one's own packages is a little daunting to start with, but it has numerous advantages. First, R provides a number of functions that will perform various checks on package contents, including things like documentation. So, if you plan to share or archive your code, writing a package is a good way to be sure that you include a minimal level of infrastructure. Packages also provide a simple interface for the user to be able to access functions and datasets.

Package construction is straightforward if only R code and objects are to be used. If compilation of source code is required (see Section 8.6) then complications ensue, and extra software may be required. The extra software that is useful to support package construction varies across the operating system supporting the build, and also depends on the target operating system. We will cover the steps for building source packages on Windows.

`package.skeleton` is an extremely useful function for constructing packages. This function checks the current workspace for functions and data, and creates a directory structure suitable to house all the existing objects, including skeletal help files and a set of instructions for how to finish creating the package. Specific objects can be selected for inclusion in the package if only a subset of the available objects are wanted. Also, it may be necessary to construct new help files for objects that are added to the package later, in which case the `prompt` function is useful, as it facilitates post-hoc help file construction for arbitrary objects. This function will work on all three major platforms.

Here we use the `package.skeleton` function to construct a draft package to accompany this book, including one dataset and one function.

```
> rm(list=ls())
> ufc <- read.csv("../data/ufc.csv")
> vol.m3 <- function(dbh.cm, height.m, multiplier=0.5) {
+     vol.m3 <- pi * (dbh.cm/200)^2 * height.m * multiplier
+     }
> package.skeleton(name = "spuRs", path = "../package", force = TRUE)
```

```
package.skeleton(name = "spuRs", path = "../package", force = TRUE)
Creating directories ...
Creating DESCRIPTION ...
Creating Read-and-delete-me ...
Saving functions and data ...
Making help files ...
Done.
Further steps are described in '../package/spuRs/Read-and-delete-me'.
```

This command creates the necessary directories and files at the path nominated by the appropriate argument. These resources are then used as the basis of package construction. Next we must complete the task begun by `package.skeleton`. The important items to fix are the help files, which are stored in the `man` directory and denoted `object.Rd`. We must add appropriate details, keywords, and simple examples of the use of the functions. The `force` argument tells R to replace the previous version if there is one.

The default installation of R upon Windows does not support package construction, at the time of writing.[1] Further tools are necessary. Specifically, R needs:

1. An installation of Perl, version 5.8.0 or later,
2. Unix-like command line tools, and
3. MinGW compilers.

These tools are all available for free download, kindly provided by Duncan Murdoch, a member of R-core, as `Rtools.exe`.[2] During the process of installation, up-to-date information is provided about any other software that might be useful, for example, LaTeX and the Microsoft HTML Help Workshop. In order for these programs to be used on your system, the PATH must be set. The installation software will optionally change your PATH appropriately, in which case a restart of your computer will be necessary.

Open a shell command prompt and change directory so that you are in the directory that contains the top-level folder for the package. For our example, we wish to be in the `../package/` directory. We can then learn about the various command-line options via

```
> R CMD --help
```

[1] Version 2.8.0.
[2] See `http://www.murdoch-sutherland.com/Rtools/`.

Then we input

```
> R CMD build spuRs
```

to build the package, and

```
> R CMD check spuRs
```

which will report any problems. Each of these commands has numerous options, which you can learn about using

```
> R CMD build --help
> R CMD check --help
```

Here is how to construct a Windows-ready binary.

```
> R CMD build --binary spuRs
```

This invocation creates a package that we can now install and load using `library` as we need. For example, *after installing* the new **spuRs** package in R, we can do the following.

```
> library(spuRs)
> data(ufc)
> str(ufc)

'data.frame':   336 obs. of  5 variables:
 $ plot    : int  2 2 3 3 3 4 4 5 5 6 ...
 $ tree    : int  1 2 2 5 8 1 2 2 4 1 ...
 $ species : Factor w/ 4 levels "DF","GF","WC",..: 1 4 2 3 3 3 1 1 2 1 ...
 $ dbh.cm  : num  39 48 52 36 38 46 25 54.9 51.8 40.9 ...
 $ height.m: num  20.5 33 30 20.7 22.5 18 17 29.3 29 26 ...
```

8.2 Frames and environments

In order to interact with R effectively in more complex settings, it is important to know something about how R organises the objects that it creates and contains. The following description provides enough background for the rest of the chapter, but we have brushed over some of the deeper details.

R uses frames and environments to organise the objects created within it. A *frame* is a device that ties the names of objects to their R representations. An *environment* is a frame combined with a reference to another environment, its *parent environment*. Environments are nested, and the parent environment is the environment that directly contains the current environment. That is, following the analogy developed in Section 2.2, if a variable is like a folder with a name on the front, the frame is the catalogue of folder names and locations of the folders in memory, and an environment is a catalogue plus the address of another catalogue that contains the current catalogue.

When R is started, a workspace is created. This workspace is called the global environment, and is the default container for the objects that are subsequently created. When packages are loaded they may have their own environment (*namespace*) associated with them, which is added to the R search path.

The contents of the search path can be found by

```
> search()

 [1] ".GlobalEnv"        "package:lattice"   "package:stats"
 [4] "package:graphics"  "package:grDevices" "package:utils"
 [7] "package:datasets"  "package:methods"   "Autoloads"
[10] "package:base"
```

When a function is called, R creates a new environment, which is enclosed in the current environment. The objects that are named in the function arguments are passed from the current environment to the new environment. The expressions within the function are then applied to the objects within the new environment. Objects that are created in the new environment are not available in the parent environment. Likewise, if a function is called from within a function, another new environment is created, enclosed within the recently created environment, and so on. When objects are created in the code, R will create them in the current environment, unless instructed to do otherwise.

When an expression is evaluated, R looks in the current environment for all the objects that the expression includes. If any objects cannot be located in the frame of the current environment, then R searches the frame of the parent environment. This search for objects continues up through the environments until it reaches the global environment, and then if necessary, the search path. If an object that is passed to the function is subsequently changed within the function, then a copy is made and the changes are actually made to a copy of the object. Otherwise the object itself is made available to the function. This makes a noticeable difference when the object is complicated, as we show in the code below.

```
# program spuRs/resources/scripts/scoping.r
# Script to demonstrate the difference between passing and copying
# arguments.
# We use an artificial example with no real-world utility.

require(nlme)

fm1 <- lme(distance ~ age, data = Orthodont)
fm1$numIter <- 1

fm2 <- fm1

nochange <- function(x) {
  2 * x$numIter
```

```
  return(x)
}

change <- function(x) {
  x$numIter <- integer(2)
  return(x)
}
```

```
> source("../scripts/scoping.r")
> system.time(for (i in 1:10000) change(fm1))

   user  system elapsed
  0.515   0.030   0.584

> system.time(for (i in 1:10000) nochange(fm1))

   user  system elapsed
  0.049   0.000   0.050
```

This execution shows the extra time taken to copy the object to the environment of the function when we wish to alter it within the function.

It is possible to evaluate expressions in arbitrary environments, as we will show in the next section.

8.3 Debugging again

In some cases, it is useful to be able to examine objects in parent environments. We can do so using the `eval` and `expression` functions. We can use

```
> eval(expression( {R expression here} ),
+      envir = sys.frame(n))
```

to evaluate `{R expression here}` in relative environment `n`, which might, for example, be `-1`, denoting the environment from which your function has been called. We wrap the R expression in the `expression` function to ensure that it is passed to `eval` without being evaluated. We tell `eval` the relative environment in which to evaluate the expression using the `envir` argument.

For example, to list the objects that are defined in the environment that is one level up, you might do the following.

```
> rm(list = ls())
> ls()

character(0)
```

```
> x <- 2
> ls.up <- function() {
+     eval(expression(ls()), envir = sys.frame(-1))
+ }
> ls.up()
```

```
[1] "ls.up" "x"
```

This example is not particularly useful by itself, but it shows how to construct commands that operate in different environments, and such commands are very useful when debugging inside a browser. Often the root cause of problems with a function is located in the environment from which the function was called, rather than in the function itself. It is convenient to be able to show or manipulate the values of the calling environment, in order to be able to efficiently detect a bug.

Calling `ls` in the parent environment is useful to see what objects are there, and calling `print` in the parent environment is useful for examining them.

The `recover` command allows the user to select an environment (but not the global environment), then calls the `browser` function to browse it. When `recover` is executed it presents a menu with the available environments. `recover` is invoked automatically upon an error if `options(error = recover)` is set. The only limitation is that it does not allow stepping through the code, as `browser` does when called directly.

Here is a very trivial example of how we might use `recover` to locate an error. We start with the default error response, which is documented in `?stop`.

```
> broken <- function(x) {
+    broken2 <- function(x) {
+    y <- x * z
+    return(y)
+    }
+  y <- broken2(x)
+  y2 <- y^2
+  return(y2)
+ }
> broken(0:2)
```

```
Error in broken2(x) : object "z" not found
```

Now we change the error response option.

```
> options(error = recover)  # Change the response to an error.
> broken(0:2)
```

```
Error in broken2(x) : object "z" not found

Enter a frame number, or 0 to exit
```

```
1: broken(0:2)
2: broken2(x)

Selection:
```

We are offered two environments to browse; we choose the second.

```
Selection: 2
Called from: eval(expr, envir, enclos)
Browse[1]>
```

We now have the browser prompt, at which we can evaluate R expressions.

```
Browse[1]> ls()

[1] "x"

Browse[1]> x

[1]  0 1 2
```

We have found that the `broken2(x)` frame lacks `z` (which is as the error reported!) We now back out of the current environment and inspect the other environment.

```
Browse[1]> c


Enter a frame number, or 0 to exit

1: broken(0:2)
2: broken2(x)

Selection: 1
Called from: eval(expr, envir, enclos)

Browse[1]> ls()

[1] "broken2" "x"

Browse[1]> broken2

function(x) {
y <- x * z
return(y)
 }
<environment: 0x87c9c14>

Browse[1]> Q

>
```

Clearly the function is looking for an object that is unavailable. Note that we were able to drop in and out of the different available environments, to try to track down where the problem was originating.

8.4 Object-oriented programming: S3

Object-oriented programming, OOP for short, is a style of programming that can simplify many problems. It is based on defining classes, and creating and manipulating objects of those classes. By an object we mean a variable with a particular structure, such as a Boolean variable, vector, dataframe, or a list. The *class* of an object is a label that describes the category of the object.

R supports OOP through so-called old-style (S3) and new-style (S4) classes. This support allows the user to define new classes of objects (variables with specific structures), and functions that apply to such objects. Also, existing functions can be augmented to apply to new classes. It is not essential to apply the principles of OOP when writing R code, but doing so can have advantages, especially when writing code that is to be used by other people.

In this section we will cover the S3 classes. The key facilities that we need to discuss are: classes, generic functions, and attributes.

Note that the class of an object should not be confused with the mode, which is similar but more primitive, and refers to how R stores the object in memory.

Below, we develop an example that demonstrates the application of S3 classes. In order to follow the development, it is helpful to understand the definition and use of generic functions.

8.4.1 Generic functions

In R, a generic function is a function that examines the class of its first argument, and chooses another function appropriate to that class. For example, if we look at the innards of the mean function,

```
> mean

function (x, ...)
UseMethod("mean")
<environment: namespace:base>
```

we see that it merely passes the string 'mean' to a function called `UseMethod`. The `UseMethod` function then calls the version of `mean` that is designated for the class of the object named in the first argument. If the class of the object is, for example, widget, then `UseMethod` calls `mean.widget` with the same arguments that were supplied to `mean` by the user.

Any generic function can be extended by adding a version that acts differently depending on the class. Not all functions are generic; we can find out if a function is generic using the `methods` function.[3]

[3] Observe that in the response to `methods(var)`, R noted that some of the functions are non-visible. R located the names of the methods even though they are not available on the search path. Examine such objects using the `getAnywhere` function.

```
> methods(mean)

[1] mean.Date       mean.POSIXct    mean.POSIXlt    mean.data.frame
[5] mean.default    mean.difftime

> methods(var)

[1] var.test          var.test.default* var.test.formula*

   Non-visible functions are asterisked
Warning message:
In methods(var) : function 'var' appears not to be generic
```

In brief, to write a function that will be *automatically* invoked to replace an existing generic function `fu` when applied to an object of class `bar`, we need merely name the function `fu.bar`. Of course, non-generic functions can also be written specifically for objects of a given class, but do not require the trailing object name.

Note that the `function.class` approach that we have described here is only for S3 classes; S4 classes have a different and more formal approach.

8.4.2 Example: seed dispersal

We will illustrate OOP in R through an example. We will create a new class, in other words a new type of object, and then write generic functions that apply to that class. Specifically we write new versions of the generic functions `print` and `mean`.

Our example comes from plant biology. In studying the spread of plants (in particular weeds), it is useful to know how their seeds are dispersed. Seed dispersal is measured using seed traps, which are containers of fixed size that are situated at a known distance and bearing from a parent plant for a set amount of time. The number of seeds landing in each container during that time are counted, and the number that can be attributed to the parent plant are identified.

For our model, we assume that the seed traps are laid out in a straight line anchored at the parent plant. The seeds in each trap are then counted after a certain time (see Figure 8.1).

Let us imagine that the data available to an analyst for such a setup is the distance of the centre of each trap from the plant, the trap area, and the count of seeds in each trap. Presently there is no `trapTransect` class, inasmuch as there are no special methods for objects of that have class `trapTransect`, viz:

```
> methods(class = "trapTransect")

no methods were found
```

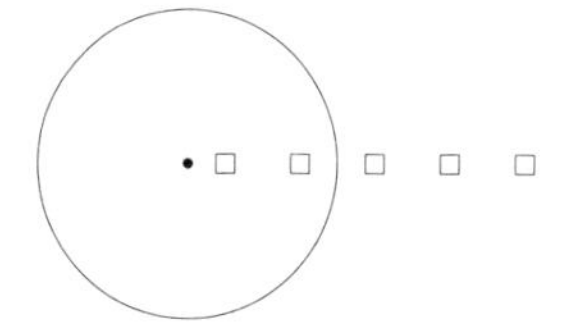

Figure 8.1 *Transect of seed traps from plant; squares represent seed traps, the circle represents the median of the overall seed shadow, the black dot is the focal parent plant.*

We can invent one. First, we have to decide what attributes we would like objects of our new class to have. Each object will contain the data from a single transect of traps, so we want to store the trap distances from the parent plant and the seed counts. The trap sizes will be assumed to be constant. The basic structure of an object of class `trapTransect` will therefore be a list with three components: trap distances, trap seed counts, and trap size.

In R, we invent S3 classes by writing a constructor function that creates objects of the appropriate structure, sets the `class` attribute, and returns the object. We then write functions that manipulate objects of that class. We start with a constructor function for this class, which applies simple checks to the inputs, constructs the list, sets the class attribute, and returns the list, now considered an object of class `trapTransect`.

```
> trapTransect <- function(distances, seed.counts, trap.area = 0.0001) {
+   if (length(distances) != length(seed.counts))
+     stop("Lengths of distances and counts differ.")
+   if (length(trap.area) != 1) stop("Ambiguous trap area.")
+   trapTransect <- list(distances = distances,
+                        seed.counts = seed.counts,
+                        trap.area = trap.area)
+   class(trapTransect) <- "trapTransect"
+   return(trapTransect)
+ }
```

In the interests of brevity, we have omitted checks for missing values, inputs that are not numbers, etc., although an operational solution would require those. Also, our function assumes that the trap area is 0.0001, although this can be overridden in the arguments when the function is called. We do not need to specify the units; a more complete solution would do so.

We now create a function that prints out relevant information about the trap-Transect data when invoked, `print.trapTransect`, using the very handy `str` function. Note that our goal is to provide a compact example, and our use of `str` reflects an austere rather than an aesthetic choice!

```
> print.trapTransect <- function(x, ...) {
```

```
+     str(x)
+ }
```

Also note that we have chosen the arguments x and ...; these are the same arguments as for the generic function print.

We now write a specific function for the mean that uses the structure of the trapTransect object to compute the mean dispersal distance from the plant along the transect. We indicate that this version of the mean should be used only for trapTransect objects by post-fixing '.trapTransect' to the function name.

```
> mean.trapTransect <- function(x, ...) {
+   return(weighted.mean(x$distances, w = x$seed.counts))
+ }
```

R now knows about the trapTransect class, in the sense that it recognises specific versions of generic methods that are suitable to the class.

```
> methods(class = "trapTransect")

[1] mean.trapTransect  print.trapTransect
```

Finally, we demonstrate the use of the class methods, print.trapTransect and mean.trapTransect:

```
> s1 <- trapTransect(distances = 1:4, seed.counts = c(4, 3, 2, 0))
> s1

List of 3
 $ distances   : int [1:4] 1 2 3 4
 $ seed.counts: num [1:4] 4 3 2 0
 $ trap.area   : num 1e-04
 - attr(*, "class")= chr "trapTransect"

> mean(s1)

[1] 1.777778
```

It is important to note that even though s1 is an object of class trapTransect, it is still a list. We say that objects of class trapTransect *inherit* the characteristics of objects of class list.

```
> is.list(s1)

[1] TRUE
```

This means that we can still manipulate the object using its list structure. For example, if we would like to know about the first element, we can examine it using

```
> s1[[1]]

[1] 1 2 3 4
```

We continue the development of this example in Section 21.4.2.

Creating all of this infrastructure seems like a lot of work. The advantages become apparent when we need to construct, manipulate, and analyse models of complex systems. OOP supports easy and rapid prototyping, and provides a pathway for adding complexity as complexity becomes necessary. For example, it is straightforward to drop in a new `mean.trapTransect` function should we deem it necessary.

There is a final consideration. It is possible that functions provided by different packages could be incompatible, in the sense that they have the same name but different effects. One solution is the use of namespaces (not covered here, but see L. Tierney, 2003. 'Name Space Management for R', R-news 3/1 2–5). Accessing a specific object in a namespace requires the use of explicit calls such as `package::object`, see `?::` for more information.

8.5 Object-oriented programming: S4

In this section we briefly cover the infrastructure that is provided by S4 classes. S4 classes provide a formal object-method framework for object-oriented programming. Here we reinvent the seed trap example of Section 8.4 using S4 classes, with an explanatory commentary. We assume that the structure of the class and objects will remain as above.

First, we tell R about the class itself, using the `setClass` function, which takes the proposed class name and the proposed structure of the objects of the class as arguments.

```
> setClass("trapTransect",
+          representation(distances = "numeric",
+                         seed.counts = "numeric",
+                         trap.area = "numeric"))

[1] "trapTransect"
```

Writing an object constructor is a little more involved than for S3 classes. The constructor function is called `new`, but if we wish to do any processing of the arguments, including validity checks, then we need to add a specific `initialize` function, which will be called by `new`.

```
> setMethod("initialize",
+           "trapTransect",
+           function(.Object,
+                    distances = numeric(0),
```

```
+                   seed.counts = numeric(0),
+                   trap.area = numeric(0)) {
+           if (length(distances) != length(seed.counts))
+             stop("Lengths of distances and counts differ.")
+           if (length(trap.area) != 1)
+             stop("Ambiguous trap area.")
+           .Object@distances <- distances
+           .Object@seed.counts <- seed.counts
+           .Object@trap.area <- trap.area
+           .Object
+         })
```

```
[1] "initialize"
```

`new` creates an empty object and passes it to `initialize`, along with the arguments that were provided to it. `initialize` then returns the updated object, if the evaluations are successful.

```
> s1 <- new("trapTransect",
+           distances = 1:4,
+           seed.counts = c(4, 3, 2, 0),
+           trap.area = 0.0001)
```

Objects from S4 classes differ from objects of S3 classes in a few important ways. The elements that comprise the object, as defined in the `setClass` function, are called *slots*. The names of the slots can be found by

```
> slotNames(s1)
```

```
[1] "distances"   "seed.counts" "trap.area"
```

The values in the slots are accessed by either the `slot` function or the "`@`" operator, which takes the place of the `$` operator used previously.

```
> s1@distances
```

```
[1] 1 2 3 4
```

We now add two methods for the class: `show`, to print objects of the class when just the object name is input, and `mean`, to compute and return the mean seed distance from the object. In each case we use the `setMethod` function, which requires the method name, the pattern of expected formal arguments (called the *signature*), and the function itself.

```
> setMethod("mean",
+           signature(x = "trapTransect"),
+           function(x, ...) weighted.mean(x@distances,
+                                          w = x@seed.counts))
```

```
[1] "mean"
```

```
> setMethod("show",
+           signature(object = "trapTransect"),
+           function(object) str(object))

[1] "show"
```

We demonstrate the application of the new methods to the object.

```
> s1

Formal class 'trapTransect' [package ".GlobalEnv"] with 3 slots
  ..@ distances   : int [1:4] 1 2 3 4
  ..@ seed.counts: num [1:4] 4 3 2 0
  ..@ trap.area   : num 1e-04

> mean(s1)

[1] 1.777778
```

We list the S4 methods for the `trapTransect` class by

```
> showMethods(classes = "trapTransect")

Function: initialize (package methods)
.Object="trapTransect"

Function: mean (package base)
x="trapTransect"

Function: show (package methods)
object="trapTransect"
```

To display the code for a particular S4 method, we use

```
> getMethod("mean", "trapTransect")

Method Definition:

function (x, ...)
weighted.mean(x@distances, w = x@seed.counts)

Signatures:
        x
target  "trapTransect"
defined "trapTransect"
```

See `?Classes` and `?Methods` for more information.

8.6 Compiled code

R can provide straightforward links to any compiled C or FORTRAN code that has been appropriately structured. The ease with which R calls and executes such code is first-rate compensation for any disadvantage in speed compared with its competition. Some operations that take considerable time in R can be executed vastly more quickly in C or FORTRAN, sometimes three orders of magnitude more quickly. More details and examples can be found in 'S Programming' by W.N. Venables and B.D. Ripley (Springer, 2000), and in the 'Writing R Extensions' manual, which is installed with R by default, and is also available on CRAN.

There are four steps required to link compiled C code to R:

8.6.1 Writing

We brush over the details of writing C code. It is beyond the scope of this book to provide any great detail on this topic. Instead, we provide a simple example, and point out the important elements of working in C with R. We write a function to sum the elements of a vector in C, and call the function from R. Note that we do not advocate this function as a replacement for the built-in `sum` function, but present it as an instructive example.

Here's the example written in C:

```
void csum(double *data, int *ndata, double *sum)
{
  int i;
  sum[0] = 0;
  for (i = 0; i < *ndata; i++) {
      sum[0] += data[i];
  }
}
```

The compiled code has to be modular, and written so that all communication is through the passed arguments, which must be pointers to the objects. In practical terms this means that the arguments must be declared as pointers in the function definition, and that care must be taken in the code so that the pointers are handled correctly. In C, the code must be always be a function of type void. Also note the following points:

- C indexes vectors differently than R: the first element of an array is element number 0, *not* element 1.
- We pass the `data` object and the object length via the arguments, rather than determining the length within C. In general, we try to do in R what we can do well in R.

8.6.2 Compiling

This code is subsequently compiled into a binary object that can be called from R. The nature and specificity of the binary object depends on the operating system upon which it was compiled; unlike R code, such objects usually cannot be shared across operating systems. We will refer to the binary object as a *shared library*. As far as we are concerned, it is a binary blob that R can communicate with. For the following development, we will restrict ourselves to the Windows environment.

Compilation is straightforward if the appropriate software is available. For Windows the necessary software is the same as that required for package construction (Section 8.1.2), namely Perl, command line tools and MinGW compilers. We create the program above as a text file using a text editor, and name it something useful, for example `sum.c`. We then open the built-in command line tool, traverse to the directory that contains the file, and type:

```
> R CMD SHLIB csum.c
```

The MinGW compiler will compile and link the code, and create an object in the same directory called `sum.dll`. The next step is to link the object to R (see below). If compilation fails, then the compiler will present reasonably helpful error messages.

8.6.3 Attaching

In R, we load the binary object using the `dyn.load` function, and write a function to call it, as follows:

```
> mySum <- function(data) {
+   if (!(is.loaded("csum"))) dyn.load("../src/csum.dll")
+   .C("csum",
+      as.double(data),
+      as.integer(length(data)),
+      sum = double(1))$sum
+ }
```

This code tells R where to find the shared library, and what to do with it. Note that in the declaration of the function we name the C subroutine and the arguments. We tell R that when we call `mySum(data)`, `data` will point to a double-precision array, that R should compute the length of `data`, that the length should be of mode integer, and that sum is an empty, double-precision array of size one for the output. Then the `$sum` at the end tells R to return that array, hopefully no longer empty.

8.6.4 Call

We can now call the function like any other:

```
> mySum(1)

[1] 1

> mySum(0.5^(1:1000))

[1] 1
```

In practice linking compiled C code into R can save a lot of time if you need to make intensive numerical calculations involving lots of loops. By combining C with R it is possible to use each for what it is best at: R for manipulating objects interactively and C for heavy-duty number crunching (in particular if vectorisation within R is not feasible). As an example, some time ago the third author was working on calculating signed-rank equivalence tests (S. Wellek, Testing statistical hypotheses of equivalence, Chapman & Hall, 2003). Each test required a triple loop over the data, which in this case comprised many thousands of data points, requiring approximately 27 trillion operations to compute the test statistic. The entire run took three hours in C, but was estimated to have gone for a year in R.

8.7 Further reading

For more information on programming in R you should consult the documentation included with the R distribution: 'An Introduction to R' (R-intro), 'Writing R Extensions' (R-exts), 'R Data Import/Export' (R-data), 'The R Language Definition' (R-lang), and 'R Installation and Administration' (R-admin). These documents are invaluable in describing the contemporary requirements and functionality for R.

There are also numerous books about R (and S) available. Of particular note relating to this chapter is 'S Programming' by W.N. Venables and B.D. Ripley (Springer, 2000). Also, for the use of R, we recommend 'Modern Applied Statistics with S. Fourth Edition' by W.N. Venables and B.D. Ripley (Springer, 2002), 'Statistical Models in S', edited by J. Chambers and T. Hastie (Wadsworth & Brooks/Cole, 1992), and 'Data Analysis and Graphics Using R: An Example-Based Approach. Second Edition' by J. Maindonald and J. Braun (Cambridge University Press, 2006).

8.8 Exercises

1. Student records.

Create an S3 class `studentRecord` for objects that are a list with the named elements 'name', 'subjects completed', 'grades', and 'credit'.

Write a `studentRecord` method for the generic function `mean`, which returns a weighted GPA, with subjects weighted by credit. Also write a `studentRecord` method for `print`, which employs some nice formatting, perhaps arranging subjects by year code.

Finally create a further class for a cohort of students, and write methods for `mean` and `print` which, when applied to a cohort, apply `mean` or `print` to each student in the cohort.

2. Let `Omega` be an ordered vector of numbers and define a subset of `Omega` to be an ordered subvector. Define a class `set` for subsets of `Omega` and write functions that perform union, intersection, and complementation on objects of class `set`.

   Do not use R's built-in functions `union`, `intersect`, `setdiff`, or `setequal`.

3. Continued fractions.

   A continued fraction has the form

$$a_0 + \cfrac{1}{a_1 + \cfrac{1}{a_2 + \cfrac{1}{a_3 + \cfrac{1}{\ddots}}}}.$$

   The representation is finite if all the a_k are zero for $k \geq k_0$.

   Write a class for continued fractions (with a finite representation). Now write functions to convert numbers from decimal to continued fraction form, and back.

PART II

Numerical techniques

CHAPTER 9

Numerical accuracy and program efficiency

When using a computer to perform intensive numerical calculations, there are two important issues we should bear in mind: accuracy and speed.

In this chapter we will consider technical details about how computers operate, and their ramifications for programming practice, particularly within R. We look at how computers represent numbers, and the effect that this has on the accuracy of computation results. We also discuss the time it takes to perform a computation and programming techniques for speeding things up. Finally we consider the effects of memory limitations on computation efficiency.

9.1 Machine representation of numbers

Computers use switches to encode information. A single ON/OFF indicator is called a bit; a group of eight bits is called a byte. Although it is quite arbitrary, it is usual to associate 1 with ON and 0 with OFF.

9.1.1 Integers

A fixed number of bytes is used to represent a single integer, usually four or eight. Let k be the number of bits we have to work with (usually 32 or 64). There are a number of schema used to encode integers. We describe three of them: the *sign-and-magnitude*, *biased*, and *two's complement* schema.

In the sign-and-magnitude scheme, we use one bit to represent the sign $+/-$ and the remainder to give a binary representation of the magnitude. Using the sequence $\pm b_{k-2}\cdots b_2 b_1 b_0$, where each b_i is 0 or 1, we represent the decimal number $\pm(2^0 b_0 + 2^1 b_1 + 2^2 b_2 + \cdots + 2^{k-2} b_{k-2})$. For example, taking $k = 8$, -0100101 is interpreted as $-(2^5 + 2^2 + 2^0) = -37$. The smallest and largest integers we can represent under this scheme are $-(2^{k-1} - 1)$ and $2^{k-1} - 1$.

The disadvantage of this scheme is that there are two representations of 0.

In the biased scheme, we use the sequence $b_{k-1}\cdots b_1 b_0$ to represent the decimal number $2^0 b_0 + 2^1 b_1 + \cdots 2^{k-1} b_{k-1} - (2^{k-1} - 1)$. For example, taking $k = 8$,

00100101 is interpreted as $37 - 255 = -218$. The smallest and largest integers represented under this scheme are $-(2^{k-1} - 1)$ and 2^{k-1}.

The disadvantage of this scheme is that addition becomes a little more complex.

The most popular scheme for representing integers on a computer is the *two's complement* scheme. Given k bits, the numbers $0, 1, \ldots, 2^{k-1} - 1$ are represented in the usual way, using a binary expansion, but the numbers $-1, -2, \ldots, -2^{k-1}$ are represented by $2^k - 1, 2^k - 2, \ldots, 2^k - 2^{k-1}$. We will not go into the details, but it turns out that addition under this scheme is equivalent to addition modulo 2^k, and can be implemented very efficiently.

The representation of integers on your computer happens at a fundamental level, and R has no control over it. The largest integer you can represent on your computer (whatever encoding scheme is in use) is known as *maxint*; R records the value of maxint on your computer in the variable `.Machine`.

```
> .Machine$integer.max
[1] 2147483647
```

If you know a number is integer valued then it is efficient to store it as such. However in practice R almost always treats integers the same way it does real numbers, for which it uses *floating point representation.*

9.1.2 Real numbers

Floating point representation is based on binary scientific notation. In decimal scientific notation, we write $x = d_0.d_1d_2 \cdots \times 10^m$, where $d_0, d_1, \ldots$ are digits, with $d_0 \neq 0$ unless $x = 0$. In binary scientific notation, we write $x = b_0.b_1b_2 \cdots \times 2^m$, where $b_0, b_1, \ldots$ are all 0 or 1, with $b_0 = 1$ unless $x = 0$. The sequence $d_0.d_1d_2 \cdots$ or $b_0.b_1b_2 \cdots$ is called the mantissa and m the exponent.

R can use scientific `e` notation to represent numbers:

```
> 1.2e3
[1] 1200
```

The `e` should be read as 'ten raised to the power' and should not be confused with the exponential.

In practice we must limit the size of the mantissa and the exponent; that is, we limit the precision and range of the real numbers we work with. In *double precision* eight bytes are used to represent floating point numbers: 1 bit is used for the sign, 52 bits for the mantissa, and 11 bits for the exponent. The biased scheme is used to represent the exponent, which thus takes on values from -1023 to 1024. For the mantissa, 52 bits are used for $b_1, \ldots, b_{52}$ while the value of b_0 depends on m:

- If $m = -1023$ then $b_0 = 0$, which allows us to represent 0, using $b_1 = \cdots = b_{52} = 0$, or numbers smaller in size than 2^{-1023} otherwise (these are called denormalised numbers).
- If $-1023 < m < 1024$ then $b_0 = 1$.
- If $m = 1024$ then we use $b_1 = \cdots = b_{52} = 0$ to represent $\pm\infty$, which R writes as `-Inf` and `+Inf`. If one of the $b_i \neq 0$ then we interpret the representation as `NaN`, which stands for Not a Number.

```
> 1/0

[1] Inf

> 0/0

[1] NaN
```

In double precision, the smallest non-zero positive number is 2^{-1074} and the largest number is $2^{1023}(2 - 2^{-53})$ (sometimes called realmax). More importantly, the smallest number x such that $1 + x$ can be distinguished from 1 is $2^{-52} \approx 2.220446 \times 10^{-16}$, which is called *machine epsilon*. Thus, in base 10, double precision is roughly equivalent to 16 significant figures, with exponents of size up to ± 308.

```
> 2^-1074 == 0

[1] FALSE

> 1/(2^-1074)

[1] Inf

> 2^1023 + 2^1022 + 2^1021

[1] 1.572981e+308

> 2^1023 + 2^1022 + 2^1022

[1] Inf

> x <- 1 + 2^-52
> x - 1

[1] 2.220446e-16

> y <- 1 + 2^-53
> y - 1

[1] 0
```

When arithmetic operations on double precision floating point numbers produce a result smaller in magnitude than 2^{-1074} or larger in magnitude than realmax, then the result is 0 or $\pm\infty$, respectively. We call this *underflow* or *overflow*.

The double precision scheme we have described here is part of the IEEE Standard for Binary Floating-Point Arithmetic IEEE 754-1985. This standard is used in practically all contemporary computers, though compliance cannot be guaranteed. The implementation of floating point arithmetic happens at a fundamental level of the computer and is not something R can control. It is something R is aware of however, and the constant `.Machine` will give you details about your local numerical environment.

9.2 Significant digits

Using double precision numbers is roughly equivalent to working with 16 significant digits in base 10. Arithmetic with integers will be exact for values from $-(2^{53}-1)$ to $2^{53}-1$ (roughly -10^{16} to 10^{16}), but as soon as you start using numbers outside this range, or fractions, you can expect to lose some accuracy due to *roundoff error*. For example, 1.1 does not have a finite binary expansion, so in double precision its binary expansion is rounded to $1.00011001100\cdots001$, with an error of roughly 2^{-53}.

To allow for roundoff error when comparing numbers, we can use `all.equal(x, y, tol)`, which returns `TRUE` if `x` and `y` are within `tol` of each other, with default given by the square root of machine epsilon (roughly 10^{-8}).

Let $\tilde{a}$ be an approximation of a, then the *absolute error* is $|\tilde{a}-a|$ and the *relative error* is $|\tilde{a}-a|/a$. Restricting $\tilde{a}$ to 16 significant digits is equivalent to allowing a relative error of 10^{-16}. When adding two approximations we add the absolute errors to get (a bound on) the absolute error of the result. When multiplying two approximations we add the relative errors to get (an approximation of) the relative error of result: suppose ϵ and δ are the (small) relative errors of a and b, then

$$\tilde{a}\tilde{b} = a(1+\epsilon)b(1+\delta) = ab(1+\epsilon+\delta+\epsilon\delta) \approx ab(1+\epsilon+\delta).$$

Suppose we add 1,000 numbers each of size around 1,000,000 with relative errors of up to 10^{-16}. Each thus has an absolute error of up to 10^{-10}, so adding them all we would have a number of size around 1,000,000,000 with an absolute error of up to 10^{-7}. That is, the relative error remains much the same at 10^{-16}. However, things can look very different when you start subtracting numbers of a similar size. For example, consider

$$1,234,567,812,345,678 - 1,234,567,800,000,000 = 12,345,678.$$

If the two numbers on the left-hand side have relative errors of 10^{-16}, then the right-hand side has an absolute error of about 1, which is a relative error of around 10^{-8}: a dramatic loss in accuracy, which we call *catastrophic cancellation* error.

Catastrophic cancellation is an inherent problem when you have a finite number of significant digits. However if you are aware of it, it can sometimes be avoided.

9.2.1 Example: $\sin(x) - x$ *near 0*

Since $\lim_{x\to 0} \sin(x)/x = 1$, we have that $\sin(x) \approx x$ near 0. Thus if we wish to know $\sin(x) - x$ near 0, then we expect catastrophic cancellation to reduce the accuracy of our calculation.

The Taylor expansion of $\sin(x)$ about 0 is $\sum_{n=0}^{\infty}(-1)^n x^{2n+1}/(2n+1)!$, thus

$$\sin(x) - x = \sum_{n=1}^{\infty}(-1)^n \frac{x^{2n+1}}{(2n+1)!}.$$

If we truncate this expansion to N terms, then the error is at most $|x^{2N+1}/(2N+1)!|$ (this can be proved using the fact that the summands oscillate in sign while decreasing in magnitude). Suppose we approximate $\sin(x)-x$ using two terms, namely

$$\sin(x) - x \approx -\frac{x^3}{6} + \frac{x^5}{120} = -\frac{x^3}{6}\left(1 - \frac{x^2}{20}\right).$$

If $|x| < 0.001$ then the error in the approximation is less than $0.001^5/120 < 10^{-17}$ in magnitude. If $|x| < 0.000001$ then the error is less than 10^{-302}. Since this formula does not involve subtraction of similarly sized numbers, it does not suffer from catastrophic cancellation.

```
> x <- 2^-seq(from = 10, to = 40, by = 10)
> x

[1] 9.765625e-04 9.536743e-07 9.313226e-10 9.094947e-13

> sin(x) - x

[1] -1.552204e-10 -1.445250e-19  0.000000e+00  0.000000e+00

> -x^3/6 * (1 - x^2/20)

[1] -1.552204e-10 -1.445603e-19 -1.346323e-28 -1.253861e-37
```

We see that for $x = 2^{-20} \approx 10^{-6}$, catastrophic cancellation when calculating $\sin(x) - x$ naively has resulted in an absolute error of around 10^{-23}, which may sound alright until we reflect that this is a relative error of around 10^{-4}. For $x = 2^{-30}$ the relative error is 1!

9.2.2 Example: range reduction

When approximating $\sin(x)$ using a Taylor expansion about 0, the further x is from 0, the more terms we need in the expansion to get a reasonable approximation. But $\sin(x)$ is periodic, so to avoid this, for large x we can just take k such that $|x - 2k\pi| \leq \pi$, then use $\sin(x) = \sin(x - 2k\pi)$.

Unfortunately this procedure can cause problems because of catastrophic cancellation.

Suppose we start with 16 significant digits. If x is large, say around 10^8, then the absolute error of x will be around 10^{-8} and thus the absolute error of $x - 2k\pi$ will be around 10^{-8}. This means the relative error of $x - 2k\pi$ has increased to (at least) 10^{-8} (more if $x - 2k\pi$ is close to 0).

9.3 Time

Ultimately we measure how efficient a program is by how long it takes to run, which will depend on the language it is written in and the computer it is run on. Also, computers typically are doing a number of things at once, such as playing music, watching the mouse or the mailbox, so the time taken to run a program will also depend on what else the computer is doing at the same time. To measure how many CPU (Computer Processing Unit) seconds are spent evaluating an expression, we use `system.time(expression)`. More generally, the expression `proc.time()` will tell you how much time you have used on the CPU since the current R session began.

```
> system.time(source("primedensity.r"))
   user  system elapsed
   0.08    0.03    0.19
```

The sum of the user and system time gives the CPU seconds spent evaluating the expression. The elapsed time also includes time spent on tasks unrelated to your R session.

In most cases of interest, the time taken to run a program depends on the nature of the inputs. For example, the time taken to sum a vector or sort a vector will clearly depend on the length of the vector, n say. Alternatively we may want a solution to the equation $f(x) = 0$, accurate to within some tolerance ϵ, in which case the time taken will depend on ϵ. Because we cannot test a program with all possible inputs, we need a theoretical method of assessing efficiency, which would depend on n or ϵ, and which will give us a measure of how long the program should take to run. We do this by counting the number of operations executed in running a program, where operations are tasks such as addition, multiplication, logical comparison, variable assignment, and calling built-in functions.

For example, the following program will sum the elements of a vector `x`:

```
S <- 0
for (a in x) S <- S + a
```

Let n be the length of `x`, then when we run this program we carry out n addition operations and $2n + 1$ variable assignments (we assign a value to `a` and to `S` each time we go around the `for` loop).

For a second example, suppose we are using the Taylor series $\sum_{n=1}^{N}(-1)^{n+1}x^n/n$ to approximate $\log(1+x)$, for $0 \le x \le 1$, and we want an error of at most $\pm\epsilon$. It can be shown that the approximation error is no greater in magnitude than the last term in the sum, which suggests the following code:

```
eps <- 1e-12
x <- 0.5

n <- 0
log1x <- 0
while (n == 0 || abs(last.term) > eps) {
  n <- n + 1
  last.term <- (-1)^(n+1)*x^n/n
  log1x <- log1x + last.term
}
```

How many arithmetic operations are performed when running this program? When we go around the loop the n-th time we perform three additions and $2n+3$ multiplications/divisions, noting that x^n requires n multiplications. We loop until $x^n/n < \epsilon$. Putting $x = 1$ we get $n = \lceil 1/\epsilon \rceil$, which is an upper bound on n for all $x \in (0, 1]$. (Here $\lceil 1/\epsilon \rceil$ is the *ceiling* of $1/\epsilon$, obtained by rounding up to the nearest integer.) Thus the total number of additions will be bounded by $3\lceil 1/\epsilon \rceil$ and the total number of multiplications/divisions bounded by

$$\sum_{n=1}^{\lceil 1/\epsilon \rceil}(2n+3) = \lceil 1/\epsilon \rceil(\lceil 1/\epsilon \rceil + 1) + 3\lceil 1/\epsilon \rceil = \lceil 1/\epsilon \rceil^2 + 4\lceil 1/\epsilon \rceil.$$

In this example, a simple modification to the program will improve the efficiency. Change the line `last.term <- (-1)^(n+1)*x^n/n` to

```
last.term <- -last.term*x*(1 - 1/n)
```

We now have just three multiplications/divisions each time we go around the loop, so the total number will be bounded by $3\lceil 1/\epsilon \rceil$. (Multiplying by -1 does not count as a multiplication.)

In practice, if we know the number of operations grows like an^b where n is the problem size (the length of the vector or the inverse tolerance in our examples), then the value of b is *much* more important than the value of a. For this reason rather than count operations exactly, it is usually sufficient to ascertain how fast they grow as a function of n. Let f and g be functions of n, then we say that $f(n) = O(g(n))$ as $n \to \infty$ if $\lim_{n\to\infty} f(n)/g(n) < \infty$, and $f(n) = o(g(n))$ as $n \to \infty$ if $\lim_{n\to\infty} f(n)/g(n) = 0$. Our first example required $O(n)$ operations to sum a vector of length n. In its initial form, our second example required $O(1/\epsilon^2)$ operations to calculate $\log(1+x)$ to within tolerance ϵ.

Some operations take much longer than others. Variable assignment (to an existing variable), addition and subtraction are quick. Multiplication and division take a bit longer, and powers take longer still. Transcendental functions such as sin and log have to be calculated and so take even longer, but not as long as user-defined functions.

As we have already seen in Example 3.3.4, creating or changing the size of a vector (also called redimensioning an array) is relatively slow, which is why, when we know how big a vector is going to be, it is better to initialise it fully grown (but full of zeros) than to increase it incrementally. We can compare the relative speeds using `system.time`:

```
> n <- 10000
> x <- rep(0, n)
> system.time(for (i in 1:n) x[i] <- i^2)

   user  system elapsed
  0.023   0.000   0.024

> x <- c()
> system.time(for (i in 1:n) x[i] <- i^2)

   user  system elapsed
  0.515   0.044   0.621
```

In practice what we do is identify the longest or most important operation in a program, and count how many times it is performed. For example, for numerical integration, root-finding, and optimisation, we are working with a user-defined function f, and we count how many times $f(x)$ is evaluated, for different x. For numerical sorting algorithms (see Exercise 7), we count how many comparisons of the form $x < y$ are made. For advanced uses, the function `Rprof` can be used to capture a lot of information about a program as it runs.

9.4 Loops versus vectors

In R, vector operations are generally faster than equivalent loops. However, if you just count operations there appears to be no reason for this. R is a very high-level language in which it is relatively easy to create and manipulate variables. The price we pay for this flexibility is speed. When you evaluate an expression in R, it is 'translated' into a faster lower-level language before being evaluated, then the result is translated back into R. It is the translation that takes much of the time, and vectorisation saves on the amount of translation required.[1]

For example, take the following code to square each element of x:

```
for (i in 1:length(x)) x[i] <- x[i]^2
```

Each time we evaluate the expression `x[i] <- x[i]^2`, we have to translate `x[i]` into our lower-level language, and then translate the result back. In contrast, to evaluate the expression `x <- x^2`, we translate `x` all at once and then square it, before translating the answer back: all the work takes place in our faster lower-level language.

Many of R's functions are vectorised, which means that if the first argument is a vector, then the output will be a vector of the same length, computed by applying the function elementwise to the input vector. Vectorisation allows for faster executing code that is easier to read. User-defined functions can also be vectorised if they comprise functions that are innately vectorised, or are invoked using one of the `apply` family of functions (see Sections 5.4 and 6.4).

When we have a numerically intensive algorithm that uses lots of loops and cannot be vectorised, then R allows us to encode the algorithm in C or Fortran (which are faster lower-level languages) and then access this as a function. Section 8.6 gives some pointers as to how this is done.

9.4.1 Example: column sums of a matrix

We demonstrate a collection of different approaches to solving the problem of summing across the columns of a matrix. We confine ourselves to R code, and we order the solutions from the least to the most efficient.

```
> big.matrix <- matrix(1:1e+06, nrow = 1000)
> colsums <- rep(NA, dim(big.matrix)[2])
```

We compare

1. A double loop of summations,

```
> system.time({
+     for (i in 1:dim(big.matrix)[2]) {
+         s <- 0
+         for (j in 1:dim(big.matrix)[1]) {
+             s <- s + big.matrix[j, i]
+         }
+         colsums[i] <- s
+     }
+ })
```

[1] This is a rather simplified view of what is going on. Nonetheless, it provides us with a workable cognitive model in which we can express the problem.

```
   user  system elapsed
  1.727   0.000   1.903
```

2. The use of apply,

```
> system.time(colsums <- apply(big.matrix, 2, sum))
```

```
   user  system elapsed
  0.035   0.008   0.044
```

3. A single loop of sums, and

```
> system.time(for (i in 1:dim(big.matrix)[2]) {
+     colsums[i] <- sum(big.matrix[, i])
+ })
```

```
   user  system elapsed
  0.029   0.001   0.030
```

4. Using the dedicated R function:

```
> system.time(colsums <- colSums(big.matrix))
```

```
   user  system elapsed
  0.004   0.000   0.003
```

We note that using apply is not faster than using a for loop. This is because, in R, apply creates its own for loop.

9.5 Memory

Computer memory comes in a variety of forms. For most purposes it is sufficient to think in terms of RAM (random access memory), which is fast, and the hard disk, which is slow.

Variables require memory. By default they are stored in RAM, but if you have too many they will be stored on the hard disk then swapped into RAM when needed, which takes time. Historically RAM was expensive and in short supply, and keeping memory use to a minimum was important if you wanted your programs to run quickly. In the present day however, RAM is relatively cheap, and programmers seldom have to worry about how many variables they use. Moreover, because accessing an existing variable is invariably quicker than recalculating it, it is usual to store commonly used quantities for reuse.

For example, consider the function prime from Example 5.4.1:

```
# program spuRs/resources/scripts/prime.r

prime <- function(n) {
    # returns TRUE if n is prime
    # assumes n is a positive integer
```

```
    if (n == 1) {
        is.prime <- FALSE
    } else if (n == 2) {
        is.prime <- TRUE
    } else {
        is.prime <- TRUE
        m <- 2
        m.max <- sqrt(n)  # only want to calculate this once
        while (is.prime && m <= m.max) {
            if (n %% m == 0) is.prime <- FALSE
            m <- m + 1
        }
    }
    return(is.prime)
}
```

Calculating $\sqrt{n}$ is relatively slow, so we do this once and store the result. An alternative is for the main while loop to start as follows:

```
while (is.prime && m <= sqrt(n))
```

Coding the loop this way would require us to recalculate $\sqrt{n}$ each time we check the loop condition, which is inefficient.

Because R works much more quickly with vectors than loops, it is usual to try to vectorise R programs. This will occasionally produce very large vectors. As soon as a vector (or list) is too large to store in RAM all at once, the speed at which you can use it will drop dramatically. If it is sufficiently large, then it may not be possible to store it at all, in which case you are said to have run out of memory.

R has an absolute limit on the length of a vector of $2^{31} - 1 = 2{,}147{,}483{,}647$ (the result of using signed 32-bit integers to index vectors), however, if you run out of memory it is more likely that the problem is that you have reached the limits of your computing environment. If you find this happening then you will need to break your vectors down into smaller subvectors and deal with each in turn. In extreme cases it may be necessary to save a variable and then delete it from the workspace, using `save` and `rm`, to free up enough memory for your program to keep running.

9.6 Caveat

In this chapter we have considered programming efficiency only from the point of view of code execution. A more useful approach is to consider programming efficiency from the point of view of code creation as well as execution; that is, to include the cost of code development. It may well be that judicious refining can trim an hour off the execution of your code, but if it takes two hours to do so, then perhaps there is no real gain. This is the evaluation that a programmer must make: what are the short-term and long-term benefits of code optimisation against the short-term cost of programming time?

For large projects involving more than one programmer other considerations become important, such as the clarity and stability of your code. That is, can others easily understand what the code does, and does it produce sensible answers no matter what sort of input is provided (even if that is just an informative error message). The practice of systematically developing and maintaining large complicated programs is often referred to as *software engineering*.

9.7 Exercises

1. In single precision four bytes are used to represent a floating point number: 1 bit is used for the sign, 8 for the exponent, and 23 for the mantissa.

   What are the largest and smallest non-zero positive numbers in single precision (including denormalised numbers)?

   In base 10, how many significant digits do you get using single precision?
2. What is the relative error in approximating π by 22/7? What about 355/113?
3. Suppose x and y can be represented without error in double precision. Can the same be said for x^2 and y^2?

   Which would be more accurate, $x^2 - y^2$ or $(x - y)(x + y)$?
4. To calculate $\log(x)$ we use the expansion

   $$\log(1 + x) = x - \frac{x^2}{2} + \frac{x^3}{3} - \frac{x^4}{4} + \cdots.$$

   Truncating to n terms, the error is no greater in magnitude than the last term in the sum. How many terms in the expansion are required to calculate $\log 1.5$ with an error of at most 10^{-16}? How many terms are required to calculate $\log 2$ with an error of at most 10^{-16}?

   Using the fact that $\log 2 = 2 \log \sqrt{2}$, suggest a better way of calculating $\log 2$.
5. The sample variance of a set of observations $x_1, \ldots, x_n$ is given by $S^2 = \sum_{i=1}^n (x_i - \overline{x})^2/(n - 1) = (\sum_{i=1}^n x_i^2 - n\overline{x}^2)/(n - 1)$, where $\overline{x} = \sum_{i=1}^n x_i/n$ is the sample mean.

   Show that the second formula is more efficient (requires fewer operations) but can suffer from catastrophic cancellation. Demonstrate catastrophic cancellation with an example sample of size $n = 2$.
6. Horner's algorithm for evaluating the polynomial $p(x) = a_0 + a_1x + a_2x^2 + \cdots + a_nx^n$ consists of re-expressing it as

   $$a_0 + x(a_1 + x(a_2 + \cdots + x(a_{n-1} + xa_n) \cdots)).$$

How many operations are required to evaluate $p(x)$ in each form?

7. This exercise is based on the problem of sorting a list of numbers, which is one of the classic computing problems. Note that R has an excellent sorting function, `sort(x)`, which we will not be using.

   To judge the effectiveness of a sorting algorithm, we count the number of *comparisons* that are required to sort a vector `x` of length `n`. That is, we count the number of times we evaluate logical expressions of the form `x[i] < x[j]`. The fewer comparisons required, the more efficient the algorithm.

   Selection sort The simplest but least efficient sorting algorithm is selection sort. The selection sort algorithm uses two vectors, an unsorted vector and a sorted vector, where all the elements of the sorted vector are less than or equal to the elements of the unsorted vector. The algorithm proceeds as follows:

   1. Given a vector x, let the initial unsorted vector u be equal to x, and the initial sorted vector s be a vector of length 0.
   2. Find the smallest element of u then remove it from u and add it to the end of s.
   3. If u is not empty then go back to step 2.

   Write an implementation of the selection sort algorithm. To do this you may find it convenient to write a function that returns the *index* of the smallest element of a vector.

   How many comparisons are required to sort a vector of length n using the selection sort algorithm?

   Insertion sort Like selection sort, insertion sort uses an unsorted vector and a sorted vector, moving elements from the unsorted to the sorted vector one at a time. The algorithm is as follows:

   1. Given a vector x, let the initial unsorted vector u be equal to x, and the initial sorted vector s be a vector of length 0.
   2. Remove the last element of u and insert it into s so that s is still sorted.
   3. If u is not empty then go back to step 2.

   Write an implementation of the insertion sort algorithm. To insert an element a into a sorted vector $s = (b_1, \ldots, b_k)$ (as per step 2 above), you do not usually have to look at every element of the vector. Instead, if you start searching from the end, you just need to find the first i such that $a \geq b_i$, then the new sorted vector is $(b_1, \ldots, b_i, a, b_{i+1}, \ldots, b_k)$.

   Because inserting an element into a sorted vector is usually quicker than finding the minimum of a vector, insertion sort is usually quicker than selection sort, but the actual number of comparisons required depends on the initial vector x. What are the worst and best types of vector x, with respect to sorting using insertion sort, and how many comparisons are required in each case?

Bubble sort Bubble sort is quite different from selection sort and insertion sort. It works by repeatedly comparing adjacent elements of the vector $x = (a_1, \ldots, a_n)$, as follows:

1. For $i = 1, \ldots, n-1$, if $a_i > a_{i+1}$ then swap a_i and a_{i+1}.
2. If you did any swaps in step 1, then go back and do step 1 again.

Write an implementation of the bubble sort algorithm and work out the minimum and maximum number of comparisons required to sort a vector of length n.

Bubble sort is not often used in practice. Its main claim to fame is that it does not require an extra vector to store the sorted values. There was a time when the available memory was an important programming consideration, and so people worried about how much storage an algorithm required, and bubble sort is excellent in this regard. However at present computing speed is more of a bottleneck than memory, so people worry more about how many operations an algorithm requires.

If you like bubble sort then you should look up the related algorithm *gnome sort*, which was named after the garden gnomes of Holland and their habit of rearranging flower pots.

Quick sort The quick sort algorithm is (on average) one of the fastest sorting algorithms currently available and is widely used. It was first described by C.A.R. Hoare in 1960. Quick sort uses a 'divide-and-conquer' strategy: it is a recursive algorithm that divides the problem into two smaller (and thus easier) problems. Given a vector $x = (a_1, \ldots, a_n)$, the algorithm works as follows:

1. If $n = 0$ or 1 then x is sorted so stop.
2. If $n > 1$ then split $(a_2, \ldots, a_n)$ into two subvectors, l and g, where l consists of all the elements of x less than a_1, and g consists of all the elements of x greater than a_1 (ties can go in either l or g).
3. Sort l and g. Call the sorted subvectors $(b_1, \ldots, b_i)$ and $(c_1, \ldots, c_j)$, respectively, then the sorted vector x is given by $(b_1, \ldots, b_i, a_1, c_1, \ldots, c_j)$.

Implement the quick sort algorithm using a recursive function.

It can be shown that on average the quick sort algorithm requires $O(n \log n)$ comparisons to sort a vector of length n, though its worst-case performance is $O(n^2)$. Also, it is possible to implement quick sort so that it uses memory efficiently while remaining quick.

Two other sorting algorithms that also require on average $O(n \log n)$ comparisons are *heap sort* and *merge sort*.

8. Use the `system.time` function to compare the programs `primedensity.r` and `primesieve.r`, given in Chapter 5.

9. For $\mathbf{x} = (x_1, \ldots, x_n)^T$ and $\mathbf{y} = (y_1, \ldots, y_n)^T$, the *convolution* of $\mathbf{x}$ and $\mathbf{y}$ is

the vector $\mathbf{z} = (z_1, \ldots, z_{2n})^T$ given by

$$z_k = \sum_{i=\max\{1,k-n\}}^{\min\{k,n\}} x_i \cdot y_{k-i}.$$

Write two programs to convolve a pair of vectors, one using loops and the other using vector operations, then use `system.time` to compare their speed.

10. Use the `system.time` function to compare the relative time that it takes to perform addition, multiplication, powers, and other simple operations. You may wish to perform each more than once!

CHAPTER 10

Root-finding

10.1 Introduction

The next few chapters introduce numerical algorithms for solving some common applied mathematical problems. In each case we present motivating examples, some underpinning theory, and applications in R. This chapter focuses on root-finding, and covers fixed-point iteration, the Newton-Raphson method, the secant method, and the bisection method.

Suppose that $f : \mathbb{R} \to \mathbb{R}$ is a continuous function. A *root* of f is a solution to the equation $f(x) = 0$ (see Figure 10.1 for example). That is, a root is a number $a \in \mathbb{R}$ such that $f(a) = 0$. If we draw the graph of our function, say $y = f(x)$, which is a curve in the plane, a solution of $f(x) = 0$ is the x-coordinate of a point at which the curve crosses the x-axis.

The roots of a function are important algebraically, for example, we use the

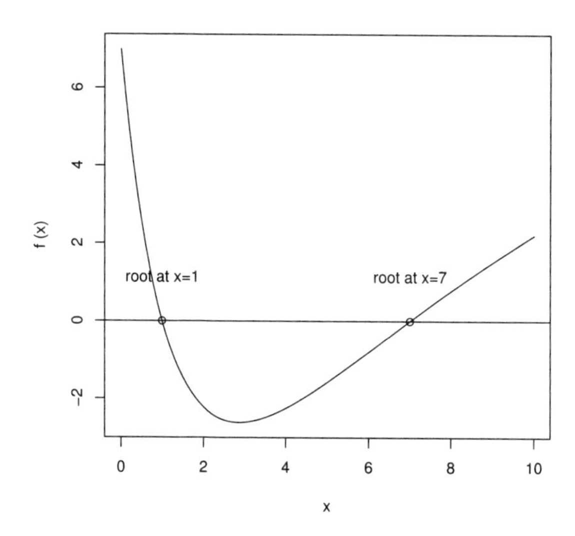

Figure 10.1 *The roots of the function f.*

roots of a polynomial to factorise it. Moreover the solution to a physical problem can often be expressed as the root of a suitable function. Root-finding is also a classical numerical or computational problem, and provides a good introduction to important issues in numerical mathematics.

10.1.1 Example: loan repayments

Suppose that a loan has an initial amount P, a monthly interest rate r, a duration of N months, and a monthly repayment of A. The remaining debt after n months is given by P_n, where

$$\begin{aligned} P_0 &= P; \\ P_{n+1} &= P_n(1+r) - A. \end{aligned}$$

That is, each month you pay interest on the previous balance, then reduce the balance of the loan by amount A. This is a first-order recurrence equation, and has the following solution (check that it works):

$$P_n = P(1+r)^n - A((1+r)^n - 1)/r.$$

Putting $P_N = 0$, we get

$$\frac{A}{P} = \frac{r(1+r)^N}{(1+r)^N - 1}.$$

Suppose that we know P, N, and A, then we can find r by finding the root(s) of the following function:

$$f(x) = \frac{A}{P} - \frac{x(1+x)^N}{(1+x)^N - 1}.$$

Choose some values for P, N, and A and then try finding r analytically. If/when you decide this is too hard, you can try doing it numerically with one of the techniques below.

10.2 Fixed-point iteration

Let $g : \mathbb{R} \to \mathbb{R}$ be a continuous function. A *fixed point* of g is a number a such that $g(a) = a$. That is, a is a solution of the equation $g(x) = x$. Graphically, a fixed point is where the graph $y = g(x)$ of the function crosses the line $y = x$.

The computational problem of finding fixed points of a function is easily reduced to the problem of finding roots. To see this define the function $f(x)$ by the equation $f(x) = c(g(x) - x)$, where c is a non-zero constant, then one clearly has $f(a) = 0$ if and only if $g(a) = a$. So to find the fixed points of the function g we need only find the roots of the associated function f, that is solutions of the equation $f(x) = 0$. Conversely, the problem of finding roots

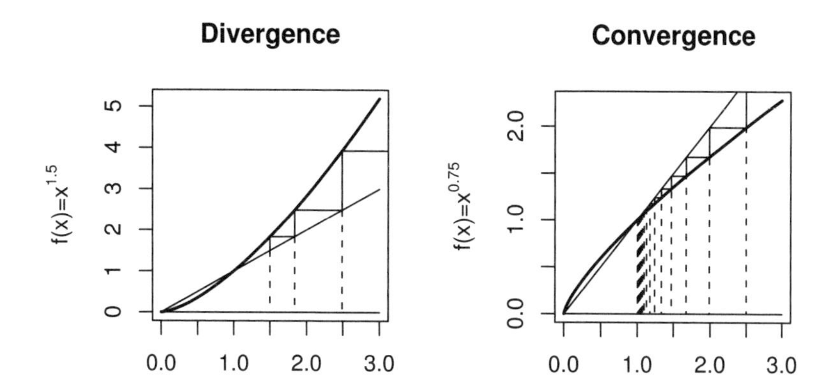

Figure 10.2 *The fixed-point algorithm applied to the function* $y = x^{1.5}$*, starting at* $x_0 = 1.5$*, and to the function* $y = x^{0.75}$*, starting at* $x_0 = 2.5$.

of $f(x) = 0$ is equivalent to the problem of finding fixed points of the function $g(x) = c \cdot f(x) + x$.

Although this is one way to convert from one form of problem to the other, it is not the only way, and in practice some ways are better than others.

The 'fixed-point method' is an iterative method for solving $g(x) = x$. That is, it generates a sequence of points $x_0, x_1, x_2, \ldots$ that (hopefully) converges to some point a such that $g(a) = a$. Starting with our initial guess x_0, we generate the next guess using $x_1 = g(x_0)$ and repeat. This gives the following first-order recurrence relation (also called a difference equation):

Fixed-point method

$$x_{n+1} = g(x_n).$$

If $x_n \to a$ then, given g is continuous, we have

$$a = \lim_{n\to\infty} x_{n+1} = \lim_{n\to\infty} g(x_n) = g(\lim_{n\to\infty} x_n) = g(a).$$

So a is a fixed point of g. But does the sequence $\{x_n\}_{n=0}^{\infty}$ always converge? The answer, sadly, is no.

In Figure 10.2 we illustrate the application of the fixed-point method to the functions $g_1(x) = x^{1.5}$ and $g_2(x) = x^{0.75}$, starting to the right of 1 in each case. The dotted lines give the successive values of x_n. The solid lines show how we obtain x_{n+1} from x_n. Both functions have fixed points at 1, but the algorithm diverges when applied to g_1 and converges when applied to g_2. The crucial difference is the value of g' at the fixed point: if $|g'(a)| < 1$ at the fixed point a, then the algorithm will converge, otherwise it diverges. It is also necessary to start relatively 'close' to the fixed point to guarantee that you will converge to it. We will not prove this result here, though Exercise 2 gives an outline of how it is done. You should be able to convince yourself that it is true by looking at the figure.

Even when the method does converge, we still have a (small) problem: $\{x_n\}_{n=0}^{\infty}$ may converge to a, but never actually reach it. We can never avoid this problem, rather we have to accommodate it. The best we can do is ask for an x_n within distance δ of a, for some small $\delta > 0$.

Practically, to avoid iterating forever, we stop when $|x_n - x_{n-1}| \leq \epsilon$ for some (user-specified) tolerance ϵ. Noting that $g(a) = a$, and that $g(x) - g(a) \approx g'(a)(x - a)$ for x close to a, we have the following:

$$\begin{aligned}
|x_n - x_{n-1}| \leq \epsilon \quad &\Leftrightarrow \quad |g(x_{n-1}) - x_{n-1}| \leq \epsilon \\
&\Leftrightarrow \quad |g(x_{n-1}) - g(a) - (x_{n-1} - a)| \leq \epsilon \\
&\Rightarrow \quad |x_{n-1} - a| \leq \epsilon + |g(x_{n-1}) - g(a)| \approx \epsilon + g'(a)|x_{n-1} - a| \\
&\Rightarrow \quad |x_{n-1} - a| \leq \epsilon/(1 - g'(a)).
\end{aligned}$$

Thus, to ensure $|x_n - a| \leq \delta$ we need to choose $\epsilon \leq \delta(1 - g'(a))$ (approximately). Of course, until we know a we can't find $g'(a)$, so in practice we just choose ϵ to be small.

Note that the fixed-point method can still be used if g' does not exist, provided g is continuous. However, the convergence properties of the method are harder to describe in such a case.

The code below implements the fixed-point algorithm in a function `fixedpoint`. To use it you first need to create a function, `ftn(x)` say, that returns $g(x)$. `fixedpoint(ftn, x0, tol = 1e-9, max.iter = 100)` has four inputs:

`ftn` is the name of a function that takes a single numeric input and returns a single numeric result.

`x0` is the starting point for the algorithm.

`tol` is such that the algorithm will stop if $|x_n - x_{n-1}| \leq$ tol, with default 10^{-9}.

`max.iter` is such that the algorithm will stop when $n =$ max.iter, with default 100.

We remark that, because the fixed-point method is not guaranteed to converge, our coding of the algorithm counts how many iterations have been performed, and stops if they exceed some specified maximum. This prevents the function running on forever.

```
# program spuRs/resources/scripts/fixedpoint.r
# loadable spuRs function

fixedpoint <- function(ftn, x0, tol = 1e-9, max.iter = 100) {
```

```
  # applies the fixed-point algorithm to find x such that ftn(x) == x
  # we assume that ftn is a function of a single variable
  #
  # x0 is the initial guess at the fixed point
  # the algorithm terminates when successive iterations are
  # within distance tol of each other,
  # or the number of iterations exceeds max.iter

  # do first iteration
  xold <- x0
  xnew <- ftn(xold)
  iter <- 1
  cat("At iteration 1 value of x is:", xnew, "\n")

  # continue iterating until stopping conditions are met
  while ((abs(xnew-xold) > tol) && (iter < max.iter)) {
    xold <- xnew;
    xnew <- ftn(xold);
    iter <- iter + 1
    cat("At iteration", iter, "value of x is:", xnew, "\n")
  }

  # output depends on success of algorithm
  if (abs(xnew-xold) > tol) {
    cat("Algorithm failed to converge\n")
    return(NULL)
  } else {
    cat("Algorithm converged\n")
    return(xnew)
  }
}
```

10.2.1 Example: finding the root of $f(x) = \log(x) - \exp(-x)$

We consider three approaches to solving the equation $f(x) = \log(x) - \exp(-x) = 0$. First, we put one term on each side of the equation and exponentiate both sides to get

$$x = \exp(\exp(-x)) = g_1(x).$$

Second, we subtract each side from x to get

$$x = x - \log x + \exp(-x) = g_2(x).$$

Finally, we add x to both sides to get

$$x = x + \log x - \exp(-x) = g_3(x).$$

1. Applying the fixed-point method to g_1, we find the sequence appears to con-

verge but it takes 14 iterations for successive guesses to agree to 6 decimal places.

```
> source("../scripts/fixedpoint.r")
> ftn1 <- function(x) return(exp(exp(-x)))
> fixedpoint(ftn1, 2, tol = 1e-06)
```

```
At iteration 1 value of x is: 1.144921
At iteration 2 value of x is: 1.374719
At iteration 3 value of x is: 1.287768
At iteration 4 value of x is: 1.317697
At iteration 5 value of x is: 1.307022
At iteration 6 value of x is: 1.310783
At iteration 7 value of x is: 1.309452
At iteration 8 value of x is: 1.309922
At iteration 9 value of x is: 1.309756
At iteration 10 value of x is: 1.309815
At iteration 11 value of x is: 1.309794
At iteration 12 value of x is: 1.309802
At iteration 13 value of x is: 1.309799
At iteration 14 value of x is: 1.309800
Algorithm converged
[1] 1.309800
```

2. Using g_2, we find the sequence appears to converge and it takes only 6 iterations.

```
> ftn2 <- function(x) return(x - log(x) + exp(-x))
> fixedpoint(ftn2, 2, tol = 1e-06)
```

```
At iteration 1 value of x is: 1.442188
At iteration 2 value of x is: 1.312437
At iteration 3 value of x is: 1.309715
At iteration 4 value of x is: 1.309802
At iteration 5 value of x is: 1.309799
At iteration 6 value of x is: 1.309800
Algorithm converged
[1] 1.309800
```

3. Using g_3, we find the sequence does not appear to converge at all.

```
> ftn3 <- function(x) return(x + log(x) - exp(-x))
> fixedpoint(ftn3, 2, tol = 1e-06, max.iter = 20)
```

```
At iteration 1 value of x is: 2.557812
At iteration 2 value of x is: 3.41949
At iteration 3 value of x is: 4.616252
At iteration 4 value of x is: 6.135946
At iteration 5 value of x is: 7.947946
At iteration 6 value of x is: 10.02051
At iteration 7 value of x is: 12.32510
At iteration 8 value of x is: 14.83673
At iteration 9 value of x is: 17.53383
```

```
At iteration 10 value of x is: 20.39797
At iteration 11 value of x is: 23.4134
At iteration 12 value of x is: 26.56671
At iteration 13 value of x is: 29.84637
At iteration 14 value of x is: 33.24243
At iteration 15 value of x is: 36.74626
At iteration 16 value of x is: 40.35030
At iteration 17 value of x is: 44.04789
At iteration 18 value of x is: 47.83317
At iteration 19 value of x is: 51.70089
At iteration 20 value of x is: 55.64637
Algorithm failed to converge
NULL
```

This example illustrates that as a method for finding roots, the fixed-point method has some disadvantages. One needs to convert the problem into fixed-point form, but there are many ways to do this, each of which will have different convergence properties and some of which will not converge at all. We consider the question of the best way of converting a root-finding problem to a fixed-point problem in Exercise 7.

It also turns out that the fixed-point method is relatively slow, in that the error is usually divided by a constant factor at each iteration. Both of our next two algorithms, the Newton-Raphson method and the secant method, converge more quickly because they make informed guesses as to where to find a better approximation to the root.

10.3 The Newton-Raphson method

Suppose our function f is differentiable with continuous derivative f' and a root a. Let $x_0 \in \mathbb{R}$ and think of x_0 as our current 'guess' at a. Now the straight line through the point $(x_0, f(x_0))$ with slope $f'(x_0)$ is the best straight line approximation to the function $f(x)$ at the point x_0 (this is the *meaning* of the derivative). The equation of this straight line is given by

$$f'(x_0) = \frac{f(x_0) - y}{x_0 - x}.$$

Now this straight line crosses the x-axis at a point x_1, which should be a better approximation than x_0 to a. To find x_1 we observe

$$f'(x_0) = \frac{f(x_0) - 0}{x_0 - x_1} \quad \text{and so} \quad x_1 = x_0 - \frac{f(x_0)}{f'(x_0)}.$$

In other words, the next best guess x_1 is obtained from the current guess x_0 by subtracting a correction term $f(x_0)/f'(x_0)$ (Figure 10.3).

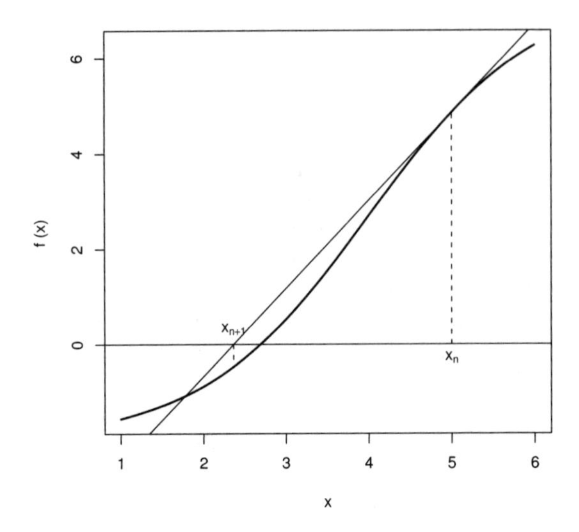

Figure 10.3 *A step in the Newton-Raphson root-finding method.*

Now that we have x_1, we use the same procedure to get the next guess

$$x_2 = x_1 - \frac{f(x_1)}{f'(x_1)}$$

or in general:

Newton-Raphson method

$$x_{n+1} = x_n - \frac{f(x_n)}{f'(x_n)}.$$

Like the fixed-point method, this is a first-order recurrence relation. It can be shown that if f is 'well behaved' at a (which means $f'(a) \neq 0$ and f'' is finite and continuous at a)[1] and you start with x_0 'close enough' to a, then x_n will converge to a quickly. Unfortunately, like the fixed-point method, we don't know if f is well behaved at a until we know a, and we don't know beforehand how close is close enough.

So, we cannot guarantee convergence of the Newton-Raphson algorithm. However, if $x_n \to a$ then, since f and f' are continuous, we have

$$a = \lim_{n\to\infty} x_{n+1} = \lim_{n\to\infty} \left(x_n - \frac{f(x_n)}{f'(x_n)} \right)$$

[1] In fact, we can get away with the slightly less restrictive but more technical condition that $f'(a) \neq 0$ and f' is Lipschitz-continuous in a neighbourhood of a, which means that for some constant c, $|f'(x) - f'(y)| \leq c|x - y|$.

$$= \lim_{n\to\infty} x_n - \frac{f(\lim_{n\to\infty} x_n)}{f'(\lim_{n\to\infty} x_n)} = a - \frac{f(a)}{f'(a)}.$$

Thus, provided $f'(a) \neq \pm\infty$, we must have $f(a) = 0$.

Since we are expecting $f(x_n) \to 0$, a good stopping condition for the Newton-Raphson algorithm is $|f(x_n)| \leq \epsilon$ for some tolerance ϵ. If the sequence $\{x_n\}_{n=0}^{\infty}$ is converging to a root a, then for x close to a we have $f(x) \approx f'(a)(x - a)$. So if $|f(x_n)| \leq \epsilon$ we have $|x - a| \leq \epsilon / f'(a)$ (approximately).

The code below implements the Newton-Raphson algorithm in a function `newtonraphson`. To use it you first need to create a function, `ftn(x)` say, which returns the vector $(f(x), f'(x))$. `newtonraphson(ftn, x0, tol = 1e-9, max.iter = 100)` has four inputs:

`ftn` is the name of a function that takes a single numeric input and returns a numeric vector of length two. If x is the input then the output must be $(f(x), f'(x))$.

`x0` is the starting point for the algorithm.

`tol` is such that the algorithm will stop if $|f(x_n)| \leq$ tol, with default 10^{-9}.

`max.iter` is such that the algorithm will stop when $n =$ max.iter, with default 100.

As for the fixed-point method, because we cannot guarantee convergence, we count the number of iterations and stop if this gets too large. This prevents the program running indefinitely, though of course you have to make sure that you do not stop it too soon, in case it is converging more slowly than you expected. Note that, because our stopping condition only depends on $|f(x_n)|$, and not $|x_n - x_{n-1}|$, we do not have to store the previous iteration, as we did with function `fixedpoint`.

```
# program spuRs/resources/scripts/newtonraphson.r
# loadable spuRs function

newtonraphson <- function(ftn, x0, tol = 1e-9, max.iter = 100) {
  # Newton_Raphson algorithm for solving ftn(x)[1] == 0
  # we assume that ftn is a function of a single variable that returns
  # the function value and the first derivative as a vector of length 2
  #
  # x0 is the initial guess at the root
  # the algorithm terminates when the function value is within distance
  # tol of 0, or the number of iterations exceeds max.iter

  # initialise
  x <- x0
  fx <- ftn(x)
  iter <-  0

  # continue iterating until stopping conditions are met
```

```
  while ((abs(fx[1]) > tol) && (iter < max.iter)) {
    x <- x - fx[1]/fx[2]
    fx <- ftn(x)
    iter <-  iter + 1
    cat("At iteration", iter, "value of x is:", x, "\n")
  }

  # output depends on success of algorithm
  if (abs(fx[1]) > tol) {
    cat("Algorithm failed to converge\n")
    return(NULL)
  } else {
    cat("Algorithm converged\n")
    return(x)
  }
}
```

When applied to the function $\log x - \exp(-x)$ with derivative $1/x + \exp(-x)$, we get impressively fast convergence

```
> source("../scripts/newtonraphson.r")
> ftn4 <- function(x) {
+     fx <- log(x) - exp(-x)
+     dfx <- 1/x + exp(-x)
+     return(c(fx, dfx))
+ }
> newtonraphson(ftn4, 2, 1e-06)

At iteration 1 value of x is: 1.122020
At iteration 2 value of x is: 1.294997
At iteration 3 value of x is: 1.309709
At iteration 4 value of x is: 1.309800
Algorithm converged
[1] 1.309800
```

10.4 The secant method

A problem with the Newton-Raphson algorithm is that it needs the derivative f'. If the derivative is hard to compute or does not exist, then we can use the secant method, which only requires that the function f is continuous.

Like the Newton-Raphson method, the secant method is based on a linear approximation to the function f. Suppose that f has a root at a. For this method we assume that we have two current 'guesses', x_0 and x_1, for the value of a. We will think of x_0 as an older guess and we want to replace the pair x_0, x_1 by the pair x_1, x_2, where x_2 is a new guess.

To find a good new guess x_2 we first draw the straight line from $(x_0, f(x_0))$ to

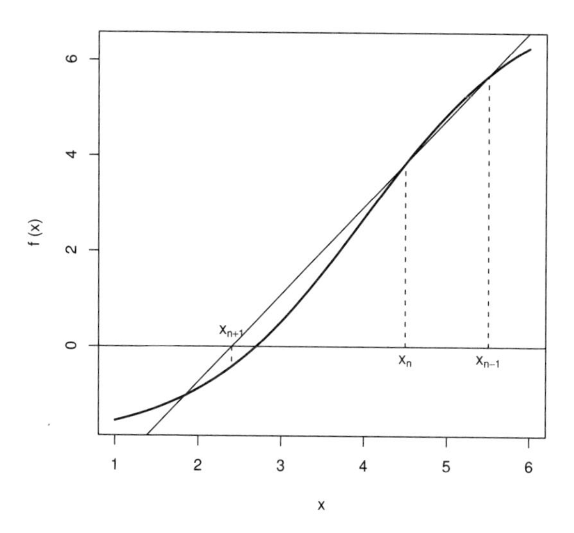

Figure 10.4 *A step in the secant root-finding method.*

$(x_1, f(x_1))$, which is called a secant of the curve $y = f(x)$. Like the tangent, the secant is a linear approximation of the behaviour of $y = f(x)$, in the region of the points x_0 and x_1. As the new guess we will use the x-coordinate x_2 of the point at which the secant crosses the x-axis (Figure 10.4). Now the equation of the secant is given by

$$\frac{y - f(x_1)}{x - x_1} = \frac{f(x_0) - f(x_1)}{x_0 - x_1}$$

and so x_2 can be found from

$$\frac{0 - f(x_1)}{x_2 - x_1} = \frac{f(x_0) - f(x_1)}{x_0 - x_1}$$

which gives

$$x_2 = x_1 - f(x_1)\frac{x_0 - x_1}{f(x_0) - f(x_1)}.$$

Repeating this we get a second-order recurrence relation (each new value depends on the previous two):

Secant method

$$x_{n+1} = x_n - f(x_n)\frac{x_n - x_{n-1}}{f(x_n) - f(x_{n-1})}.$$

Note that if x_n and x_{n-1} are close together, then

$$f'(x_n) \approx \frac{f(x_n) - f(x_{n-1})}{x_n - x_{n-1}}$$

so we can view the secant method as an approximation of the Newton-Raphson method, where we substitute $(f(x_n) - f(x_{n-1}))/(x_n - x_{n-1})$ for $f'(x_n)$.

The convergence properties of the secant method are similar to those of the Newton-Raphson method. If f is well behaved at a and you start with x_0 and x_1 sufficiently close to a, then x_n will converge to a quickly, though not quite as fast as the Newton-Raphson method. As for the Newton-Raphson method, we cannot guarantee convergence. Comparing the secant method to the Newton-Raphson method, we see a trade-off: we no longer need to know f' but in return we give up some speed and have to provide two initial points, x_0 and x_1.

The problem of implementing the secant method appears as Exercise 6.

10.5 The bisection method

The Newton-Raphson and secant root-finding methods work by producing a sequence of guesses to the root and, under favourable circumstances, converge rapidly to the root from an initial guess. Unfortunately they cannot be guaranteed to work. A more reliable but slower approach is root-bracketing, which works by first isolating an interval in which the root must lie, and then successively refining the bounding interval in such a way that the root is guaranteed to always lie inside the interval. The canonical example is the bisection method, in which the width of the bounding interval is successively halved.

Suppose that f is a continuous function, then it is easy to see that f has a root in the interval (x_l, x_r) if either $f(x_l) < 0$ and $f(x_r) > 0$ or $f(x_l) > 0$ and $f(x_r) < 0$. A convenient way to verify this condition is to check if $f(x_l)f(x_r) < 0$. The bisection method works by taking an interval (x_l, x_r) that contains a root, then successively refining x_l and x_r until $x_r - x_l \le \epsilon$, where ϵ is some predefined tolerance. The algorithm is as follows:

Bisection method Start with $x_l < x_r$ such that $f(x_l)f(x_r) < 0$.

1. if $x_r - x_l \le \epsilon$ then stop.
2. put $x_m = (x_l + x_r)/2$; if $f(x_m) = 0$ then stop.
3. if $f(x_l)f(x_m) < 0$ then put $x_r = x_m$ otherwise put $x_l = x_m$.
4. go back to step 1.

Note that at every iteration of the algorithm, we know that there is root in the interval (x_l, x_r). Provided we start with $f(x_l)f(x_r) < 0$, the algorithm is guaranteed to converge, with the approximation error reducing by a constant factor $1/2$ at each iteration. If we stop when $x_r - x_l \le \epsilon$, then we know that both x_l and x_r are within distance ϵ of a root.

Note that the bisection method cannot find a root a if the function f just touches the x-axis at a, that is, if the x-axis is a tangent to the function at a. The Newton-Raphson method will still work in this case. The most popular current root-finding methods use root-bracketing to get close to a root, then switch over to the Newton-Raphson or secant method when it seems safe to do so. This strategy combines the safety of bisection with the speed of the secant method.

Here is an implementation of the bisection method in R. Because this algorithm makes certain assumptions about x_l and x_r, we check that these assumptions hold before the algorithm runs. Also, because the algorithm is guaranteed to converge (provided the initial conditions are met), we do not need to put a bound on the maximum number of iterations. Note that the code has a number of `return` statements. Recall that a function terminates the first time a `return` is executed. In this function, if we detect a problem with the inputs then we print an error message and immediately `return(NULL)`, so that the remainder of the function is not executed.

In Exercise 12 you are asked to generalise `bisection` so that it can deal with the case $f(x_l)f(x_r) > 0$.

```
# program spuRs/resources/scripts/bisection.r
# loadable spuRs function

bisection <- function(ftn, x.l, x.r, tol = 1e-9) {
  # applies the bisection algorithm to find x such that ftn(x) == 0
  # we assume that ftn is a function of a single variable
  #
  # x.l and x.r must bracket the fixed point, that is
  # x.l < x.r and ftn(x.l) * ftn(x.r) < 0
  #
  # the algorithm iteratively refines x.l and x.r and terminates when
  # x.r - x.l <= tol

  # check inputs
  if (x.l >= x.r) {
    cat("error: x.l >= x.r \n")
    return(NULL)
  }
  f.l <- ftn(x.l)
  f.r <- ftn(x.r)
  if (f.l == 0) {
    return(x.l)
  } else if (f.r == 0) {
    return(x.r)
  } else if (f.l * f.r > 0) {
    cat("error: ftn(x.l) * ftn(x.r) > 0 \n")
    return(NULL)
  }
```

```
  # successively refine x.l and x.r
  n <- 0
  while ((x.r - x.l) > tol) {
    x.m <- (x.l + x.r)/2
    f.m <- ftn(x.m)
    if (f.m == 0) {
      return(x.m)
    } else if (f.l * f.m < 0) {
      x.r <- x.m
      f.r <- f.m
    } else {
      x.l <- x.m
      f.l <- f.m
    }
    n <- n + 1
    cat("at iteration", n, "the root lies between", x.l, "and", x.r, "\n")
  }

  # return (approximate) root
  return((x.l + x.r)/2)
}
```

Here it is in action. Observe how slow the method is compared to the Newton-Raphson method.

```
> source("../scripts/bisection.r")
> ftn5 <- function(x) return(log(x) - exp(-x))
> bisection(ftn5, 1, 2, tol = 1e-06)

at iteration 1 the root lies between 1 and 1.5
at iteration 2 the root lies between 1.25 and 1.5
at iteration 3 the root lies between 1.25 and 1.375
at iteration 4 the root lies between 1.25 and 1.3125
at iteration 5 the root lies between 1.28125 and 1.3125
at iteration 6 the root lies between 1.296875 and 1.3125
at iteration 7 the root lies between 1.304688 and 1.3125
at iteration 8 the root lies between 1.308594 and 1.3125
at iteration 9 the root lies between 1.308594 and 1.310547
at iteration 10 the root lies between 1.309570 and 1.310547
at iteration 11 the root lies between 1.309570 and 1.310059
at iteration 12 the root lies between 1.309570 and 1.309814
at iteration 13 the root lies between 1.309692 and 1.309814
at iteration 14 the root lies between 1.309753 and 1.309814
at iteration 15 the root lies between 1.309784 and 1.309814
at iteration 16 the root lies between 1.309799 and 1.309814
at iteration 17 the root lies between 1.309799 and 1.309807
at iteration 18 the root lies between 1.309799 and 1.309803
at iteration 19 the root lies between 1.309799 and 1.309801
at iteration 20 the root lies between 1.309799 and 1.309800
[1] 1.309800
```

10.6 Exercises

1. Draw a function $g(x)$ for which the fixed-point algorithm produces the oscillating sequence $1, 7, 1, 7, \ldots$ when started with $x_0 = 7$.

2. (a). Suppose that $x_0 = 1$ and that for $n \geq 0$

$$x_{n+1} = \alpha x_n.$$

Find a formula for x_n. For which values of α does x_n converge, and to what?

(b). Consider the fixed-point algorithm for finding x such that $g(x) = x$:

$$x_{n+1} = g(x_n).$$

Let c be the fixed point, so $g(c) = c$. The first-order Taylor approximation of g about the point c is

$$g(x) \approx g(c) + (x - c)g'(c).$$

Apply this Taylor approximation to the fixed-point algorithm to give a recurrence relation for $x_n - c$.

What condition on the function g at the point c will result in the convergence of x_n to c?

3. Use `fixedpoint` to find the fixed point of $\cos x$. Start with $x_0 = 0$ as your initial guess (the answer is 0.73908513).

Now use `newtonraphson` to find the root of $\cos x - x$, starting with $x_0 = 0$ as your initial guess. Is it faster than the fixed-point method?

4. A picture is worth a thousand words.

The function `fixedpoint_show.r` below is a modification of `fixedpoint` that plots intermediate results. Instead of using the variables `tol` and `max.iter` to determine when the algorithm stops, at each step you will be prompted to enter `"y"` at the keyboard if you want to continue. There are also two new inputs, `xmin` and `xmax`, which are used to determine the range of the plot. `xmin` and `xmax` have defaults `x0 - 1` and `x0 + 1`, respectively.

```
# program spuRs/resources/scripts/fixedpoint_show.r
# loadable spuRs function

fixedpoint_show <- function(ftn, x0, xmin = x0-1, xmax = x0+1) {
  # applies fixed-point method to find x such that ftn(x) == x
  # x0 is the starting point
  # subsequent iterations are plotted in the range [xmin, xmax]

  # plot the function
  x <- seq(xmin, xmax, (xmax - xmin)/200)
  fx <- sapply(x, ftn)
  plot(x, fx, type = "l", xlab = "x", ylab = "f(x)",
```

```
    main = "fixed point f(x) = x", col = "blue", lwd = 2)
  lines(c(xmin, xmax), c(xmin, xmax), col = "blue")

  # do first iteration
  xold <- x0
  xnew <- ftn(xold)
  lines(c(xold, xold, xnew), c(xold, xnew, xnew), col = "red")
  lines(c(xnew, xnew), c(xnew, 0), lty = 2, col = "red")

  # continue iterating while user types "y"
  cat("last x value", xnew, " ")
  continue <- readline("continue (y or n)? ") == "y"
  while (continue) {
    xold <- xnew;
    xnew <- ftn(xold);
    lines(c(xold, xold, xnew), c(xold, xnew, xnew), col = "red")
    lines(c(xnew, xnew), c(xnew, 0), lty = 2, col = "red")
    cat("last x value", xnew, " ")
    continue <- readline("continue (y or n)? ") == "y"
  }

  return(xnew)
}
```

Use `fixedpoint_show` to investigate the fixed points of the following functions:

(a). $\cos(x)$ using $x_0 = 1, 3, 6$

(b). $\exp(\exp(-x))$ using $x_0 = 2$

(c). $x - \log(x) + \exp(-x)$ using $x_0 = 2$

(d). $x + \log(x) - \exp(-x)$ using $x_0 = 2$

5. Below is a modification of `newtonraphson` that plots intermediate results, analogous to `fixedpoint_show` above. Use it to investigate the roots of the following functions:

(a). $\cos(x) - x$ using $x_0 = 1, 3, 6$

(b). $\log(x) - \exp(-x)$ using $x_0 = 2$

(c). $x^3 - x - 3$ using $x_0 = 0$

(d). $x^3 - 7x^2 + 14x - 8$ using $x_0 = 1.1, 1.2, \ldots, 1.9$

(e). $\log(x)\exp(-x)$ using $x_0 = 2$.

```
# program spuRs/resources/scripts/newtonraphson_show.r
# loadable spuRs function

newtonraphson_show <- function(ftn, x0, xmin = x0-1, xmax = x0+1) {
  # applies Newton-Raphson to find x such that ftn(x)[1] == 0
  # x0 is the starting point
  # subsequent iterations are plotted in the range [xmin, xmax]
```

```
  # plot the function
  x <- seq(xmin, xmax, (xmax - xmin)/200)
  fx <- c()
  for (i in 1:length(x)) {
    fx[i] <- ftn(x[i])[1]
  }
  plot(x, fx, type = "l", xlab = "x", ylab = "f(x)",
    main = "zero f(x) = 0", col = "blue", lwd = 2)
  lines(c(xmin, xmax), c(0, 0), col = "blue")

  # do first iteration
  xold <- x0
  f.xold <- ftn(xold)
  xnew <- xold - f.xold[1]/f.xold[2]
  lines(c(xold, xold, xnew), c(0, f.xold[1], 0), col = "red")

  # continue iterating while user types "y"
  cat("last x value", xnew, " ")
  continue <- readline("continue (y or n)? ") == "y"
  while (continue) {
    xold <- xnew;
    f.xold <- ftn(xold)
    xnew <- xold - f.xold[1]/f.xold[2]
    lines(c(xold, xold, xnew), c(0, f.xold[1], 0), col = "red")
    cat("last x value", xnew, " ")
    continue <- readline("continue (y or n)? ") == "y"
  }

  return(xnew)
}
```

6. Write a program, using both `newtonraphson.r` and `fixedpoint.r` for guidance, to implement the secant root-finding method:

$$x_{n+1} = x_n - f(x_n)\frac{x_n - x_{n-1}}{f(x_n) - f(x_{n-1})}.$$

First test your program by finding the root of the function $\cos x - x$. Next see how the secant method performs in finding the root of $\log x - \exp(-x)$ using $x_0 = 1$ and $x_1 = 2$. Compare its performance with that of the other two methods.

Write a function `secant_show.r` that plots the sequence of iterates produced by the secant algorithm.

7. Adaptive fixed-point iteration.

To find a root a of f we can apply the fixed-point method to $g(x) = x + cf(x)$, where c is some non-zero constant. That is, given x_0 we put $x_{n+1} = g(x_n) = x_n + cf(x_n)$.

From Taylor's theorem we have

$$\begin{aligned} g(x) &\approx g(a) + (x-a)g'(a) \\ &= a + (x-a)(1 + cf'(a)) \end{aligned}$$

so

$$g(x) - a \approx (x-a)(1 + cf'(a)).$$

Based on this approximation, explain why $-1/f'(a)$ would be a good choice for c.

In practice we don't know a so cannot find $-1/f'(a)$. At step n of the iteration, what would be your best guess at $-1/f'(a)$? Using this guess for c, what happens to the fixed-point method? (You can allow your guess to change at each step.)

8. The iterative method for finding the fixed point of a function works in very general situations. Suppose $A \subset \mathbb{R}^d$ and $f : A \to A$ is such that for some $0 \le c < 1$ and all vectors $\mathbf{x}, \mathbf{y} \in A$,

$$\|f(\mathbf{x}) - f(\mathbf{y})\|_d \le c\|\mathbf{x} - \mathbf{y}\|_d.$$

It can be shown that for such an f there is a unique point $\mathbf{x}_* \in A$ such that $f(\mathbf{x}_*) = \mathbf{x}_*$. Moreover for any $\mathbf{x}_0 \in A$, the sequence defined by $\mathbf{x}_{n+1} = f(\mathbf{x}_n)$ converges to $\mathbf{x}_*$. Such a function is called a *contraction mapping*, and this result is called the *contraction mapping theorem*, which is one of the fundamental results in the field of *functional analysis*.

Modify the function `fixedpoint(ftn, x0, tol, max.iter)` given in Section 10.2, so that it works for any function `ftn(x)` that takes as input a numerical vector of length $d \ge 1$ and returns a numerical vector of length d. Use your modified function to find the fixed points of the function f below, in the region $[0, 2] \times [0, 2]$.

$$f(x_1, x_2) = (\log(1 + x_1 + x_2), \log(5 - x_1 - x_2)).$$

9. For $f : \mathbb{R} \to \mathbb{R}$, the Newton-Raphson algorithm uses a sequence of linear approximations to f to find a root. What happens if we use quadratic approximations instead?

Suppose that x_n is our current approximation to f, then a quadratic approximation to f at x_n is given by the second-order Taylor expansion:

$$f(x) \approx g_n(x) = f(x_n) + (x - x_n)f'(x_n) + \tfrac{1}{2}(x - x_n)^2 f''(x_n).$$

Let x_{n+1} be the solution of $g_n(x) = 0$ that is closest to x_n, assuming a solution exists. If $g_n(x) = 0$ has no solution, then let x_{n+1} be the point at which g_n attains either its minimum or maximum. Figure 10.5 illustrates the two cases.

Implement this algorithm in R and use it to find the fixed points of the following functions:

(a). $\cos(x) - x$ using $x_0 = 1, 3, 6$.

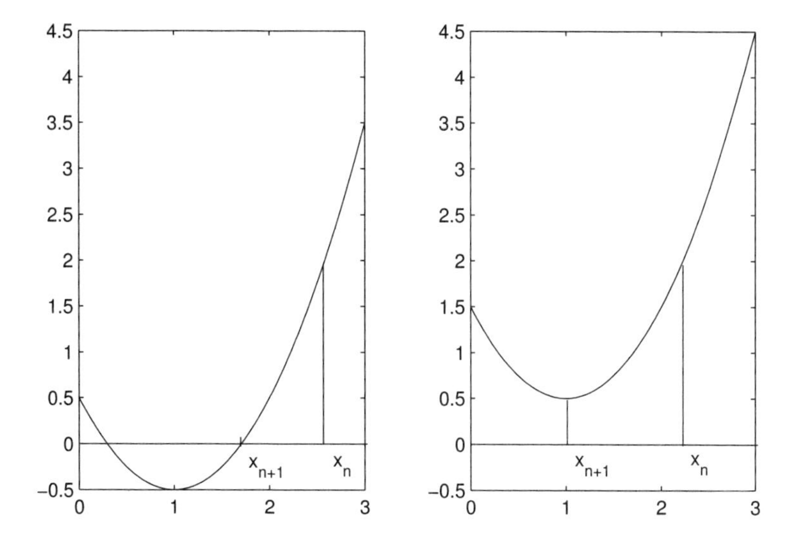

Figure 10.5 *The iterative root-finding scheme of Exercise 9*

(b). $\log(x) - \exp(-x)$ using $x_0 = 2$.

(c). $x^3 - x - 3$ using $x_0 = 0$.

(d). $x^3 - 7x^2 + 14x - 8$ using $x_0 = 1.1, 1.2, \ldots, 1.9$.

(e). $\log(x)\exp(-x)$ using $x_0 = 2$.

For your implementation, assume that you are given a function `ftn(x)` that returns the vector $(f(x), f'(x), f''(x))$. Given x_n, if you rewrite g_n as $g_n(x) = a_2x^2 + a_1x + a_0$ then you can use the formula $(-a_1 \pm \sqrt{a_1^2 - 4a_2a_0})/2a_2$ to find the roots of g_n and thus x_{n+1}. If g_n has no roots then the min/max occurs at the point $g_n'(x) = 0$.

How does this algorithm compare to the Newton-Raphson algorithm?

10. How do we know $\pi = 3.1415926$ (to 7 decimal places)? One way of finding π is to solve $\sin(x) = 0$. By definition the solutions to $\sin(x) = 0$ are $k\pi$ for $k = 0, \pm 1, \pm 2, \ldots$, so the root closest to 3 should be π.

(a). Use a root-finding algorithm, such as the Newton-Raphson algorithm, to find the root of $\sin(x)$ near 3. How close can you get to π? (You may use the function `sin(x)` provided by R.)

The function $\sin(x)$ is *transcendental*, which means that it cannot be written as a rational function of x. Instead we have to write it as an infinite sum:

$$\sin(x) = \sum_{k=0}^{\infty} (-1)^k \frac{x^{2k+1}}{(2k+1)!}.$$

(This is the infinite order Taylor expansion of $\sin(x)$ about 0.) In practice, to calculate $\sin(x)$ numerically we have to truncate this sum, so any numerical calculation of $\sin(x)$ is an approximation. In particular the function `sin(x)` provided by R is only an approximation of $\sin(x)$ (though a very good one).

(b). Put

$$f_n(x) = \sum_{k=0}^{n} (-1)^k \frac{x^{2k+1}}{(2k+1)!}.$$

Write a function in R to calculate $f_n(x)$. Plot $f_n(x)$ over the range $[0, 7]$ for a number of values of n, and verify that it looks like $\sin(x)$ for large n.

(c). Choose a large value of n, then find an approximation to π by solving $f_n(x) = 0$ near 3. Can you get an approximation that is correct up to 6 decimal places? Can you think of a better way of calculating π?

11. The astronomer Edmund Halley devised a root-finding method faster than the Newton-Raphson method, but which requires second derivative information. If x_n is our current solution then

$$x_{n+1} = x_n - \frac{f(x_n)}{f'(x_n) - (f(x_n)f''(x_n)/2f'(x_n))}.$$

Let m be a positive integer. Show that applying Halley's method to the function $f(x) = x^m - k$ gives

$$x_{n+1} = \left(\frac{(m-1)x_n^m + (m+1)k}{(m+1)x_n^m + (m-1)k} \right) x_n.$$

Use this to show that, to 9 decimal places, $59^{1/7} = 1.790518691$.

12. The bisection method can be generalised to deal with the case $f(x_l)f(x_r) > 0$, by *broadening* the bracket. That is, we reduce x_l and/or increase x_r, and try again. A reasonable choice for broadening the bracket is to double the width of the interval $[x_l, x_r]$, that is (in pseudo-code)

$$\begin{aligned} m &\leftarrow (x_l + x_r)/2 \\ w &\leftarrow x_r - x_l \\ x_l &\leftarrow m - w \\ x_r &\leftarrow m + w \end{aligned}$$

Incorporate bracket broadening into the function `bisection` given in Section 10.5. Note that broadening is *not guaranteed* to find x_l and x_r such that $f(x_l)f(x_r) \leq 0$, so you should include a limit on the number of times it can be tried.

Use your modified function to find a root of

$$f(x) = (x-1)^3 - 2x^2 + 10 - \sin(x),$$

starting with $x_l = 1$ and $x_r = 2$.

CHAPTER 11

Numerical integration

It is frequently necessary to compute definite integrals $\int_a^b f(x)dx$ of a given function f. From the Fundamental Theorem of Calculus we know that if we can find an antiderivative or indefinite integral F, such that $F'(x) = \frac{d}{dx}F(x) = f(x)$, then $\int_a^b f(x)dx = F(b) - F(a)$. However for many functions f it is impossible to write down an antiderivative in closed form. That is, we have no finite formula for F. In such cases we can use numerical integration to approximate the definite integral.

For example, in statistics we often use definite integrals of the standard normal density, that is, integrals of the form

$$\Phi(z) = \int_{-\infty}^{z} \frac{1}{\sqrt{2\pi}} e^{-x^2/2}\, dx.$$

We know that $\Phi(0) = 1/2$ and $\Phi(\infty) = 1$, but for all other z numerical integration is used.

In this chapter we consider three numerical integration techniques: the trapezoid rule, Simpson's rule, and adaptive quadrature. In each case we suppose that we are given an integrable[1] function $f(x)$ and an interval $[a, b]$ and the object is to approximate

$$\int_a^b f(x)\, dx.$$

We subdivide the interval $[a, b]$ into n equal subintervals each of length $h = (b - a)/n$. The endpoints of these subintervals are labelled

$$a = x_0, x_1, x_2, \ldots, x_{n-1}, x_n = b.$$

We approximate the integral on each of these small intervals, then add all the small approximations to give a total approximation to the original integral.

11.1 Trapezoidal rule

The trapezoidal approximation is obtained by approximating the area under $y = f(x)$ over the subinterval $[x_i, x_{i+1}]$ by a trapezoid. That is, the function

[1] All our examples deal with integrable functions, but we are not concerned here with formal proofs of integrability.

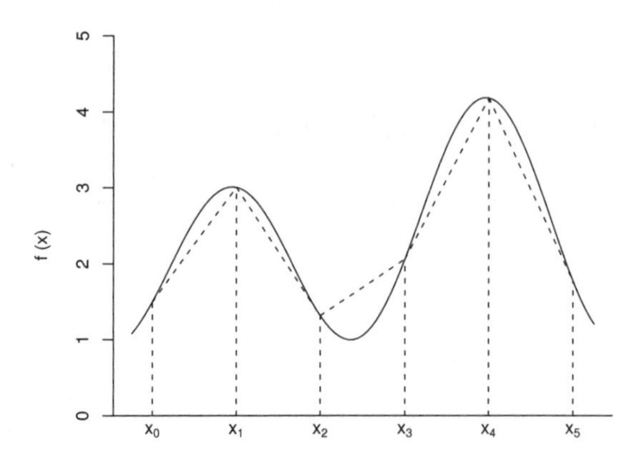

Figure 11.1 *The approximation of f used by the trapezoidal rule.*

$f(x)$ is approximated by a straight line over the subinterval $[x_i, x_{i+1}]$ (Figure 11.1). The width of the trapezoid is h, the left side of the trapezoid has height $f(x_i)$ and the right side has height $f(x_{i+1})$. The area of the trapezoid is thus

$$\frac{h}{2}(f(x_i) + f(x_{i+1})).$$

Now we add the areas for all of the subintervals together to get our trapezoidal approximation T to the integral $\int_a^b f(x)dx$:

Trapezoidal rule

$$T = \frac{h}{2}(f(x_0) + 2f(x_1) + 2f(x_2) + \cdots + 2f(x_{n-1}) + f(x_n)).$$

Notice that for $i = 1, \ldots, n-1$, $f(x_i)$ contributes to the area of the trapezoid to the left of x_i and to the right of x_i and so appears multiplied by 2 in the formula above. In contrast $f(x_0)$ and $f(x_n)$ contribute only to the area of the first and last trapezoid, respectively.

Here is an implementation in R. We use it to estimate $\int_0^1 4x^3\,dx = 1$.

```
# program spuRs/resources/scripts/trapezoid.r

trapezoid <- function(ftn, a, b, n = 100) {
  # numerical integral of ftn from a to b
  # using the trapezoid rule with n subdivisions
  #
  # ftn is a function of a single variable
  # we assume a < b and n is a positive integer
```

```
  h <- (b-a)/n
  x.vec <- seq(a, b, by = h)
  f.vec <- sapply(x.vec, ftn)
  T <- h*(f.vec[1]/2 + sum(f.vec[2:n]) + f.vec[n+1]/2)
  return(T)
}
```

```
> source("../scripts/trapezoid.r")
> ftn6 <- function(x) return(4 * x^3)
> trapezoid(ftn6, 0, 1, n = 20)

[1] 1.0025

> trapezoid(ftn6, 0, 1, n = 40)

[1] 1.000625

> trapezoid(ftn6, 0, 1, n = 60)

[1] 1.000278
```

Note that as defined, the function `ftn6` is vectorised (given a vector as input, it will return a vector as output). Thus, in `trapezoid`, the command `sapply(x.vec, ftn)` could be replaced by `ftn(x.vec)`. The advantage of using `sapply` is that it will work even if `ftn` is not vectorised.

The trapezoid rule gives exact results if f is a constant or a linear function, otherwise there will be an error, corresponding to the extent that our trapezoidal approximation overshoots or undershoots the actual graph of f.

11.2 Simpson's rule

Simpon's rule subdivides the interval $[a, b]$ into n subintervals, where n is even, then on each consecutive pair of subintervals, it approximates the behaviour of $f(x)$ by a parabola (polynomial of degree 2) rather than by the straight lines used in the trapezoidal rule.

Let $u < v < w$ be any three points distance h apart. For $x \in [u, w]$ we want to approximate $f(x)$ by a parabola which passes through the points $(u, f(u))$, $(v, f(v))$, and $(w, f(w))$. There is exactly one such parabola $p(x)$ and it is given by the formula

$$p(x) = f(u)\frac{(x-v)(x-w)}{(u-v)(u-w)} + f(v)\frac{(x-u)(x-w)}{(v-u)(v-w)} + f(w)\frac{(x-u)(x-v)}{(w-u)(w-v)}.$$

As an approximation to the area under the curve $y = f(x)$, we use $\int_u^w p(x)dx$. A rather lengthy but elementary calculation shows

$$\int_u^w p(x)dx = \frac{h}{3}(f(u) + 4f(v) + f(w)).$$

Now, assuming that n is even, we add up the approximations for the subintervals $[x_{2i}, x_{2i+2}]$ to obtain Simpson's approximation S to the integral $\int_a^b f(x)dx$.

Simpson's rule

$$S = \frac{h}{3}(f(x_0) + 4f(x_1) + 2f(x_2) + 4f(x_3) + \cdots + 4f(x_{n-1}) + f(x_n)).$$

Notice that the $f(x_i)$ for i odd are all weighted 4, while the $f(x_i)$ for i even (except 0 and n) are weighted 2 as they each appear in two subintervals.

Obviously Simpson's rule gives exact results if $f(x)$ is a quadratic function since it is based on approximating each piece of $f(x)$ by a parabola. Surprisingly, it also gives exact results if $f(x)$ is a cubic function. In general it gives better results than the trapezoid rule.

Here is an implementation in R:

```
#program spuRs/resources/scripts/simpson_n.r

simpson_n <- function(ftn, a, b, n = 100) {
  # numerical integral of ftn from a to b
  # using Simpson's rule with n subdivisions
  #
  # ftn is a function of a single variable
  # we assume a < b and n is a positive even integer

  n <- max(c(2*(n %/% 2), 4))
  h <- (b-a)/n
  x.vec1 <- seq(a+h, b-h, by = 2*h)
  x.vec2 <- seq(a+2*h, b-2*h, by = 2*h)
  f.vec1 <- sapply(x.vec1, ftn)
  f.vec2 <- sapply(x.vec2, ftn)
  S <- h/3*(ftn(a) + ftn(b) + 4*sum(f.vec1) + 2*sum(f.vec2))
  return(S)
}

> source("../scripts/simpson_n.r")
> ftn6 <- function(x) return(4 * x^3)
> simpson_n(ftn6, 0, 1, 20)

[1] 1
```

11.2.1 *Example:* $\Phi(z)$ `Phi.r`

One of Gauss' many prodigious acts was to compile by hand tables of $\Phi(z) = \int_{-\infty}^{z} \frac{1}{\sqrt{2\pi}} e^{-x^2/2}\, dx$, estimated to several decimal places. (This is the distribution function of a normal or Gaussian random variable; see Section 16.5.1.) Thankfully we can now do this using a computer, as follows.

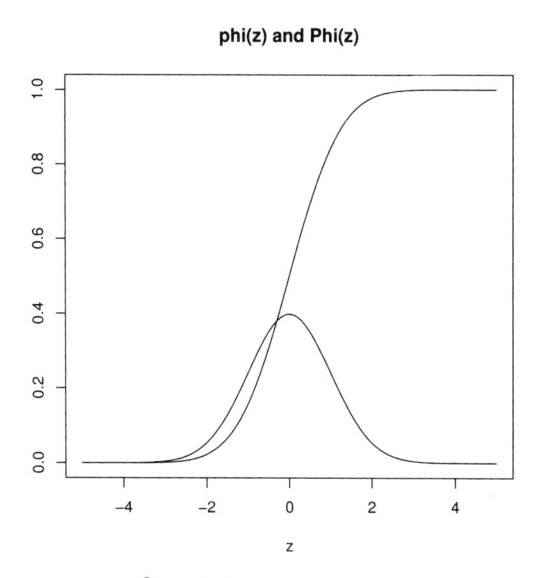

Figure 11.2 $\phi(z) = e^{-x^2/2}/\sqrt{2\pi}$ *and its integral* Φ*; see Example 11.2.1.*

```
# program spuRs/resources/scripts/Phi.r
# estimate and plot the normal cdf Phi

rm(list = ls()) # clear the workspace
source("../scripts/simpson_n.r")
phi <- function(x) return(exp(-x^2/2)/sqrt(2*pi))
Phi <- function(z) {
  if (z < 0) {
    return(0.5 - simpson_n(phi, z, 0))
  } else {
    return(0.5 + simpson_n(phi, 0, z))
  }
}

z <- seq(-5, 5, by = 0.1)
phi.z <- sapply(z, phi)
Phi.z <- sapply(z, Phi)
plot(z, Phi.z, type = "l", ylab = "", main = "phi(z) and Phi(z)")
lines(z, phi.z)
```

Running the command `source("../scripts/Phi.r")` we get the output given in Figure 11.2. We will see in Section 16.1 that R actually has a built-in function for calculating $\Phi(z)$, namely `pnorm`.

11.2.2 Example: convergence of Simpson's rule `simpson_test.r`

To test the accuracy of Simpson's rule we estimated $\int_{0.01}^{1}(1/x)\,dx = -\log(0.01)$ for a sequence of increasing values of n, the number of partitions. A plot of log(error) against $\log(n)$ appears to have a slope of roughly -4 for large values of n, indicating that the error decays like n^{-4}. This can in fact be shown to hold in general for functions f with a continuous fourth derivative.

```
# program simpson_test.r
# test the accuracy of Simpson's rule
# using the integral of 1/x from 0.01 to 1

rm(list = ls()) # clear the workspace
source("../scripts/simpson_n.r")
ftn <- function(x) return(1/x)
S <- function(n) simpson_n(ftn, 0.01, 1, n)

n.vec <- seq(10, 1000, by = 10)
S.vec <- sapply(n.vec, S)

opar <- par(mfrow = c(1, 2), pty="s", mar=c(4,4,2,1), las=1)
plot(n.vec, S.vec + log(0.01), type = "l",
  xlab = "n", ylab = "error")
plot(log(n.vec), log(S.vec + log(0.01)), type = "l",
  xlab = "log(n)", ylab = "log(error)")
par(opar)
```

Running the command `source("../scripts/simpson_test.r")` we get the output given in Figure 11.3.

11.2.3 Achieving a set tolerance

When using `simpson_n` in practice we need some rule for choosing n which results in a reasonably accurate approximation. Suppose that f is continuous and we wish to estimate $I = \int_a^b f(x)dx$. Let $S(n)$ be the value of the approximation when we use a partition of size n, then $S(n) \to I$ as $n \to \infty$. Thus $S(2n) - S(n) \to 0$ and we can use a stopping rule of the form n large enough that $|S(2n) - S(n)| \le \epsilon$, where $\epsilon > 0$ is some small tolerance. Unfortunately, given a $\delta > 0$ we cannot in general find an ϵ such that $|S(2n) - S(n)| \le \epsilon \Rightarrow |S(2n) - I| \le \delta$. (If we know something about f' then it is possible to bound the error, but we will not pursue this here.) As a rule of thumb, the square root of machine epsilon is a good place to start when choosing ϵ, that is, around 10^{-8} if you are working in double precision.

Here is a modification of `simpson_n` such that instead of specifying the partition size n, we specify a tolerance ϵ. That is, we automatically increase n until $|S(2n) - S(n)| \le \epsilon$. Observe that we increase n by a factor of 2 each time, so

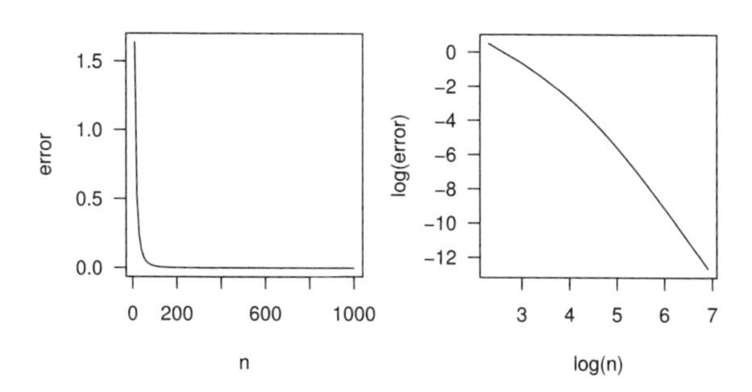

Figure 11.3 *Error using Simpson's method with a partition of size n; see Example 11.2.2.*

that we can reuse previous function evaluations. In practice evaluating f is the most expensive operation we perform when doing numerical integration. If we increased n by just 2 (n has to be even), then the points at which we evaluate f would all change, except a and b, so we would be calculating f at $n-1$ new points. If we double n then we can reuse all our existing f values, so we only need to calculate f at n new points. Our modified function `simpson` is in the script `simpson.r`:

```
# program spuRs/resources/scripts/simpson.r

simpson <- function(ftn, a, b, tol = 1e-8, verbose = FALSE) {
  # numerical integral of ftn from a to b
  # using Simpson's rule with tolerance tol
  #
  # ftn is a function of a single variable and a < b
  # if verbose is TRUE then n is printed to the screen

  # initialise
  n <- 4
  h <- (b - a)/4
  fx <- sapply(seq(a, b, by = h), ftn)
  S <- sum(fx*c(1, 4, 2, 4, 1))*h/3
  S.diff <- tol + 1  # ensures we loop at least once

  # increase n until S changes by less than tol
  while (S.diff > tol) {
    # cat('n =', n, 'S =', S, '\n')  # diagnostic
    S.old <- S
```

```
    n <- 2*n
    h <- h/2
    fx[seq(1, n+1, by = 2)] <- fx  # reuse old ftn values
    fx[seq(2, n, by = 2)] <- sapply(seq(a+h, b-h, by = 2*h), ftn)
    S <- h/3*(fx[1] + fx[n+1] + 4*sum(fx[seq(2, n, by = 2)]) +
         2*sum(fx[seq(3, n-1, by = 2)]))
    S.diff <- abs(S - S.old)
  }
  if (verbose) cat('partition size', n, '\n')
  return(S)
}
```

11.3 Adaptive quadrature

In this section we present a program that does *adaptive quadrature* using Simpson's rule as its basic method. In adaptive quadrature, the subinterval width h is not constant over the interval $[a, b]$, but instead adapts to the function. The key observation is that h only needs to be small where the integrand f is steep.

To see why it is a good idea to change the size of h to suit the behaviour of the function, we will consider the integral $\int_0^1 (k+1)x^k dx = 1$. How large a partition is required to estimate $\int_0^1 (k+1)x^k dx = 1$ using a tolerance of 10^{-9}?

```
> source("../scripts/simpson.r")
> options(digits = 16)
> f4 <- function(x) 5 * x^4
> simpson(f4, 0, 1, tol = 1e-09, verbose = TRUE)

partition size 512
[1] 1.000000000009701

> f8 <- function(x) 9 * x^8
> simpson(f8, 0, 1, tol = 1e-09, verbose = TRUE)

partition size 1024
[1] 1.000000000015280

> f12 <- function(x) 13 * x^12
> simpson(f12, 0, 1, tol = 1e-09, verbose = TRUE)

partition size 2048
[1] 1.000000000005419
```

Clearly as k increases $(k+1)x^k$ gets steeper and we need a smaller h to achieve a given tolerance. But $(k+1)x^k$ is much steeper over the interval $[0.5, 1]$ than the interval $[0, 0.5]$. Thus if we split the integral into two bits, we should find that we need a much smaller h for the interval $[0.5, 1]$ than for $[0, 0.5]$:

```
> S1 <- simpson(f12, 0, 0.5, tol = 5e-10, verbose = TRUE)

partition size 256

> S2 <- simpson(f12, 0.5, 1, tol = 5e-10, verbose = TRUE)

partition size 1024

> S1 + S2

[1] 1.000000000008118
```

Note that when we split the integral into two, we halved the tolerance for each part. That way, the tolerance for the recombined integral `S1 + S2` is guaranteed to remain less than 10^{-9}.

Adaptive quadrature automatically allows the interval width h to vary over the range of integration, using a recursive algorithm. The basic idea is to apply Simpson's rule using some initial h and $h/2$. If the difference between the two estimates is less than some given tolerance ϵ, then we are done. If not then we split the range of integration $[a, b]$ into two parts ($[a, c]$ and $[c, b]$ where $c = (a + b)/2$) and on each part we apply Simpson's rule using interval widths $h/2$ and $h/4$ and a tolerance of $\epsilon/2$. (If the error on each subinterval is less than $\epsilon/2$, then the error of the combined estimates will be less than ϵ.) By decreasing h we improve the accuracy. If the desired tolerance is not met on a given subinterval then we split it further, but we only do this for subintervals that do not achieve the desired tolerance. Thus a small h is used only where needed. The method only subdivides intervals when it needs the greater resolution (generally where the function is spiky) and thereby saves a lot of work (measured by the number of function evaluations required).

In the implementation below we also keep track of how often we have subdivided (called the *level of recursion*) since if the function has a singularity (a point where it heads off to infinity) then it could recursively call itself forever!

```
# program spuRs/resources/scripts/quadrature.r
# numerical integration using adaptive quadrature

quadrature <- function(ftn, a, b, tol = 1e-8, trace = FALSE) {
  # numerical integral of ftn from a to b
  # ftn is a function of one variable
  # the partition used is recursively refined until the
  # estimate on successive partitions differs by at most tol
  # if trace is TRUE then intermediate results are printed
  #
  # the main purpose of this function is to call function q.recursion
  #
  # the function returns a vector of length 2 whose first element
  # is the integral and whose second element is the number of
  # function evaluations required
```

```
  c = (a + b)/2
  fa <- ftn(a)
  fb <- ftn(b)
  fc <- ftn(c)
  h <- (b - a)/2
  I.start <- h*(fa + 4*fc + fb)/3 # Simpson's rule
  q.out <- q.recursion(ftn,a,b,c,fa,fb,fc,I.start,tol,1,trace)
  q.out[2] <- q.out[2] + 3
  if (trace) {
    cat("final value is", q.out[1], "in",
        q.out[2], "function evaluations\n")
  }
  return(q.out)
}

q.recursion <- function(ftn,a,b,c,fa,fb,fc,I.old,tol,level,trace) {
  # refinement of the numerical integral of ftn from a to b
  # ftn is a function of one variable
  # the current partition is [a, c, b]
  # fi == ftn(i)
  # I.old is the value of the integral I using the current partition
  # if trace is TRUE then intermediate results are printed
  # level is the current level of refinement/nesting
  #
  # the function returns a vector of length 2 whose first element
  # is the integral and whose second element is the number of
  # function evaluations required
  #
  # I.left and I.right are estimates of I over [a, c] and [c, b]
  # if |I.old - I.left - I.right| <= tol then we are done, otherwise
  # I.left and I.right are recursively refined

  level.max <- 64
  if (level > level.max) {
    cat("recursion limit reached: singularity likely\n")
    return(NULL)
  } else {
    h <- (b - a)/4
    f1 <- ftn(a + h)
    f2 <- ftn(b - h)
    I.left <- h*(fa + 4*f1 + fc)/3  # Simpson's rule for left half
    I.right <- h*(fc + 4*f2 + fb)/3 # Simpson's rule for right half
    I.new <- I.left + I.right       # new estimate for the integral
    f.count <- 2

    if (abs(I.new - I.old) > tol) { # I.new not accurate enough
      q.left <- q.recursion(ftn, a, c, a + h, fa, fc, f1, I.left,
                            tol/2, level + 1, trace)
```

```
      q.right <- q.recursion(ftn, c, b, b - h, fc, fb, f2, I.right,
                             tol/2, level + 1, trace)
      I.new <- q.left[1] + q.right[1]
      f.count <-  f.count + q.left[2] + q.right[2];
    } else { # we have achieved the desired tolerance
      if (trace) {
        cat("integral over [", a, ", ", b, "] is ", I.new,
            " (at level ", level, ")\n", sep = "")
      }
    }

    return(c(I.new, f.count))
  }
}
```

We apply `quadrature` to the function $f(x) = 1.5\sqrt{x}$ over the range $[0, 1]$. f is only steep near $x = 0$, so the method keeps subdividing until it handles the leftmost subinterval correctly, then doesn't have to subdivide again as it comes back up.

```
> rm(list = ls())
> source("../scripts/quadrature.r")
> ftn <- function(x) return(1.5 * sqrt(x))
> quadrature(ftn, 0, 1, tol = 0.001, trace = TRUE)

integral over [0, 0.0009765625] is 3.005339e-05 (at level 11)
integral over [0.0009765625, 0.001953125] is 5.579888e-05 (at level 11)
integral over [0.001953125, 0.00390625] is 0.0001578231 (at level 10)
integral over [0.00390625, 0.0078125] is 0.0004463910 (at level 9)
integral over [0.0078125, 0.015625] is 0.001262585 (at level 8)
integral over [0.015625, 0.03125] is 0.003571128 (at level 7)
integral over [0.03125, 0.0625] is 0.01010068 (at level 6)
integral over [0.0625, 0.125] is 0.02856903 (at level 5)
integral over [0.125, 0.25] is 0.08080541 (at level 4)
integral over [0.25, 0.5] is 0.2285522 (at level 3)
integral over [0.5, 1] is 0.6464433 (at level 2)
final value is 0.9999944 in 45 function evaluations
[1]  0.9999944 45.0000000
```

Using a more realistic tolerance, we compare adaptive quadrature to the standard Simpson's rule:

```
> quadrature(ftn, 0, 1, 1e-09, trace = FALSE)

[1]    1 1205

> source("../scripts/simpson.r")
> simpson(ftn, 0, 1, 1e-09, verbose = TRUE)

partition size 524288
[1] 1
```

In this case the standard Simpson's rule used more than 400 times as many function calls than the adaptive quadrature approach (524,288 + 1 versus 1,205). We can also use the `system.time` function to compare the efficiency of the algorithms, and see that adaptive quadrature is substantially faster:

```
> rm(list = ls())
> ftn <- function(x) return(1.5 * sqrt(x))
> source("../scripts/quadrature.r")
> system.time(quadrature(ftn, 0, 1, 1e-09, trace = FALSE))

   user  system elapsed
  0.015   0.000   0.015

> source("../scripts/simpson.r")
> system.time(simpson(ftn, 0, 1, 1e-09, verbose = FALSE))

   user  system elapsed
  3.132   0.059   3.729
```

We end this section by noting that R has a built-in function `integrate`, which performs adaptive quadrature, and for multivariate integration we can use the function `adapt` from the `adapt` package.

11.4 Exercises

1. Let p be the quadratic $p(x) = c_0 + c_1 x + c_2 x^2$. Simpson's rule uses a quadratic to approximate a given function f over two adjacent intervals, then uses the integral of the quadratic to approximate the integral of the function.

   (a). Show that

   $$\int_{-h}^{h} p(x)\,dx = 2hc_0 + \frac{2}{3}c_2 h^3;$$

   (b). Write down three equations that constrain the quadratic to pass through the points $(-h, f(-h))$, $(0, f(0))$, and $(h, f(h))$, then solve them for c_0 and c_2;

   (c). Show that

   $$\int_{-h}^{h} p(x)\,dx = \frac{h}{3}(f(-h) + 4f(0) + f(h)).$$

2. Suppose $f : [0, 2\pi] \to [0, \infty)$ is continuous and $f(0) = f(2\pi)$. For $(x, y) \in \mathbb{R}^2$ let (R, θ) be the polar coordinates of (x, y), so $x = R\cos\theta$ and $y = R\sin\theta$. Define the set $A \subset \mathbb{R}^2$ by

   $$(x, y) \in A \text{ if } R \leq f(\theta).$$

   We consider the problem of calculating the area of A.

   We approximate the area of A using triangles. For small ϵ, the area of

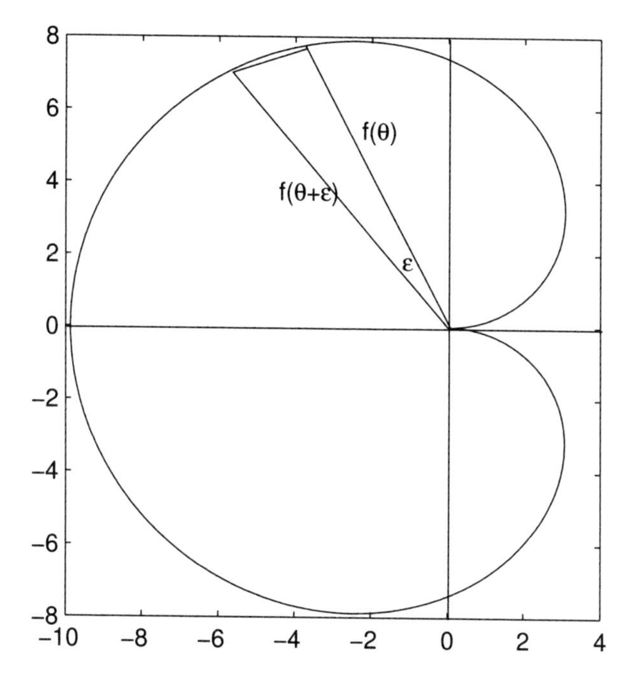

Figure 11.4 *Integration using polar coordinates, as per Exercise 2.*

the triangle with vertices $(0,0)$, $(f(\theta)\cos\theta, f(\theta)\sin\theta)$ and $(f(\theta+\epsilon)\cos(\theta+\epsilon), f(\theta+\epsilon)\sin(\theta+\epsilon))$ is $\sin(\epsilon)f(\theta)f(\theta+\epsilon)/2 \approx \epsilon f(\theta)f(\theta+\epsilon)$ (since $\sin(x) \approx x$ near 0). Thus the area of A is approximately

$$\sum_{k=0}^{n-1} \sin(2\pi/n)f(2\pi k/n)f(2\pi(k+1)/n)/2$$
$$\approx \sum_{k=0}^{n-1} \pi f(2\pi k/n)f(2\pi(k+1)/n)/n. \quad (11.1)$$

See, for example, Figure 11.4.

Write a program to numerically calculate this polar-integral using the summation formula (11.1).

Check numerically (or otherwise) that as $n \to \infty$ the polar-integral (11.1) converges to $\frac{1}{2}\int_0^{2\pi} f^2(x)\,dx$. Use $f_1(x) = 2$ and $f_2(x) = 4\pi^2 - (x-2\pi)^2$ as test cases.

3. The standard normal distribution function is given by

$$\Phi(z) = \int_{-\infty}^{z} \frac{1}{\sqrt{2\pi}} e^{-x^2/2}\,dx.$$

For $p \in [0,1]$, the $100p$ standard normal percentage point is defined as that

z_p for which

$$\Phi(z_p) = p.$$

Using the function `Phi(z)` from Example 11.2.1, calculate z_p for $p = 0.5$, 0.95, 0.975, and 0.99.

Hint: express the problem as a root-finding problem.

4. Consider

$$I = \int_0^1 \frac{3}{2}\sqrt{x}\,dx = 1.$$

Let T_n be the approximation to I given by the trapezoid rule with a partition of size n and let S_n be the approximation given by Simpson's rule with a partition of size n.

Let $n_T(\epsilon)$ be the smallest value of n for which $|T_n - I| \leq \epsilon$ and let $n_S(\epsilon)$ be the smallest value of n for which $|S_n - I| \leq \epsilon$. Plot $n_T(\epsilon)$ and $n_S(\epsilon)$ against ϵ for $\epsilon = 2^{-k}$, $k = 2, \ldots, 16$.

5. Let $T(n)$ be the estimate of $I = \int_a^b f(x)dx$ obtained using the trapezoidal method with a partition of size n. If f has a continuous second derivative, then using Taylor's theorem one can show that the error $E(n) = |I - T(n)| = O(1/n^2)$. This suggests a method for improving the trapezoid method: if $T(n) \approx I + c/n^2$ and $T(2n) \approx I + c/(2n)^2$, for some constant c, then

$$R(2n) = (4T(2n) - T(n))/3 \approx (4I + c/n^2 - I - c/n^2)/3 = I.$$

That is, the errors cancel.

This is called Richardson's deferred approach to the mean. Show that $R(2n)$ is precisely $S(2n)$, that is, Simpson's rule using a partition of size $2n$.

CHAPTER 12

Optimisation

This chapter concerns the problem of optimisation, that is, finding the maximum or minimum of a function. The search for efficient optimisation techniques is one of the major endeavours of modern mathematics. We will consider this problem first in one dimension, then in higher dimensions. We will restrict our attention to maxima, but everything we say can be equally well applied to minima, by the simple expedient of multiplying the function by -1. For univariate functions we will consider Newton's method and the golden-section method. For multivariate functions we will consider Newton's method (again) and steepest ascent. We also provide some basic information about the optimisation tools that are available in R.

In one dimension we suppose that we have a function $f : \mathbb{R} \to \mathbb{R}$ with continuous first and second derivatives. f has a *global maximum* at x^* if $f(x) \leq f(x^*)$ for all x. f has a *local maximum* at x^* if $f(x) \leq f(x^*)$ for all x in a neighbourhood of x^* (that is, all x such that $|x - x^*| < \epsilon$ for some $\epsilon > 0$). A necessary condition for x^* to be a local maximum is $f'(x^*) = 0$ and $f''(x^*) \leq 0$. A sufficient condition is $f'(x^*) = 0$ and $f''(x^*) < 0$.

Finding a local maximum is much easier than finding a global maximum. All of the algorithms we consider are *local search* techniques. They work by generating a sequence of points, $x(0)$, $x(1)$, $x(2)$, ..., which (hopefully) converge to a local maximum of f. Given a prospective solution $x(n)$, we look for the next prospective solution $x(n + 1)$ in some neighbourhood of $x(n)$. Because they never consider the whole space of possible solutions, local search techniques can only ever be guaranteed to find local maxima.

Let x^* be a local maximum of f. Supposing that $x(n) \to x^*$ as $n \to \infty$, we need *stopping criteria* to decide when to stop searching. We would like to be able to stop when $|x(n) - x^*| \leq \epsilon$, for some predetermined tolerance ϵ. Unfortunately this is not possible in general, and instead we use combinations of the following criteria:

- $|x(n) - x(n - 1)| \leq \epsilon$;
- $|f(x(n)) - f(x(n - 1))| \leq \epsilon$;
- $|f'(x(n))| \leq \epsilon$.

If the sequence $\{x(n)\}_{n=1}^{\infty}$ converges to a local maximum, then all three criteria will be satisfied, but the converse is not true. Thus, even when a local search technique appears to converge, we may still need to check that the final solution really is a local maximum.

Another problem with local search techniques is that they may not converge at all. For example if f is unbounded then we may find $x(n) \to \infty$. For this reason it is usual to specify a maximum number of iterations $n_{\max}$, and stop when $n = n_{\max}$.

12.1 Newton's method for optimisation

If $f : [a, b] \to \mathbb{R}$ has a continuous derivative f', then the problem of finding the maximum of f is equivalent to finding the maximum of $f(a)$, $f(b)$, and $f(x_1), \ldots, f(x_n)$, where $x_1, \ldots, x_n$ are the roots of f'. If we apply the Newton-Raphson method for root-finding to f', we get the Newton method for optimising f:

$$x(n+1) = x(n) - \frac{f'(x(n))}{f''(x(n))}.$$

By strange convention Newton usually shares credit for this algorithm when it is applied to root-finding, but not when it is used for optimisation.

In implementing Newton's method we will suppose that we have already coded up a function that takes argument x and returns the vector $(f(x), f'(x), f''(x))$. Our example will be a member of the gamma family of probability density functions (Figure 12.1). Because we are searching for a point x^* such that $f'(x^*) = 0$, we will use $|f'(x(n))| \leq \epsilon$ as our stopping condition.

```
# Code spuRs/resources/scripts/newton_gamma.r

newton <- function(f3, x0, tol = 1e-9, n.max = 100) {
    # Newton's method for optimisation, starting at x0
    # f3 is a function that given x returns the vector
    # (f(x), f'(x), f''(x)), for some f

    x <- x0
    f3.x <- f3(x)
    n <- 0
    while ((abs(f3.x[2]) > tol) & (n < n.max)) {
        x <- x - f3.x[2]/f3.x[3]
        f3.x <- f3(x)
        n <- n + 1
    }
    if (n == n.max) {
        cat('newton failed to converge\n')
    } else {
```

```
        return(x)
    }
}

gamma.2.3 <- function(x) {
    # gamma(2,3) density
    if (x < 0) return(c(0, 0, 0))
    if (x == 0) return(c(0, 0, NaN))
    y <- exp(-2*x)
    return(c(4*x^2*y, 8*x*(1-x)*y, 8*(1-2*x^2)*y))
}
```

```
> source("../scripts/newton_gamma.r")
> newton(gamma.2.3, 0.25)

[1] 1.978656e-12

> newton(gamma.2.3, 0.5)

[1] 0

> newton(gamma.2.3, 0.75)

[1] 1
```

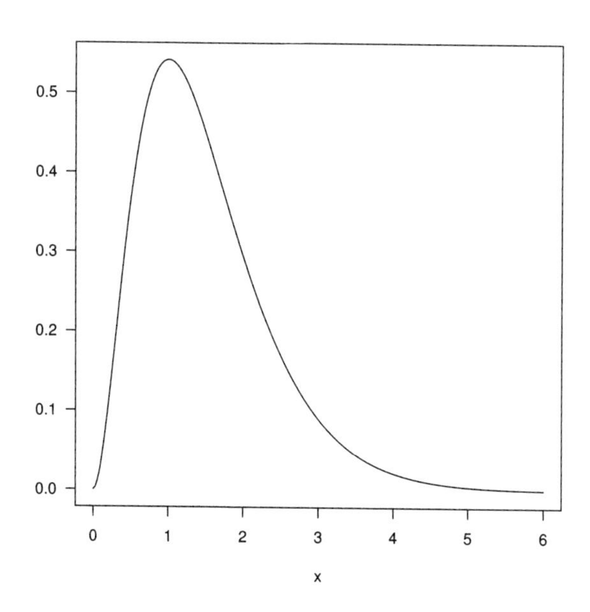

Figure 12.1 *The function* $f(x) = 4x^2e^{-2x}$ *(a* $\Gamma(2,3)$ *density), to which we apply Newton's method for optimisation.*

From this example we see that when the Newton algorithm converges, we can end up with a minimum, or indeed a 'flat spot', just as easily as a maximum. The reason is that all such stationary points satisfy $f'(x^*) = 0$. Using the corresponding root-finding theorem, it can be shown that if x^* is a local maximum, $f'(x^*) = 0$, $f''(x^*) < 0$ and f'' is Lipschitz-continuous[1] in a neighbourhood of x^*, then provided $x(0)$ is close enough to x^*, $x(n) \to x^*$ quickly, as $n \to \infty$.

We revisit Newton's method later, on a higher plane (that is, in higher dimensions).

12.2 The golden-section method

The golden-section method works in one dimension only, but does not need f'.

The golden-section method is similar to the root-bracketing technique for root-finding. Let $f : \mathbb{R} \to \mathbb{R}$ be a continuous function (note that we do not assume that we have a derivative). If we have two points $a < b$ such that $f(a)f(b) \le 0$ then we know that there is a zero in the interval $[a, b]$. To determine if we have a local maximum we need three points: if $a < c < b$ and $f(a) \le f(c)$ and $f(b) \le f(c)$ then there must be a local maximum in the interval $[a, b]$. This observation leads to the following algorithm:

Golden-section method 1 Start with $x_l < x_m < x_r$ such that $f(x_l) \le f(x_m)$ and $f(x_r) \le f(x_m)$

1. if $x_r - x_l \le \epsilon$ then stop
2. if $x_r - x_m > x_m - x_l$ then do 2a otherwise do 2b

   2a. choose a point $y \in (x_m, x_r)$
   if $f(y) \ge f(x_m)$ then put $x_l = x_m$ and $x_m = y$ otherwise put $x_r = y$

   2b. choose a point $y \in (x_l, x_m)$
   if $f(y) \ge f(x_m)$ then put $x_r = x_m$ and $x_m = y$ otherwise put $x_l = y$

3. go back to step 1

Note that so far we have not specified how to choose y other than to say it should be in the larger of the two intervals (x_l, x_m) and (x_m, x_r). Suppose that (x_m, x_r) is the larger interval, as in Figure 12.2. Let $a = x_m - x_l$, $b = x_r - x_m$, and $c = y - x_m$. The golden-section algorithm chooses y so that the ratio of the lengths of the larger to the smaller interval stays the same at each iteration.

[1] That is, there exists a k such that for all x and y, $|f''(x) - f''(y)| \le k|x - y|$.

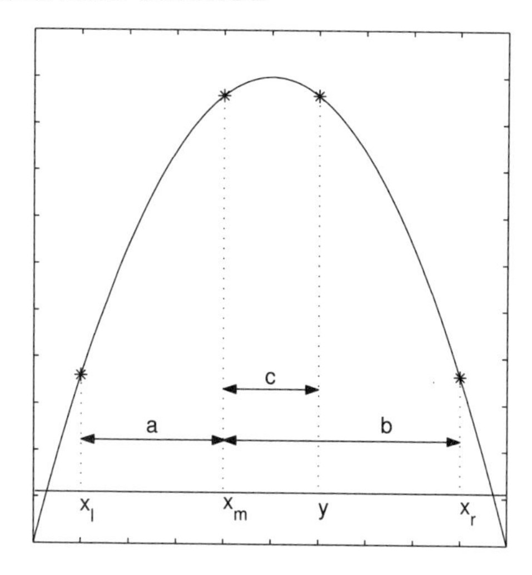

Figure 12.2 *Successive approximations to the location of the maximum using the golden-section method.*

That is, if the new bracketing interval is $[x_l, y]$ then

$$\frac{a}{c} = \frac{b}{a}$$

while if the new bracketing interval is $[x_m, x_r]$ then

$$\frac{b-c}{c} = \frac{b}{a}.$$

Put $\rho = b/a$ then solving these for c we get

$$\rho^2 - \rho - 1 = 0 \quad \text{so} \quad \rho = \frac{1+\sqrt{5}}{2}$$

which is the famous golden ratio. We also get $a = b - c$, so $c = b/(1+\rho)$ (since $(\rho-1)/\rho = 1/(1+\rho) = 3-\sqrt{5})$ and thus $y = x_m + c = x_m + (x_r - x_m)/(1+\rho)$.

The length ratio of the new interval to the old is either $b/(a+b)$ or $(a+c)/(a+b)$, which both work out as $\rho/(1+\rho)$.

An analogous argument applies if (x_l, x_m) is the larger interval. Using this method for choosing y gives the following version of the algorithm.

Golden-section method 2 Start with $x_l < x_m < x_r$ such that $f(x_l) \leq f(x_m)$ and $f(x_r) \leq f(x_m)$

1. if $x_r - x_l \leq \epsilon$ then stop
2. if $x_r - x_m > x_m - x_l$ then do 2a otherwise do 2b

   2a. let $y = x_m + (x_r - x_m)/(1+\rho)$
   if $f(y) \geq f(x_m)$ then put $x_l = x_m$ and $x_m = y$ otherwise put $x_r = y$

   2b. let $y = x_m - (x_m - x_l)/(1+\rho)$
   if $f(y) \geq f(x_m)$ then put $x_r = x_m$ and $x_m = y$ otherwise put $x_l = y$

3. go back to step 1

Here it is in R.

```
# Program spuRs/resources/scripts/gsection.r

gsection <- function(ftn, x.l, x.r, x.m, tol = 1e-9) {
  # applies the golden-section algorithm to maximise ftn
  # we assume that ftn is a function of a single variable
  # and that x.l < x.m < x.r and ftn(x.l), ftn(x.r) <= ftn(x.m)
  #
  # the algorithm iteratively refines x.l, x.r, and x.m and terminates
  # when x.r - x.l <= tol, then returns x.m

  # golden ratio plus one
  gr1 <- 1 + (1 + sqrt(5))/2

  # successively refine x.l, x.r, and x.m
  f.l <- ftn(x.l)
  f.r <- ftn(x.r)
  f.m <- ftn(x.m)
  while ((x.r - x.l) > tol) {
    if ((x.r - x.m) > (x.m - x.l)) {
      y <- x.m + (x.r - x.m)/gr1
      f.y <- ftn(y)
      if (f.y >= f.m) {
        x.l <- x.m
        f.l <- f.m
        x.m <- y
        f.m <- f.y
      } else {
        x.r <- y
        f.r <- f.y
      }
    } else {
      y <- x.m - (x.m - x.l)/gr1
      f.y <- ftn(y)
```

```
      if (f.y >= f.m) {
        x.r <- x.m
        f.r <- f.m
        x.m <- y
        f.m <- f.y
      } else {
        x.l <- y
        f.l <- f.y
      }
    }
  }
  return(x.m)
}
```

The argument above shows that if we start with x_m chosen so that the ratio $(x_r - x_m)/(x_m - x_l) = \rho$ or $1/\rho$, then at each iteration the width of the bracketing interval is reduced by a factor of $\rho/(1+\rho)$ and so must eventually go to zero. It can be shown that this rate of convergence is optimal in the sense that if you choose y any other way, then, in the worst-case, the bracketing interval will converge more slowly. Also, it is not a problem if you do not start with the ratio $(x_r - x_m)/(x_m - x_l) = \rho$ or $1/\rho$, because as soon as you have an iteration that puts $x_m = y$, this will be the case. Since $x_r - x_m \to 0$ as $n \to \infty$, to stop the golden-section algorithm it is sufficient to specify a tolerance $\epsilon > 0$ then stop when $x_r - x_l \leq \epsilon$. Moreover, provided our initial bracketing triple satisfies $f(x_l) \leq f(x_m)$ and $f(x_r) \leq f(x_m)$, the algorithm is guaranteed to converge.

The golden ratio $\rho = (1 + \sqrt{5})/2$ is usually defined as the ratio of length to breadth of a rectangle that can be decomposed into a square and a rectangle similar to the original (Figure 12.3). Given this it is no surprise that it appears here in the context of keeping a ratio constant.

12.3 Multivariate optimisation

We now consider the more useful but more difficult problem of finding local minima or maxima of a function of several variables. This is a central problem in mathematics and statistics, and continues to be the subject of active research.

Let $f : \mathbb{R}^d \to \mathbb{R}$ and suppose that all of the first- and second-order partial derivatives of f exist and are continuous everywhere. We write $\mathbf{x} = (x_1, \ldots, x_d)^T$ for an element of $\mathbb{R}^d$ and $\mathbf{e}_i$ for the i-th co-ordinate vector: $\mathbf{x} = x_1\mathbf{e}_1 + \cdots + x_d\mathbf{e}_d$. The i-th partial derivative at $\mathbf{x}$ will be denoted $f_i(\mathbf{x}) = \partial f(\mathbf{x})/\partial x_i$ and we define the *gradient*

$$\nabla f(\mathbf{x}) = (f_1(\mathbf{x}), \ldots, f_d(\mathbf{x}))^T$$

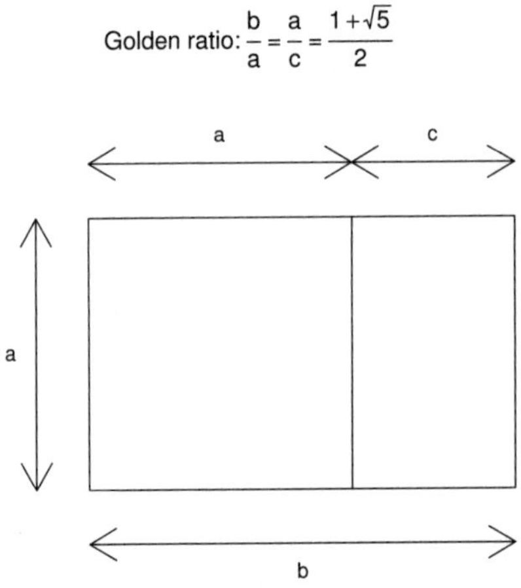

Golden ratio: $\frac{b}{a} = \frac{a}{c} = \frac{1+\sqrt{5}}{2}$

Figure 12.3 *Defining the golden ratio using similar rectangles.*

and the *Hessian*

$$\mathbf{H}(\mathbf{x}) = \begin{pmatrix} \frac{\partial^2 f(\mathbf{x})}{\partial x_1 \partial x_1} & \cdots & \frac{\partial^2 f(\mathbf{x})}{\partial x_1 \partial x_d} \\ \vdots & \ddots & \vdots \\ \frac{\partial^2 f(\mathbf{x})}{\partial x_d \partial x_1} & \cdots & \frac{\partial^2 f(\mathbf{x})}{\partial x_d \partial x_d} \end{pmatrix}.$$

For any vector $\mathbf{v} \neq \mathbf{0}$ the slope at $\mathbf{x}$ in direction $\mathbf{v}$ is given by $\mathbf{v}^T \nabla f(\mathbf{x})/\|\mathbf{v}\|$, where $\|\mathbf{v}\| = \sqrt{v_1^2 + \cdots + v_d^2}$ is the Euclidean norm. The curvature at $\mathbf{x}$ in direction $\mathbf{v}$ is given by $\mathbf{v}^T \mathbf{H}(\mathbf{x})\mathbf{v}/\|\mathbf{v}\|^2$. f has a local maximum at $\mathbf{x}$ if for all $\epsilon > 0$ small enough, $f(\mathbf{x} + \epsilon \mathbf{e}_i) \leq f(\mathbf{x})$ for $i = 1, \ldots, d$. A necessary (but not sufficient) condition for a maximum at $\mathbf{x}$ is $\nabla f(\mathbf{x}) = \mathbf{0} = (0, \ldots, 0)^T$ and for all $\mathbf{v} \neq \mathbf{0}$, the slope at $\mathbf{x}$ in direction $\mathbf{v}$ is ≤ 0 (we say that the Hessian is *negative semi-definite*). A sufficient (but not necessary) condition for f to have a local maximum at $\mathbf{x}$ is that $\nabla f(\mathbf{x}) = \mathbf{0}$ and the curvature in all directions is < 0 (in which case we say that the Hessian $\mathbf{H}(\mathbf{x})$ is *negative-definite*).

Clearly, by taking $-f$, finding a local minimum is equivalent to finding a local maximum.

As in one dimension, we will use iterative local search techniques to find local maxima. Define $\|\mathbf{x}\|_\infty = \max_i |x_i|$ (the L_∞ norm). In higher dimensions we use stopping conditions that are combinations of the following:

- $\|\mathbf{x}(n) - \mathbf{x}(n-1)\|_\infty \leq \epsilon$;
- $|f(\mathbf{x}(n)) - f(\mathbf{x}(n-1))| \leq \epsilon$;
- $\|\nabla f(\mathbf{x}(n))\|_\infty \leq \epsilon$.

To guard against non-convergence, we should also specify a maximum number of iterations $n_{\max}$, then stop when $n = n_{\max}$.

12.4 Steepest ascent

Let $f : \mathbb{R}^d \to \mathbb{R}$ be a function with continuous partial derivatives everywhere. We wish to find a local maximum of f in the vicinity of some point $\mathbf{x}(0)$.

In the steepest ascent method, we put $\mathbf{x}(n+1) = \mathbf{x}(n) + \alpha\mathbf{v}$, where α is a positive scalar and the direction $\mathbf{v}$ is the direction with largest slope. That is, $\mathbf{v}$ maximises $\mathbf{v}^T\nabla f(\mathbf{x}(n))/\|\mathbf{v}\|$. Consider

$$\frac{\partial}{\partial v_i}\frac{\mathbf{v}^T\nabla f(\mathbf{x})}{\|\mathbf{v}\|} = \frac{f_i(\mathbf{x})}{\|\mathbf{v}\|} - \frac{(\mathbf{v}^T\nabla f(\mathbf{x}))v_i}{\|\mathbf{v}\|^3}.$$

Setting this to 0 we get $v_i \propto f_i(\mathbf{x})$, from which we see that at the point $\mathbf{x}$ the direction with largest slope is $\nabla f(\mathbf{x})$. (The direction with smallest slope is $-\nabla f(\mathbf{x})$, which you use if you are searching for a local minimum.) Thus, the steepest ascent method has the form

$$\mathbf{x}(n+1) = \mathbf{x}(n) + \alpha\nabla f(\mathbf{x}(n)),$$

for some $\alpha \geq 0$. Given this form, we choose $\alpha \geq 0$ to maximise

$$g(\alpha) = f\left(\mathbf{x}(n) + \alpha\nabla f(\mathbf{x}(n))\right).$$

If $\alpha = 0$ then we have reached a local maximum, while if $\alpha > 0$ then $f(\mathbf{x}(n+1)) > f(\mathbf{x}(n))$.

If f is bounded above then, because $f(\mathbf{x}(n+1)) \geq f(\mathbf{x}(n))$, the sequence $\{f(\mathbf{x}(n))\}_{n=1}^{\infty}$ must converge. This suggests that we can use the stopping condition $f(\mathbf{x}(n)) - f(\mathbf{x}(n-1)) \leq \epsilon$, for some small tolerance ϵ. However, we need to be aware that it is still possible that $\{f(\mathbf{x}(n))\}_{n=1}^{\infty}$ converges, but $\{\mathbf{x}(n)\}_{n=1}^{\infty}$ does not. In fact, it can be shown that if f is bounded and ∇f is 'well behaved' (that is, uniformly continuous in the region of interest), then the sequence $\{\mathbf{x}(n)\}_{n=1}^{\infty}$ will converge to a local maximum.[2]

We are now in a position to sketch some code for implementing the steepest ascent method. We will assume that we have already encoded f and ∇f as functions in R. We will also assume that we have some function `line.search`, which takes arguments f, $\mathbf{x}(n)$, and $\nabla f(\mathbf{x}(n))$ and returns $\mathbf{x}(n)+\alpha_m\nabla f(\mathbf{x}(n))$ where $\alpha_m = \arg\max g(\alpha)$ (that is, the value of α that maximises $g(\alpha)$).

```
# Program spuRs/resources/scripts/ascent.r

source("../scripts/linesearch.r")

ascent <- function(f, grad.f, x0, tol = 1e-9, n.max = 100) {
    # steepest ascent algorithm
    # find a local max of f starting at x0
    # function grad.f is the gradient of f
```

[2] P. Wolfe, Convergence conditions for ascent methods. *SIAM Review*, Vol. 11, pp. 226–235, 1969.

```
    x <- x0
    x.old <- x
    x <- line.search(f, x, grad.f(x))
    n <- 1
    while ((f(x) - f(x.old) > tol) & (n < n.max)) {
        x.old <- x
        x <- line.search(f, x, grad.f(x))
        n <- n + 1
    }
    return(x)
}
```

12.4.1 Line search

To complete the steepest ascent algorithm, at each step n we need to maximise $g(\alpha) = f(\mathbf{x}(n) + \alpha \nabla f(\mathbf{x}(n)))$, over $\alpha \geq 0$. As finding a global maximum is hard, we will instead look for a local maximum, for which we will use the golden-section algorithm.

The golden-section algorithm requires three initial points $\alpha_l < \alpha_m < \alpha_r$ such that $g(\alpha_m) \geq g(\alpha_l)$ and $g(\alpha_m) \geq g(\alpha_r)$. Put $\alpha_l = 0$. In theory, if $\|\nabla f(\mathbf{x}(n))\| > 0$ then $g'(0) > 0$ and thus there must be some $\epsilon > 0$ such that $g(\epsilon) > g(0)$, so we can put $\alpha_m = \epsilon$. In practice, if $g'(0)$ is very small, then $g(\epsilon) - g(0) \approx g'(0)\epsilon$ will be very very small, and we may not be able to distinguish $g(0)$ from $g(\epsilon)$ numerically.

Unfortunately there is not even a theoretical guarantee that a suitable α_r exists, because we may have g increasing over the whole interval $[0, \infty)$. To get around this problem we specify a *maximum step size* $\alpha_{\max}$, then if we cannot find $\alpha_r \leq \alpha_{\max}$ such that $g(\alpha_r) \leq g(\alpha_m)$, we just return $\alpha_{\max}$. We can now write our line search program:

```
# Program spuRs/resources/scripts/linesearch.r

source("../scripts/gsection.r")

line.search <- function(f, x, y, tol = 1e-9, a.max = 2^5) {
    # f is a real function that takes a vector of length d
    # x and y are vectors of length d
    # line.search uses gsection to find a >= 0 such that
    #   g(a) = f(x + a*y) has a local maximum at a,
    #   within a tolerance of tol
    # if no local max is found then we use 0 or a.max for a
    # the value returned is x + a*y

    if (sum(abs(y)) == 0) return(x) # g(a) constant
```

```
    g <- function(a) return(f(x + a*y))

    # find a triple a.l < a.m < a.r such that
    # g(a.l) >= g(a.m) and g(a.m) <= g(a.r)
    # a.l
    a.l <- 0
    g.l <- g(a.l)
    # a.m
    a.m <- 1
    g.m <- g(a.m)
    while ((g.m < g.l) & (a.m > tol)) {
        a.m <- a.m/2
        g.m <- g(a.m)
    }
    # if a suitable a.m was not found then use 0 for a
    if ((a.m <= tol) & (g.m < g.l)) return(x)
    # a.r
    a.r <- 2*a.m
    g.r <- g(a.r)
    while ((g.m < g.r) & (a.r < a.max)) {
        a.m <- a.r
        g.m <- g.r
        a.r <- 2*a.m
        g.r <- g(a.r)
    }
    # if a suitable a.r was not found then use a.max for a
    if ((a.r >= a.max) & (g.m < g.r)) return(x - a.max*y)

    # apply golden-section algorithm to g to find a
    a <- gsection(g, a.l, a.r, a.m)
    return(x + a*y)
}
```

Function `line.search` uses the function `gsection` to perform a golden-section search, once a bracketing triple $a.l < a.m < a.r$ is found. The function `g` defined inside the function `line.search` makes use of R's scoping rules. Because the variables `x` and `y` are not defined within `g` or passed to it as inputs, when evaluating `g` R will look in the environment within which it was called for instances of `x` and `y`. That is, it will use the values of `x` and `y` passed to the function `line.search`.

An implementation of the golden-section algorithm for finding a local maximum is given in Section 12.2.

12.4.2 Example: $\sin(x^2/2 - y^2/4)\cos(2x - \exp(y))$

We apply the steepest ascent algorithm to the function

$$f(x, y) = \sin(x^2/2 - y^2/4)\cos(2x - \exp(y)).$$

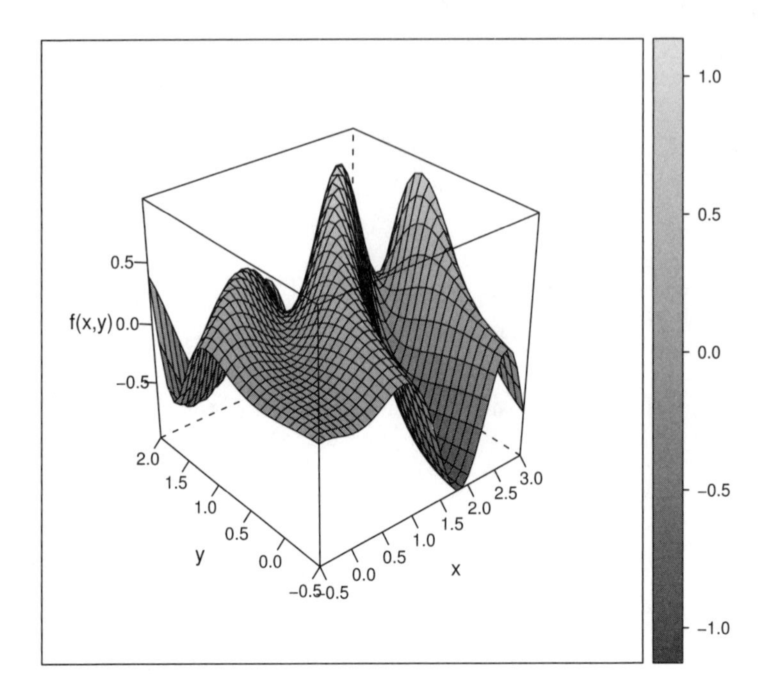

Figure 12.4 *The function* $f(x, y) = \sin(x^2/2 - y^2/4)\cos(2x - \exp(y))$.

Figure 12.4 gives a 3D-plot of the function in the region of $(0, 0)$. (The plot was produced using `wireframe`, as described in Section 7.7.) In Figure 12.5 we give a contour-plot of f, and show the sequence of points $\mathbf{x}(n)$ generated by the steepest ascent algorithm, for two different starting points: $(0.1, 0.3)$ and $(0, 0.5)$. We see that a small difference in where you start can make a big difference to where you end up. Starting at $(0.1, 0.3)$ we find the local maximum at $(2.0307, 1.4015)$, while starting at $(0, 0.5)$ we find the local maximum at $(0.3425, 1.4272)$.

Both of these search paths exhibit a well-noted characteristic of the steepest ascent algorithm: when climbing up a ridge it tends to zig-zag from one side to the other. This is because the direction $\mathbf{v}$ in which you move from $\mathbf{x}(n)$ to $\mathbf{x}(n + 1)$ is only the steepest at the point $\mathbf{x}(n)$. That is, as you move further away from $\mathbf{x}(n)$ in direction $\mathbf{v}$, you may still be moving up, but $\mathbf{v}$ will probably no longer be the direction of steepest ascent. We only stop and recalculate the gradient when there is no further gain to be made moving in direction $\mathbf{v}$.

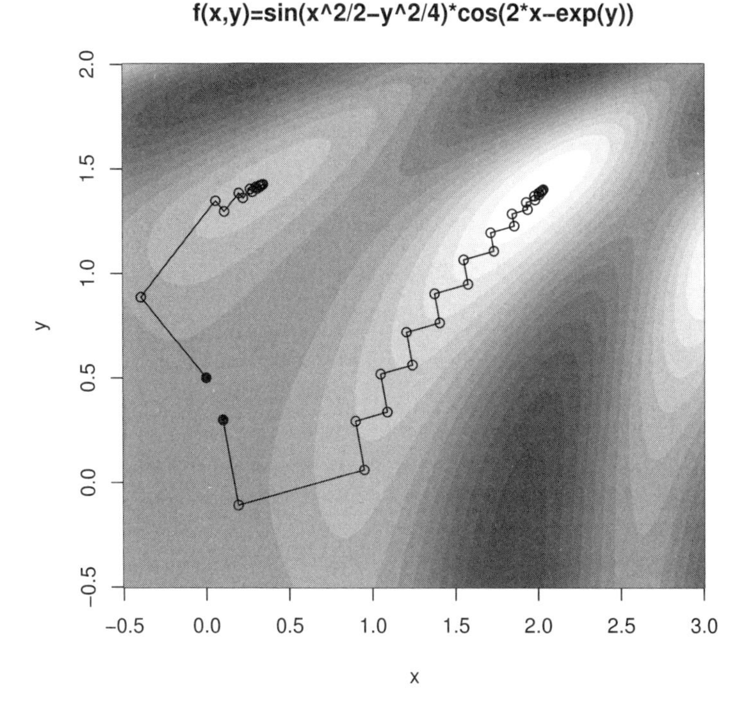

Figure 12.5 *Search paths of the steepest ascent algorithm applied to the function* $f(x,y) = \sin(x^2/2 - y^2/4)\cos(2x - \exp(y))$.

12.5 Newton's method in higher dimensions

The steepest ascent method uses information about the gradient. By making use of second-order derivatives, in other words by using the Hessian, we can construct methods that converge in fewer steps. The simplest second-order technique is Newton's method, which can be generalised from one dimension to higher dimensions relatively easily.

Newton's method looks for a point $\mathbf{x}$ such that $\nabla f(\mathbf{x}) = \mathbf{0}$. The basis of the method is a second-order Taylor expansion of f. For any $\mathbf{x}$ and $\mathbf{y}$ close together we have

$$f(\mathbf{y}) \approx f(\mathbf{x}) + (\mathbf{y}-\mathbf{x})^T \nabla f(\mathbf{x}) + \tfrac{1}{2}(\mathbf{y}-\mathbf{x})^T \mathbf{H}(\mathbf{x})(\mathbf{y}-\mathbf{x}). \tag{12.1}$$

This multidimensional approximation can be obtained from Taylor's theorem in one dimension. Put $\mathbf{v} = \mathbf{y} - \mathbf{x}$ and define $g(\alpha) = f(\mathbf{x} + \alpha\mathbf{v})$, then one has

$$g'(0) = \lim_{\alpha \to 0} \frac{g(\alpha) - g(0)}{\alpha}$$

$$
\begin{aligned}
&= \lim_{\alpha \to 0} \left(\frac{f(\mathbf{x} + \alpha v_1 \mathbf{e}_1 + \cdots + \alpha v_d \mathbf{e}_d) - f(\mathbf{x} + \alpha v_2 \mathbf{e}_2 + \cdots + \alpha v_d \mathbf{e}_d)}{\alpha} \right. \\
&\quad + \frac{f(\mathbf{x} + \alpha v_2 \mathbf{e}_2 + \cdots + \alpha v_d \mathbf{e}_d) - f(\mathbf{x} + \alpha v_3 \mathbf{e}_3 + \cdots + \alpha v_d \mathbf{e}_d)}{\alpha} \\
&\quad + \cdots \\
&\quad \left. + \frac{f(\mathbf{x} + \alpha v_d \mathbf{e}_d) - f(\mathbf{x})}{\alpha} \right) \\
&= v_1 f_1(\mathbf{x}) + v_2 f_2(\mathbf{x}) + \cdots + v_d f_d(\mathbf{x}) \ = \ \mathbf{v}^T \nabla f(\mathbf{x}).
\end{aligned}
$$

Similarly one can show from first principles that

$$g''(0) = \tfrac{1}{2}\mathbf{v}^T \mathbf{H}(\mathbf{x})\mathbf{v}.$$

Thus a second-order Taylor expansion of g about 0 gives $g(\alpha) \approx g(0) + \alpha g'(0) + \frac{1}{2}\alpha^2 g''(0) = f(\mathbf{x}) + \alpha \mathbf{v}^T \nabla f(\mathbf{x}) + \frac{1}{2}\alpha^2 \mathbf{v}^T \mathbf{H}(\mathbf{x})\mathbf{v}$. Putting $\alpha = 1$ we recover the second-order Taylor expansion of f.

Taking first-order partial derivatives on both sides of (12.1), with respect to the components of $\mathbf{y}$, we get

$$\nabla f(\mathbf{y}) \approx \nabla f(\mathbf{x}) + \mathbf{H}(\mathbf{x})(\mathbf{y} - \mathbf{x}).$$

If $\mathbf{y}$ is a local maximum then $\nabla f(\mathbf{y}) = \mathbf{0}$ and, solving the equation above, we get $\mathbf{y} = \mathbf{x} - \mathbf{H}(\mathbf{x})^{-1}\nabla f(\mathbf{x})$. This is all we need for our algorithm. Suppose $\mathbf{x}(n)$ is our current estimate, then we would like our next estimate $\mathbf{x}(n+1)$ to be a local maximum (at least approximately). Putting $\mathbf{x} = \mathbf{x}(n)$ and $\mathbf{y} = \mathbf{x}(n+1)$ does the trick:

Newton's algorithm

$$\mathbf{x}(n+1) = \mathbf{x}(n) - \mathbf{H}(\mathbf{x}(n))^{-1}\nabla f(\mathbf{x}(n)).$$

Clearly if $\mathbf{H}(\mathbf{x}(n))$ is singular (has no inverse), then Newton's method breaks down. However, as in the one-dimensional case, even if $\mathbf{H}(\mathbf{x}(n))$ is non-singular at each step, Newton's method may not converge. Despite this, if f has a local maximum at $\mathbf{x}^*$, f is 'nicely behaved' near $\mathbf{x}^*$, and if our initial point $\mathbf{x}(0)$ is 'close enough' to $\mathbf{x}^*$, then Newton's method will converge to $\mathbf{x}^*$ quickly. For our purposes f is nicely behaved near $\mathbf{x}^*$ if $\mathbf{H}(\mathbf{x}^*)$ is positive-definite and all the elements of $\mathbf{H}$ are Lipschitz-continuous.[3]

For an invertible matrix $\mathbf{A} \in \mathbb{R}^{d \times d}$ and a vector $\mathbf{b} \in \mathbb{R}^d$, $\mathbf{A}^{-1}\mathbf{b}$ is the solution to the system of equations $\mathbf{A}\mathbf{x} = \mathbf{b}$. It turns out that from a numerical point of view, it is faster and more reliable to solve the system of equations than to calculate $\mathbf{A}^{-1}$ and then multiply it by $\mathbf{b}$. The command for solving this system of linear equations in R is `solve(A, b)`. If $\mathbf{A}$ is singular then an error is generated. In our case, because $\mathbf{H}(\mathbf{x}(n))$ changes at each step, there is nothing to be gained by calculating $\mathbf{H}(\mathbf{x}(n))^{-1}$.

[3] The elements of $\mathbf{H}$ are Lipschitz-continuous if there is a constant k such that for all $\mathbf{x}$ and $\mathbf{y}$, $|f_{i,j}(\mathbf{x}) - f_{i,j}(\mathbf{y})| \le k\|\mathbf{x} - \mathbf{y}\|$, where $f_{i,j} = \partial^2 f / \partial x_i \partial x_j$.

In implementing Newton's method we will assume that we have some function `f3` that takes argument x and returns a list containing $f(x)$, $\nabla f(x)$, and $\mathbf{H}(x)$. For our stopping condition we will use $\|\nabla f(\mathbf{x}(n))\|_\infty \leq \epsilon$.

```
# program spuRs/resources/scripts/newton.r

newton <- function(f3, x0, tol = 1e-9, n.max = 100) {
    # Newton's method for optimisation, starting at x0
    # f3 is a function that given x returns the list
    # {f(x), grad f(x), Hessian f(x)}, for some f

    x <- x0
    f3.x <- f3(x)
    n <- 0
    while ((max(abs(f3.x[[2]])) > tol) & (n < n.max)) {
        x <- x - solve(f3.x[[3]], f3.x[[2]])
        f3.x <- f3(x)
        n <- n + 1
    }
    if (n == n.max) {
        cat('newton failed to converge\n')
    } else {
        return(x)
    }
}
```

We observe that this function `newton` also works if f is one dimensional.

12.5.1 Example: $\sin(x^2/2 - y^2/4)\cos(2x - \exp(y))$

We apply Newton's method to the function used in Example 12.4.2. Figure 12.6 shows the steps taken from three different starting points. We note the following:

1. Newton's method can converge to minima or saddle points, as well as maxima;
2. Newton's method is faster than steepest ascent;
3. Unless you are close to a local minimum or maximum, you can move in unexpected directions.

To emphasise the last point, we run Newton's method a number of times, with starting points clustered around $(1.5, 0.5)$. The algorithm converges each time (to a stationary point), but to very different destinations.

```
# program spuRs/resources/scripts/f3.r

f3 <- function(x) {
  a <- x[1]^2/2 - x[2]^2/4
```

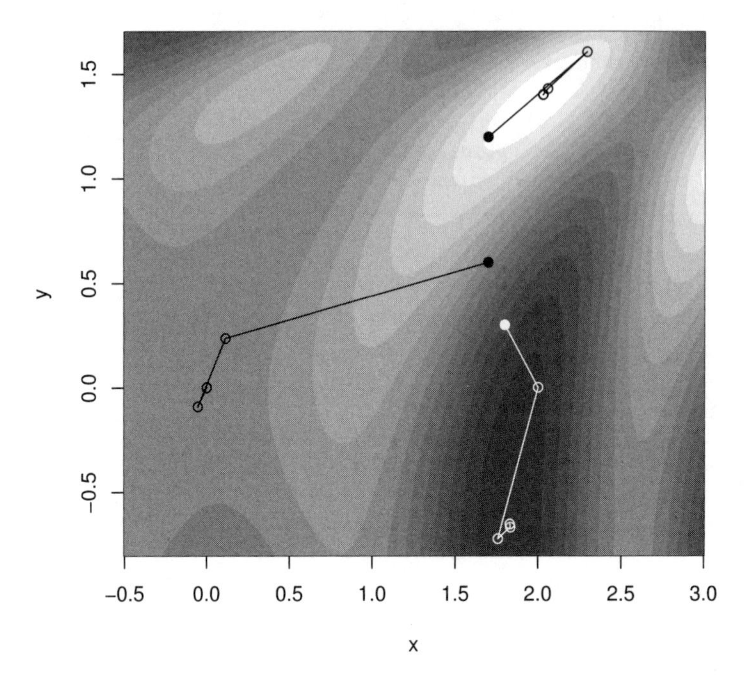

Figure 12.6 *Search paths of Newton's method applied to the function* $f(x,y) = \sin(x^2/2 - y^2/4)\cos(2x - \exp(y))$. *The starting point is indicated by a filled dot.*

```
  b <- 2*x[1] - exp(x[2])
  f <- sin(a)*cos(b)
  f1 <- cos(a)*cos(b)*x[1] - sin(a)*sin(b)*2
  f2 <- -cos(a)*cos(b)*x[2]/2 + sin(a)*sin(b)*exp(x[2])
  f11 <- -sin(a)*cos(b)*(4 + x[1]^2) + cos(a)*cos(b) -
      cos(a)*sin(b)*4*x[1]
  f12 <- sin(a)*cos(b)*(x[1]*x[2]/2 + 2*exp(x[2])) +
      cos(a)*sin(b)*(x[1]*exp(x[2]) + x[2])
  f22 <- -sin(a)*cos(b)*(x[2]^2/4 + exp(2*x[2])) - cos(a)*cos(b)/2 -
      cos(a)*sin(b)*x[2]*exp(x[2]) + sin(a)*sin(b)*exp(x[2])
  return(list(f, c(f1, f2), matrix(c(f11, f12, f12, f22), 2, 2)))
}
```

```
> source("../scripts/newton.r")
> source("../scripts/f3.r")
> for (x0 in seq(1.4, 1.6, .1)) {
+   for (y0 in seq(0.4, 0.6, .1)) {
+     cat(c(x0,y0), '-->', newton(f3, c(x0,y0)), '\n')
```

```
+   }
+ }

1.4 0.4 --> 0.04074437 -2.507290
1.4 0.5 --> 0.1179734 3.344661
1.4 0.6 --> -1.553163 6.020013
1.5 0.4 --> 2.837142 5.353982
1.5 0.5 --> 0.04074437 -2.507290
1.5 0.6 --> 9.899083e-10 1.366392e-09
1.6 0.4 --> -0.5584103 -0.7897114
1.6 0.5 --> -0.2902213 -0.2304799
1.6 0.6 --> -1.552947 -3.332638
```

12.5.2 On differentiation

A potential disadvantage of Newton's method is the need to calculate the gradient and Hessian. Steepest ascent only requires the gradient, but sometimes even this can be difficult. For functions that can be expressed in terms of polynomials and the simple transcendental functions (sin, exp, sinh, etc.), the process of calculating the gradient and Hessian is essentially mechanical and should pose no problems. Life is not always easy however, and there are plenty of situations where f is available but ∇f is not. For example f may be the result of some numerical procedure (possibly even the result of an optimisation procedure) or an approximation obtained by simulation.

In this situation there are two approaches we can take. The first assumes that even if we don't know what they are, $\mathbf{H}$ and/or ∇f do exist, in which case we can try and estimate them. Exercise 5 explores this option.

The second approach is to use an optimisation method that does not require the gradient. Such approaches tend to be relatively slow, but relatively reliable. In one dimension the golden-section algorithm is an example of a derivative-free approach. In higher dimensions there is an algorithm due to Nelder & Mead, which is well accepted and again is derivative-free. We will not describe the Nelder-Mead algorithm here, though we note that it is implemented in R (see Section 12.6).

We remarked above that for many functions, calculating the gradient and Hessian is essentially mechanical. Computers are, of course, well suited to mechanical tasks, and R provides the function `deriv` for symbolic calculation of the gradient and Hessian of a function. For example, we can reproduce the function `f3` from Example 12.5.1 as follows:

```
Df <- deriv(z ~ sin(x^2/2 - y^2/4)*cos(2*x - exp(y)),
            c('x', 'y'), func=TRUE, hessian=TRUE)

f3 <- function(x) {
```

```
  Dfx <- Df(x[1], x[2])
  f <- Dfx[1]
  gradf <- attr(Dfx, 'gradient')[1,]
  hessf <- attr(Dfx, 'hessian')[1,,]
  return(list(f, gradf, hessf))
}
```

For further details of deriv please see its help page.

12.6 Optimisation in R and the wider world

In one dimension R provides the function `optimize`, which uses a combination of the golden-section algorithm with a technique called *parabolic interpolation*.

In higher dimensions—which is where all the fun is—there are a variety of optimisation methods in current use. The R function `optim` provides implementations of three deterministic methods: the *Nelder-Mead* algorithm, a *quasi-Newton* algorithm (also called a variable metric algorithm), and the *conjugate gradient* method. Neither the steepest descent nor Newton's method are much used in practice, the first because it is relatively slow and the second because of its unpredictable behaviour.

There are also stochastic optimisation techniques, which as the name suggests, employ randomness in their search for an optimal point. `optim` implements just one of these, the *simulated annealing* method. By introducing randomness, stochastic optimisation techniques try to increase the chance of finding a *global optimum*, as opposed to a local optimum, irrespective of your starting point. With a deterministic technique, the local minimum or maximum you find depends entirely on where you start, which can be a significant drawback if you know little about the function you are trying to minimise.

The optimisation methods we have considered are for *unconstrained* problems with *continuous* variables. That is, we optimise $f(\mathbf{x})$ over all $\mathbf{x} \in \mathbb{R}^d$. Different techniques are required if we constrain $\mathbf{x}$ to lie in a connected subset of $\mathbb{R}^d$ (*constrained optimisation*) or if we constrain $\mathbf{x}$ to lie in a discrete set such as $\mathbb{Z}^d$ (*discrete optimisation*). Another practical complication, which also requires different approaches, is that we may only be able to observe $f(\mathbf{x})$ with uncertainty. In particular this will be the case if $f(\mathbf{x})$ is estimated using simulation.

For further reading on numerical optimisation we recommend *Numerical Mathematics and Computing, Sixth Edition*, W. Cheney & D. Kincaid. Thomson Brooks/Cole Publishing Co., 2008, and *Numerical Recipes 3rd Edition: The Art of Scientific Computing*, W.H. Press, S.A. Teukolsky, W.T. Vetterling, B.P. Flannery. Cambridge University Press, 2007.

12.7 A curve fitting example

Suppose we have observations $(x_1, y_1), \ldots, (x_n, y_n)$ and we want to find a function f such that $y_i \approx f(x_i)$ for $i = 1, \ldots, n$. Further suppose that f can be *parameterised* by some vector of parameters $\boldsymbol{\theta} = (\theta_1, \ldots, \theta_d)^T$. For example, if we restrict f to be a quadratic then it has the form $f(x) = ax^2+bx+c$, in which case $\boldsymbol{\theta} = (a, b, c)^T$. We write $f(x; \boldsymbol{\theta})$ for $f(x)$ to emphasise the dependence on $\boldsymbol{\theta}$.

The problem of finding the parameter $\boldsymbol{\theta}^*$, such that the fitted points $\hat{y}_i = f(x_i; \boldsymbol{\theta}^*)$ are 'closest' to the observations y_i, is called *curve fitting.*

To measure how close the fitted points are to the observed points, we use a *loss function*, which measures the distance between $\mathbf{y} = (y_1, \ldots, y_n)^T$ and $\hat{\mathbf{y}} = (\hat{y}_1, \ldots, \hat{y}_n)^T$. Two popular choices are the sum of squares

$$L_2(\boldsymbol{\theta}) = \sum_{i=1}^{n} (y_i - \hat{y}_i)^2$$

and the sum of absolute differences

$$L_1(\boldsymbol{\theta}) = \sum_{i=1}^{n} |y_i - \hat{y}_i|.$$

Note that we consider a loss function to be a function of $\boldsymbol{\theta}$, rather than a function of $\mathbf{y}$, because we are interested in how the loss changes as we change $\boldsymbol{\theta}$. Given a loss function L, we choose $\boldsymbol{\theta}^*$ to be that $\boldsymbol{\theta}$ that minimises $L(\boldsymbol{\theta})$.

As an example we consider some data on the growth of trees. In Figure 12.7 we have plotted the volume of a spruce tree at different ages.[4] Volume refers to the volume of the trunk and is measured in m^3. Age is measured by counting growth rings from a core taken at height 1.3 m, and thus is measured in years since the trunk reached a height of 1.3 m. A popular ecological model for the plant size (measured by volume) as a function of age is the Richards curve:

$$f(x) = a(1 - e^{-bx})^c.$$

Here the parameters are $\boldsymbol{\theta} = (a, b, c)^T$. Parameter a gives the maximum size of the plant and parameter b describes the speed of growth. For biological reasons parameter c is often expected to be close to 3.

We will use the R function `optim` to fit the Richards curve to the observations plotted in Figure 12.7. We will use both the sum of squares loss function and the sum of absolute differences, and compare the results. The default operation of `optim` is to minimise, which suits us, and the default method is the Nelder-Mead, which also suits us. The reason for preferring the Nelder-Mead method is that is does not require gradients, and the L_1 loss function is not everywhere

[4] Taken from A.R. von Guttenberg, Growth and yield of spruce in Hochgebirge. Franz Deuticke, Wien, 1915 (in German).

differentiable. The gradient of L_2 with respect to $\boldsymbol{\theta}$ can be calculated, so we could conceivably use other optimisation methods in that case.

Our observations are in the form of a table with three columns: `ID`, `Age`, and `Vol` (stored in a comma-separated file). The table contains data on a number of trees. Figure 12.7 is for the tree with `ID` equal to `1.3.11`. In the code below we first define the function f, called **`richards`**, and the two loss functions, `loss.L2` and `loss.L1`. We then read the data into a dataframe, extract the data relevant to tree `1.3.11`, and apply `optim`. Finally we plot the fitted functions against the observations; the result is given in Figure 12.7. We see that both loss functions have led to good fits, though slightly different.

```
> richards <- function(t, theta)
+   theta[1]*(1 - exp(-theta[2]*t))^theta[3]
> loss.L2 <- function(theta, age, vol)
+   sum((vol - richards(age, theta))^2)
> loss.L1 <- function(theta, age, vol)
+   sum(abs(vol - richards(age, theta)))
> trees <- read.csv("../data/trees.csv")
> tree <- trees[trees$ID=="1.3.11", 2:3]
> theta0 <- c(1000, 0.1, 3)
> theta.L2 <- optim(theta0, loss.L2, age=tree$Age, vol=tree$Vol)
> theta.L1 <- optim(theta0, loss.L1, age=tree$Age, vol=tree$Vol)
> par(las=1)
> plot(tree$Age, tree$Vol, type="p", xlab="Age", ylab="Volume",
+   main='Tree 1.3.11')
> lines(tree$Age, richards(tree$Age, theta.L2$par), col="blue")
> lines(tree$Age, richards(tree$Age, theta.L1$par), col="blue", lty=2)
```

12.8 Exercises

1. In the golden-section algorithm, suppose that you start with $x_l = 0$, $x_m = 0.5$, and $x_r = 1$, and that at each step if $x_r - x_m > x_m - x_l$, then $y = (x_m + x_r)/2$, while if $x_r - x_m \le x_m - x_l$, then $y = (x_l + x_m)/2$.

   In the worst-case scenario, for this choice of y, by what factor does the width of the bracketing interval reduce each time? In the worst-case scenario, is this choice of y better or worse than the usual golden-section rule?

   What about the best-case scenario?

2. Use the golden-section search algorithm to find all of the local maxima of the function

$$f(x) = \begin{cases} 0, & x = 0 \\ |x| \log(|x|/2) e^{-|x|}, & \text{o/w} \end{cases}$$

   within the interval $[-10, 10]$.

   Hint: plotting the function first will give you a good idea where to look.

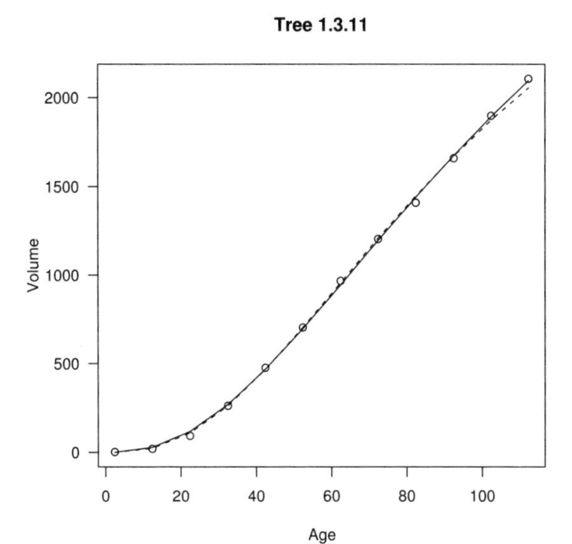

Figure 12.7 *Growth of a tree, as per Example 12.7. The points are observed values. The solid curve is a Richards curve fitted using the sum of squares loss function. The dashed curve is a Richards curve fitted using the sum of absolute differences loss function.*

3. Write a version of function `gsection` that plots intermediate results. That is, plot the function being optimised, then at each step draw a vertical line at the positions x_l, x_r, x_m, and y (with the line at y in a different colour).
4. The Rosenbrock function is a commonly used test function, given by

$$f(x, y) = (1 - x)^2 + 100(y - x^2)^2.$$

You can plot the function in the region $[-2, 2] \times [-2, 5]$ using the code below (Figure 12.8). It has a single global minimum at $(1, 1)$.

```
# program spuRs/resources/scripts/Rosenbrock.r

Rosenbrock <- function(x) {
  g <- (1 - x[1])^2 + 100*(x[2] - x[1]^2)^2
  g1 <- -2*(1 - x[1]) - 400*(x[2] - x[1]^2)*x[1]
  g2 <- 200*(x[2] - x[1]^2)
  g11 <- 2 - 400*x[2] + 1200*x[1]^2
  g12 <- -400*x[1]
  g22 <- 200
  return(list(g, c(g1, g2), matrix(c(g11, g12, g12, g22), 2, 2)))
}

x <- seq(-2, 2, .1)
y <- seq(-2, 5, .1)
```

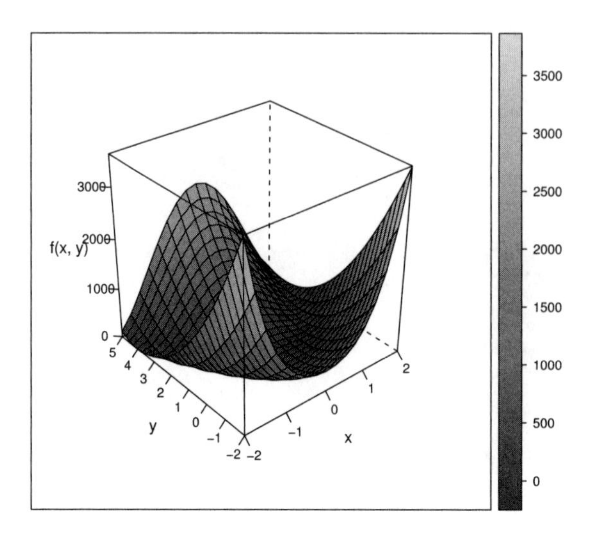

Figure 12.8 *The Rosenbrock function. See Exercise 4.*

```
xyz <- data.frame(matrix(0, length(x)*length(y), 3))
names(xyz) <- c('x', 'y', 'z')
n <- 0
for (i in 1:length(x)) {
  for (j in 1:length(y)) {
    n <- n + 1
    xyz[n,] <- c(x[i], y[j], Rosenbrock(c(x[i], y[j]))[[1]])
  }
}
library(lattice)
print(wireframe(z ~ x*y, data = xyz, scales = list(arrows = FALSE),
  zlab = 'f(x, y)', drape = T))
```

Use the function `contourplot` from the `lattice` package to form a contour-plot of the Rosenbrock function over the region $[-2,2]\times[-2,5]$. Next modify `ascent` so that it plots each step on the contour-plot, and use it to find the *minimum* of the Rosenbrock function (that is, the maximum of $-f$), starting at $(0,3)$. Use a tolerance of 10^{-9} and increase the maximum number of iterations to at least 10,000.

Now use `newton` to find the minimum (no need to use $-f$ in this case). Plot the steps made by the Newton method and compare them with those made by the steepest descent method.

5. Suppose $f : \mathbb{R}^d \to \mathbb{R}$. Since $\partial f(\mathbf{x})/\partial x_i = \lim_{\epsilon\to 0}(f(\mathbf{x}+\epsilon\mathbf{e}_i) - f(\mathbf{x}))/\epsilon$, we have for small ϵ

$$\frac{\partial f(\mathbf{x})}{\partial x_i} \approx \frac{f(\mathbf{x}+\epsilon\mathbf{e}_i) - f(\mathbf{x})}{\epsilon}.$$

In the same way, show that for $i \neq j$

$$\frac{\partial^2 f(\mathbf{x})}{\partial x_i \partial x_j} \approx \frac{f(\mathbf{x}+\epsilon\mathbf{e}_i+\epsilon\mathbf{e}_j) - f(\mathbf{x}+\epsilon\mathbf{e}_i) - f(\mathbf{x}+\epsilon\mathbf{e}_j) + f(\mathbf{x})}{\epsilon^2}$$

and

$$\frac{\partial^2 f(\mathbf{x})}{\partial x_i^2} \approx \frac{f(\mathbf{x}+2\epsilon\mathbf{e}_i) - 2f(\mathbf{x}+\epsilon\mathbf{e}_i) + f(\mathbf{x})}{\epsilon^2}.$$

(a). Test the accuracy of these approximations using the function $f(x, y) = x^3 + xy^2$ at the point $(1, 1)$. That is, for a variety of ϵ, calculate the approximate gradient and Hessian, and see by how much they differ from the true gradient and Hessian.

In R real numbers are only accurate to order 10^{-16} (try `1+10^{-16} == 1`). Thus the error in estimating $\partial f(\mathbf{x})/\partial x_i$ is of the order $10^{-16}/\epsilon$. For example, if $\epsilon = 10^{-8}$ then the error will be order 10^{-8}. It is worse for second-order derivatives: the error in estimating $\partial^2 f(\mathbf{x})/\partial x_i \partial x_j$ is of the order $10^{-16}/\epsilon^2$. Thus if $\epsilon = 10^{-8}$ then the error will be order 1. We see that we have a trade-off in our choice of ϵ: too large and we have a poor approximation of the limit; too small and we suffer rounding error.

(b). Modify the steepest ascent method, replacing the gradient with an approximation. Apply your modified algorithm to the function $f(x, y) = \sin(x^2/2 - y^2/4)\cos(2x - \exp(y))$, using the same starting points as in Example 12.4.2.

How does the algorithm's behaviour depend on your choice of ϵ? You might find it helpful to plot each step, as in Exercise 4.

6. A simple way of using local search techniques to find a global maximum is to consider several different starting points, and hope that for one of them its local maximum is in fact the global maximum. If you have no idea where to start, then randomisation can be used to choose the starting point.

   Consider the function

$$f(x, y) = -(x^2+y^2-2)(x^2+y^2-1)(x^2+y^2)(x^2+y^2+1)(x^2+y^2+2) \times \left(2 - \sin(x^2-y^2)\cos(y - \exp(y))\right).$$

   It has several local maxima in the region $[-1.5, 1.5] \times [-1.5, 1.5]$. Using several randomly chosen starting points, use steepest ascent to find all of the local maxima of f, and thus the global maximum. You can use the command `runif(2, -1.5, 1.5)` to generate a random point (x, y) in the region $[-1.5, 1.5] \times [-1.5, 1.5]$.

   A picture of f is given in Figure 12.9. Note that f has been truncated below at -3.

7. This question follows on from Example 12.7.

   The three parameters of the Richards curve give a concise summary of

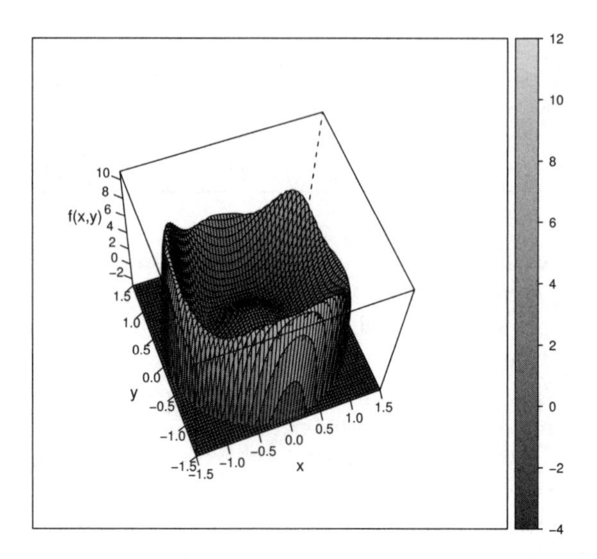

Figure 12.9 *A function with a number of local maxima: see Exercise 6.*

the growth behaviour of a tree. In practice, the optimal management of a timber plantation requires knowledge of how different trees grow in different conditions, so that you can choose which trees to put where, and how far apart.

The table `trees.csv` contains information on spruce trees grown in a number of different sites. Each tree has ID of the form $x.y.z$, where x gives the site the tree is from and y a location within that site. Fit the Richards curve to all the trees given in the table, then for each tree plot the point (a, b) on a graph, where (a, b, c) are the curve parameters. Label each point according to the site the tree comes from: can you see any relation between the site a tree is from and the parameters of its Richards curve?

Hint: to print the character `1` at point (a, b) use `text(a, b, '1')`.

PART III

Probability and statistics

CHAPTER 13

Probability

In Part III we introduce mathematical probability, which allows us to describe and think about uncertainty in a precise fashion. Probability is an essential tool for developing and interpreting stochastic simulations, which are very useful for gaining insight about situations that are too complex to analyse theoretically. Thus this part of the book launches us into Part IV, on simulation.

In this chapter we cover the probability axioms and conditional probability. We also cover the Law of Total Probability, which can be used to decompose complicated probabilities into simpler ones that are easier to compute, and Bayes' theorem, which is used to manipulate conditional probabilities in very useful ways.

13.1 The probability axioms

The first step in describing uncertainty is to consider sets of possible *outcomes*. The set of all possible outcomes is called the *sample space* and is often denoted as Ω. An *event* is defined as being any subset of Ω.

For example when rolling a die we have $\Omega = \{1, 2, 3, 4, 5, 6\}$ and the event 'an even number' is given by the subset $\{2, 4, 6\}$.

Here is some notation and useful results for manipulating sets

$$
\begin{aligned}
\emptyset &= \text{the empty set;} \\
A \backslash B &= \{x : x \in A \text{ and } x \notin B\}, A \text{ not } B; \\
\overline{A} &= \Omega \backslash A, \text{ the complement of } A; \\
A \cup B &= \{x : x \in A \text{ or } x \in B\}, \text{ the union of } A \text{ and } B; \\
A \cap B &= \{x : x \in A \text{ and } x \in B\}, \text{ the intersection of } A \text{ and } B; \\
\overline{A \cup B} &= \overline{A} \cap \overline{B}; \\
\overline{A \cap B} &= \overline{A} \cup \overline{B}; \\
|A| &= \text{size of } A \text{ (the number of elements it contains).}
\end{aligned}
$$

A and B are *disjoint* if $A \cap B = \emptyset$. A *partition* of Ω is a collection of disjoint sets whose union is Ω.

The second step in describing uncertainty is to assign a probability to events, that is, subsets of Ω. A probability measure is a function, generally denoted $\mathbb{P}$, that takes an event or set and returns a value in $[0, 1]$ that indicates how 'likely' that event is. An intuitive (Frequentist) interpretation of this value is the long-run frequency of the event in a sequence of independent identical experiments (in cases where such repetition makes sense):

$$\frac{\text{no. times } A \text{ occurs}}{\text{no. trials}} \to \mathbb{P}(A). \tag{13.1}$$

An alternative (subjective Bayesian) interpretation is to think of probability as being a personal assessment. For a subjective Bayesian, probability is not constrained to cases where a sequence of identical experiments can be identified. In either case it is clear that a probability measure should have the following properties:

$$\begin{aligned} \mathbb{P}(\Omega) &= 1; \\ \mathbb{P}(A) &\geq 0; \\ \mathbb{P}(A \cup B) &= \mathbb{P}(A) + \mathbb{P}(B) \text{ if } A \text{ and } B \text{ are disjoint.} \end{aligned}$$

These three requirements are formally adopted as *probability axioms* in the mathematical theory of probability. We *define* a probability measure to be a function on subsets of Ω that satisfies the probability axioms. Probability theory is based on these axioms: all of our proofs ultimately rely on these axioms and nothing more.[1]

It is easy to show that the following results follow from the probability axioms, for any events A and B:

$$\begin{aligned} \mathbb{P}(\emptyset) &= 0; \\ \mathbb{P}(\overline{A}) &= 1 - \mathbb{P}(A); \\ \mathbb{P}(A) &\leq 1; \\ \mathbb{P}(A) &\leq \mathbb{P}(B) \text{ if } A \subset B; \\ \mathbb{P}(A \cup B) &= \mathbb{P}(A) + \mathbb{P}(B) - \mathbb{P}(A \cap B) \text{ (addition rule).} \end{aligned}$$

A set is called *countable* if you can define a one-to-one mapping between the set and the set of positive integers. Intuitively this amounts to being able to count through all the elements of the set by successively numbering them off: one, two, three, and so on. From the addition rule we see that if we know $\mathbb{P}(\{x\})$ for all singleton sets $\{x\}$, then we can work out $\mathbb{P}(A)$ for all countable sets A.

Finite sets are clearly countable and the set of integers is countable, but it can be shown that any interval of the real line is not. If Ω is uncountably infinite, then it turns out that one consequence of adopting our probability axioms

[1] Properly speaking the third axiom should hold for a countably infinite union of disjoint sets, however this finite version is enough for our purposes.

is that it is not possible to assign probabilities to all the subsets of Ω. This leads to some fascinating theory but will not concern us here. If Ω is *countable* then we can always assign a probability to all of its subsets, so our proofs will always use a countable Ω.

A given sample space can have more than one probability measure defined on it. For example consider an experiment where we roll a die. If the die is fair then we would use the following probability:

x	1	2	3	4	5	6
$\mathbb{P}_{fair}(\{x\})$	1/6	1/6	1/6	1/6	1/6	1/6

However, if the die was weighted so that the number six was twice as likely to appear than other numbers, then we would use the following probability:

x	1	2	3	4	5	6
$\mathbb{P}_{unfair}(\{x\})$	1/7	1/7	1/7	1/7	1/7	2/7

13.1.1 Counting probability

If Ω is finite, $|\Omega| = n$, and every element of Ω is equally likely, then for all outcomes x and events A

$$\mathbb{P}(\{x\}) = \frac{1}{n} \text{ and } \mathbb{P}(A) = \frac{|A|}{n}.$$

In this situation we use techniques for counting the size of A to find the probability of A.

Counting involves permutations and combinations. When counting we need to be clear whether or not we are counting *ordered* or *unordered* sets. The number of ways of choosing r things from a set of n when order matters is $n!/(n-r)!$, a *permutation*; when order does not matter it is $\binom{n}{r} = n!/(r!(n-r)!)$, a *combination*.

If we order the suits $\clubsuit$, $\diamondsuit$, $\heartsuit$, $\spadesuit$, then we have a proper ordering of a deck of cards. Suppose we draw five cards from a well shuffled pack and we wish to know $\mathbb{P}$(exactly 2 hearts) and $\mathbb{P}$(the cards are drawn in increasing order). For the first calculation order does not matter: outcomes are unordered sets of five cards, so our state space has size

$$|\Omega_1| = \binom{52}{5}.$$

Let $A = \{\text{exactly 2 hearts}\}$ then $|A| = \binom{13}{2}\binom{39}{3}$. The elements of Ω_1 are equally likely so $\mathbb{P}(\text{exactly 2 hearts}) = \binom{13}{2}\binom{39}{3}/\binom{52}{5}$.

For the second calculation order does matter. We now suppose that outcomes are ordered sets of five cards, so our state space has size

$$|\Omega_2| = \frac{52!}{47!}.$$

Let $B = \{\text{the cards are drawn in increasing order}\}$ then $|B| = \binom{52}{5}$, since for each possible choice of five cards, in only one case are they drawn in increasing order. The elements of Ω_2 are equally likely so $\mathbb{P}(\text{the cards are drawn in increasing order}) = \binom{52}{5}47!/52!$.

13.2 Conditional probability

For events A and B the conditional probability of A given B is the probability that A occurs, assuming that B has occurred. An example of such a quantity is the probability of getting heart disease given that you are a smoker. We write this as $\mathbb{P}(A \mid B)$.

Formally, given a sequence of independent trials, to get the conditional probability of A given B, we just discard all trials with outcome not in B, then consider the frequency with which A occurs amongst the remaining trials. That is

$$\frac{\text{no. times } A \text{ and } B \text{ occur}}{\text{no. times } B \text{ occurs}} \to \mathbb{P}(A \mid B).$$

Dividing the top and bottom by the number of trials n, we get

$$\begin{aligned}\frac{\text{no. times } A \text{ and } B \text{ occur}}{\text{no. times } B \text{ occurs}} &= \frac{\text{no. times } A \text{ and } B \text{ occur}}{n} \frac{n}{\text{no. times } B \text{ occurs}} \\ &\to \mathbb{P}(A \cap B)/\mathbb{P}(B).\end{aligned}$$

For example the probability that a fair die is even (event A) is $1/2$, but if we are told that the result is less than 4 (event B), then conditional on that knowledge the probability of A changes to $1/3$. We *define* conditional probability as follows:

$$\mathbb{P}(A \mid B) = \mathbb{P}(A \cap B)/\mathbb{P}(B).$$

This is sometimes referred to as conditioning on B. Clearly the definition is only sensible if $\mathbb{P}(B) > 0$. We can rearrange the definition to obtain

$$\mathbb{P}(A \cap B) = \mathbb{P}(B)\mathbb{P}(A \mid B) = \mathbb{P}(A)\mathbb{P}(B \mid A).$$

This useful general rule for calculating the probability of the intersection of two events is called the *Multiplication rule.*

Conditional probability is equivalent to restricting our sample space to B then scaling all our probabilities by $1/\mathbb{P}(B)$. It is easy to check that $\mathbb{P}(\cdot \mid B)$ is a probability measure on B, we just need to check that the probability axioms hold. For any disjoint events $A, C \subset B$ we have:

$$\begin{aligned}\mathbb{P}(B \mid B) &= 1; \\ \mathbb{P}(A \mid B) &\geq 0;\end{aligned}$$

$$\begin{aligned}\mathbb{P}(A \cup C \,|\, B) &= \mathbb{P}((A \cup C) \cap B)/\mathbb{P}(B) \\ &= \mathbb{P}((A \cap B) \cup (C \cap B))/\mathbb{P}(B) \\ &= (\mathbb{P}(A \cap B) + \mathbb{P}(C \cap B))/\mathbb{P}(B) \\ &= \mathbb{P}(A \,|\, B) + \mathbb{P}(C \,|\, B).\end{aligned}$$

This means that all the results we derive for regular probabilities (such as the addition rule) also hold for conditional probabilities.

13.2.1 Example: life tables

From life tables, one finds that 89.935% of women live to age 60 and 57.062% live to age 80. Thus, noting that the event 'alive at 80' includes the event 'alive at 60', we have

$$\begin{aligned}&\mathbb{P}(\text{a woman lives to age 80 given that she is alive at age 60}) \\ &= \frac{0.57062}{0.89935} = 0.63448 \text{ (to 5 decimal places).}\end{aligned}$$

13.2.2 Example: indigenous deaths in custody

The following data for 1992–3 are taken from an Australian Institute of Criminology Report.[2] We consider prison deaths.

	Indigenous	Non-indigenous
Deaths in prison	4	38
Population 15+	160,000	12,926,000
Prison population	2,198	13,361

The population aged under 15 is ignored since they are not sentenced to prison, but to youth detention centres.

Let us define the following events: I, being indigenous; P, being in prison in 1992–3; D, dying in 1992–3. Let the total population over 15 (= 13,086,000) be N.

The Royal Commission into Aboriginal Deaths in Custody wanted to assess the evidence for the suspicion that indigenous people had a higher chance of dying in custody than non-indgenous people. If we compare $\mathbb{P}(D \cap P \mid I)$ and $\mathbb{P}(D \cap P \mid \bar{I})$ then from the table, we get:

$$\mathbb{P}(D \cap P \mid I) = \frac{\mathbb{P}(D \cap P \cap I)}{\mathbb{P}(I)} = \frac{4}{N} / \frac{160,000}{N} = 4/160,000 \approx 2.5 \times 10^{-5}$$

[2] `http://www.aic.gov.au/publications/dic/dic6.pdf`

but

$$\mathbb{P}(D\cap P \mid \bar{I}) = \frac{\mathbb{P}(D\cap P\cap \bar{I})}{\mathbb{P}(\bar{I})} = \frac{38}{N}\Big/\frac{12,926,000}{N} = 38/12,926,000 \approx 3.0\times 10^{-6}$$

so clearly there is a large dependence between events I and $D\cap P$.

However, looking more carefully at the data, we see that there are two components to this probability:

$$\mathbb{P}(D\cap P \mid I) = \frac{\mathbb{P}(D\cap P\cap I)}{\mathbb{P}(I)} = \frac{\mathbb{P}(D \mid P\cap I)\mathbb{P}(P\cap I)}{\mathbb{P}(I)} = \mathbb{P}(D \mid P\cap I)\mathbb{P}(P \mid I)$$

and similarly for $\bar{I}$.

The reason we write the probability this way is that $\mathbb{P}(D \mid P\cap I)$ is largely dependent on prison conditions/supervision, while $\mathbb{P}(P \mid I)$ is the incarceration rate, which depends on a much broader range of policy issues.

Using the data above, we get for the various conditional probabilities:

	$A = I$	$A = \bar{I}$
$\mathbb{P}(D \mid P\cap A)$	$\frac{4}{2198} \approx 1.8\times 10^{-3}$	$\frac{38}{13{,}361} \approx 2.8\times 10^{-3}$
$\mathbb{P}(P \mid A)$	$\frac{2198}{160{,}000} \approx 1.4\times 10^{-2}$	$\frac{13{,}361}{12{,}926{,}000} \approx 1.0\times 10^{-3}$
$\mathbb{P}(D\cap P \mid A)$	$\frac{4}{160{,}000} \approx 2.5\times 10^{-5}$	$\frac{38}{12{,}926{,}000} \approx 3.0\times 10^{-6}$

From this table, we see that the chance of dying once in prison is actually slightly higher for non-indigenous prisoners; the large discrepancy is in the incarceration rate of indigenous Australians.

13.3 Independence

We say that events A and B are *independent* if $\mathbb{P}(A\cap B) = \mathbb{P}(A)\mathbb{P}(B)$, or equivalently if $\mathbb{P}(A \mid B) = \mathbb{P}(A)$ or $\mathbb{P}(B \mid A) = \mathbb{P}(B)$. We interpret independence as the occurrence or not of B has no effect on the occurrence of A.

13.3.1 Example: disjoint events

If $\mathbb{P}(A), \mathbb{P}(B) > 0$ and A and B are disjoint, then they are dependent. This is because $\mathbb{P}(A\cap B) = \mathbb{P}(\emptyset) = 0 \neq \mathbb{P}(A)\mathbb{P}(B)$.

You should check that if $A = \emptyset$ or $A = \Omega$ then A is independent of all other events B.

13.3.2 Example: the Chevalier de Meré

The Chevalier de Meré was a seventeenth century French count who made money by betting—at even money—that he could throw at least one six in four die rolls. Eventually he couldn't find anyone to bet against him so he changed the bet to at least one double six in twenty four rolls of two dice, and started losing money.

Here's why. Using independence we have

$$\begin{aligned}
&\mathbb{P}(\text{at least one six in four rolls}) \\
&\quad = 1 - \mathbb{P}(\text{no sixes in four rolls}) \\
&\quad = 1 - \mathbb{P}(\text{no six in roll one}) \times \cdots \times \mathbb{P}(\text{no six in roll four}) \\
&\quad = 1 - (5/6)^4 \\
&\quad = 0.5177 \text{ (to four decimal places).}
\end{aligned}$$

and

$$\begin{aligned}
&\mathbb{P}(\text{at least one double six in twenty four rolls}) \\
&\quad = 1 - \mathbb{P}(\text{no double sixes in twenty four rolls}) \\
&\quad = 1 - \mathbb{P}(\text{no double six in roll one}) \times \cdots \\
&\qquad\quad \cdots \times \mathbb{P}(\text{no double six in roll twenty four}) \\
&\quad = 1 - (35/36)^{24} \\
&\quad = 0.4914 \text{ (to four decimal places).}
\end{aligned}$$

So the modified bet was not equivalent to the original as the Count had hoped, but converted a healthy 1.77% advantage into a 0.86% disadvantage.

13.4 The Law of Total Probability

Suppose that events $E_1, \ldots, E_k$ partition the sample space Ω. That is $E_i \cap E_j = \emptyset$ for all $i \neq j$ and $E_1 \cup \cdots \cup E_k = \Omega$. Then for any event A we have

$$\begin{aligned}
\mathbb{P}(A) &= \mathbb{P}(A \cap \Omega) \\
&= \mathbb{P}(A \cap (E_1 \cup \cdots \cup E_k)) \\
&= \mathbb{P}((A \cap E_1) \cup \cdots \cup (A \cap E_k)) \\
&= \mathbb{P}(A \cap E_1) + \cdots + \mathbb{P}(A \cap E_k) \text{ (disjoint events)} \\
&= \mathbb{P}(A \mid E_1)\mathbb{P}(E_1) + \cdots + \mathbb{P}(A \mid E_k)\mathbb{P}(E_k).
\end{aligned}$$

This result is called the *Law of Total Probability*. It is a very useful method for splitting a complex event A into a sequence of simpler events $A \cap E_1, \ldots, A \cap E_k$.

For example, in 2003, 53% of VCE students[3] were female, with a pass rate of 96.5%. Male students had a 95.5% pass rate. What was the overall pass rate? Let M be the event 'male', F the event 'female', and P the event 'pass', then

[3] Secondary students in the Australian state of Victoria study towards a Victorian Certificate of Education (VCE).

we have

$$\begin{aligned}\mathbb{P}(P) &= \mathbb{P}(P \mid M)\mathbb{P}(M) + \mathbb{P}(P \mid F)\mathbb{P}(F) \\ &= 0.955 \times (1 - 0.53) + 0.965 \times 0.53 \\ &= 0.960 \text{ (to three decimal places).}\end{aligned}$$

13.5 Bayes' theorem

For any events A and B we have

$$\begin{aligned}\mathbb{P}(B \mid A) &= \frac{\mathbb{P}(B \cap A)}{\mathbb{P}(A)} \\ &= \frac{\mathbb{P}(A \mid B)\mathbb{P}(B)}{\mathbb{P}(A)}.\end{aligned}$$

This result is called *Bayes' theorem.* Bayes' theorem is used to find conditional probabilities $\mathbb{P}(B \mid A)$ when you already know $\mathbb{P}(A \mid B)$.[4]

If we choose the event B to be E_1, where $E_1, \ldots, E_k$ is a partition of the sample space Ω, then we have

$$\begin{aligned}\mathbb{P}(E_1 \mid A) &= \frac{\mathbb{P}(A \mid E_1)\mathbb{P}(E_1)}{\mathbb{P}(A)} \\ &= \frac{\mathbb{P}(A \mid E_1)\mathbb{P}(E_1)}{\mathbb{P}(A \mid E_1)\mathbb{P}(E_1) + \cdots + \mathbb{P}(A \mid E_k)\mathbb{P}(E_k)}.\end{aligned}$$

The last step uses the Law of Total Probability. This is a commonly stated form of Bayes' theorem, but note that the original theorem does not require the introduction of a partition.

13.5.1 Example: prostate cancer screening

Here are data on the effectiveness of digital rectal examination (DRE) to screen for prostate cancer in men.[5]

Let P be the event 'return a positive test result' and let C be the event 'have prostate cancer', then we have

$$\begin{aligned}\mathbb{P}(P \mid C) &= 0.57; & \mathbb{P}(\overline{P} \mid C) &= 0.43; \\ \mathbb{P}(P \mid \overline{C}) &= 0.08; & \mathbb{P}(\overline{P} \mid \overline{C}) &= 0.92 \\ \mathbb{P}(C) &= 0.037; & \mathbb{P}(\overline{C}) &= 0.963.\end{aligned}$$

[4] Bayes' theorem is a fundamental part of the Bayesian theory of probability, which models how prior assumptions about the probability of events can be updated as further information becomes available.

[5] http://www.jr2.ox.ac.uk/bandolier/band74/b74-7.html

Now suppose you have no other particular reason to believe that you have cancer, but you receive a positive test result. Then what should you do? Because surgery for prostate cancer has a significant risk of complications, what you really want to know is: what is the chance of having cancer given a positive test result $\mathbb{P}(C \mid P)$?

From the Law of Total Probability we find the probability of having a positive test as

$$\mathbb{P}(P) = \mathbb{P}(P \cap C) + \mathbb{P}(P \cap \overline{C}) = \mathbb{P}(P \mid C)\mathbb{P}(C) + \mathbb{P}(P \mid \overline{C})\mathbb{P}(\overline{C}) = 0.097.$$

Thus Bayes' theorem gives us

$$\mathbb{P}(C \mid P) = \mathbb{P}(P \mid C)\mathbb{P}(C)/\mathbb{P}(P) = 0.57 \times 0.037/0.097 \approx 0.22$$

which is not very high! The reason is that the relatively high false positive rate of 8% generates many more positive test results than those arising from diseased people, as only a small proportion of people are diseased. Let us hope that better tests are developed soon!

13.6 Exercises

1. List the sample space for the following random experiment. First you toss a coin. Then, if you get a head, you throw a single die.
2. Blood is of differing types or blood groups: O, A, B, and AB. Not all are compatible for transfusion purposes. Any recipient can receive the blood from a donor with the same blood group or from a donor with type O blood. A recipient with type AB blood can receive blood of any type. No other combinations will work. Consider an experiment which consists of drawing a litre of blood and determining its type for each of the next two donors who enter a blood bank.
   (a). List the possible (ordered) outcomes of this experiment.
   (b). List the outcomes where the second donor can receive the blood of the first.
   (c). List the outcomes where each donor can receive the blood of the other.
3. (a). The number of alpha particles emitted by a radioactive sample in a fixed time interval is counted. Give the sample space for this experiment.
   (b). The elapsed time is measured until the first alpha particle is emitted. Give the sample space for this experiment.
4. An experiment is conducted to determine what fraction of a piece of metal is gold. Give the sample space for this experiment.
5. A box of n components has r ($r < n$) components which are defective. Components are tested one by one until all defective components are found, and the number of components tested is observed. Describe the sample space for this experiment.

6. Let A, B, C be three arbitrary events. Find expressions for the events that, of A, B, and C,

   (a). Only B occurs.
   (b). Both B and C, but not A, occur.
   (c). All three events occur.
   (d). At least one occurs.
   (e). None occur.

7. Using the probability axioms show that

$$\mathbb{P}(\overline{A} \cap \overline{B}) = 1 - \mathbb{P}(A \cup B).$$

   You may find it helpful to draw a Venn diagram of A and B.

8. Is it possible to have an assignment of probabilities such that $\mathbb{P}(A) = 2/3$, $\mathbb{P}(B) = 1/5$, and $\mathbb{P}(A \cap B) = 1/4$?

9. When an experiment is performed, one and only one of the events A, B, or C will occur. Find $\mathbb{P}(A)$, $\mathbb{P}(B)$, and $\mathbb{P}(C)$ under each of the following assumptions:

   (a). $\mathbb{P}(A) = \mathbb{P}(B) = \mathbb{P}(C)$.
   (b). $\mathbb{P}(A) = \mathbb{P}(B)$ and $\mathbb{P}(C) = 1/4$.
   (c). $\mathbb{P}(A) = 2\mathbb{P}(B) = 3\mathbb{P}(C)$.

10. Consider a sample space $\Omega = \{a, b, c, d, e\}$ in which the following events are defined $A = \{a\}$, $B = \{b\}$, $C = \{c\}$, $D = \{d\}$, $E = \{e\}$. We are given a number of alternative probability measures on this sample space. It seems that in some of them an error has been made with the figures. Find those cases in which an error has been made, indicating why it must be an error. In those cases where there are no apparent errors, find $\mathbb{P}(E)$.

    (a). $\mathbb{P}(A \cup B \cup C \cup D) = 0.5$, $\mathbb{P}(B \cup C \cup D) = 0.6$.
    (b). $\mathbb{P}(A \cup B) = 0.3$, $\mathbb{P}(C \cup D) = 0.5$.
    (c). $\mathbb{P}(A \cup B) = 0.6$, $\mathbb{P}(C) = 0.4$.
    (d). $\mathbb{P}(A \cup B \cup C) = 0.7$, $\mathbb{P}(A \cup B) = \mathbb{P}(B \cup C) = 0.3$.

11. Let A and B be events in a sample space such that $\mathbb{P}(A) = \alpha$, $\mathbb{P}(B) = \beta$, and $\mathbb{P}(A \cap B) = \gamma$. Find an expression for the probabilities of the following events in terms of α, β, and γ.

    (a). $\overline{A} \cap B$.
    (b). $A \cap \overline{B}$.
    (c). $\overline{A} \cap \overline{B}$.

12. If the occurrence of B makes A more likely, does the occurrence of A make B more likely?

13. Suppose that $\mathbb{P}(A) = 0.6$. What can you say about $\mathbb{P}(A|B)$ when

(a). A and B are mutually exclusive?

(b). A is a subset of B?

(c). B is a subset of A?

14. If A and B are events such that $\mathbb{P}(A) = 0.4$ and $\mathbb{P}(A \cup B) = 0.7$, find $\mathbb{P}(B)$ if A and B are

    (a). Mutually exclusive.

    (b). Independent.

15. Determine the conditions under which an event A is independent of its subset B.

16. How many times should a fair coin be tossed in order that the probability of observing at least one head is at least 0.99?

17. A random sample of size n is taken from a population of size N. Write down the number of distinct samples when sampling is:

    (a). Ordered, with replacement.

    (b). Ordered, without replacement.

    (c). Unordered, without replacement.

    (d). Unordered, with replacement.

    (Note that (d) is hard; it may help to write down a few special cases first.)

18. Suppose that both a mother and father carry genes for blood types A and B. They each pass one of these genes to a child and each gene is equally likely to be passed. We assume they pass genes independently. The child will have blood type A if both parents pass their A genes, type B if both pass their B genes, and type AB if one A and one B gene are passed. What are the probabilities that a child of these parents has type A blood? Type B? Type AB?

19. An individual plays roulette using the following system. He bets \$1 that the roulette wheel will come up black. If he wins, he quits. If he loses he makes the same bet a second time but now he bets \$2. Then irrespective of the result, he quits. What is the sample space for this experiment? Assuming he has a probability of $^1/_2$ of winning each bet, what is the probability that he goes home a winner?

20. Two dice are rolled. What is the probability that at least one is a six? If the two faces are different, what is the probability that one is a six? If the sum is seven what is the probability that one die shows a six?

21. A woman has two children. Assume that all possible outcomes for the sexes are equally likely. What is the probability that she has two boys given that

    (a). The eldest is a boy.

    (b). At least one is a boy.

22. Two archers A and B shoot at the same target. Suppose A hits the target with probability 0.65 and independently B hits the target with probability 0.5.

    (a). Given only one of the archers hits the target, what is the probability it was A?

    (b). Given at least one of them hits, what is the probability that B hits?

23. A diagnostic test is used to determine whether or not a person has a certain disease. If the test is positive, then it is assumed the person has the disease, if negative that they don't have it. However the test is not 100% accurate. If a diseased person is tested, it still gives a negative result 5% of the time (a false negative) and when testing a person free of the disease, it gives a false positive 10% of the time. Suppose we choose someone at random from a population in which only 1 person in 50 has the disease.

    (a). Find the probability that their test result is positive.

    (b). Find the probability that their test result is misleading.

    (c). Find the probability that they actually have the disease if they test positive.

24. There are two bus lines which travel between towns A and B. Bus line A runs late 20% of the time, while bus line B runs late 50% of the time. You travel three times as often by line A as you do by line B. On a certain day you arrive late. What is the probability that you used bus line B that day?

25. An electronic system receives signals as input and sends out appropriate coded messages as output.

    The system consists of 3 converters (C_1, C_2, and C_3), 2 monitors (M_1 and M_2), and a perfectly reliable three-way switch for connecting the input to the converters. The incoming signal is encoded by one or more of the converters and the monitors check whether the conversion is correct.

    Initially the signal is fed into C_1. If M_1 passes the conversion, the coded message is sent out. If M_1 rejects the conversion, the input is switched to C_2 and the conversion is checked by M_2. If M_2 passes the conversion, the coded message is sent out. If M_2 rejects the conversion, the input is switched to C_3 and the coded message is sent out without any further checks.

    Each of the converters has probability 0.9 of correctly coding the incoming message. Each of the monitors has probability 0.8 of rejecting a wrongly coded message and also probability 0.8 of passing a correctly coded message.

    Show that the probability of a correct output from the system is about 0.968.

26. The dice game craps is played as follows. The player throws two dice, and if the sum is seven or eleven, then he wins. If the sum is two, three, or twelve, then he loses. If the sum is anything else, then he continues throwing until he either throws that number again (in which case he wins) or he throws

a seven (in which case he loses). Calculate the probability that the player wins.

27. If you toss a coin four times, the probability of getting four heads in a row is $(0.5)^4 = 0.0625$. Suppose that we toss a coin twenty times; what is the probability that we get a sequence of four heads in a row at some point?

    Write a program to estimate this probability. Your answer should be greater than $5 \times 0.0625 = 0.3125$. (Why?)

    Your program should have the following structure:

    (a). A function `four.n.twenty()` that simulates twenty coin tosses, then checks to see if there are four heads in a row.

    (b). A function `four.n.twenty.prob(N)` that executes `four.n.twenty()` `N` times and returns the proportion of times there were four heads in a row.

    Using `four.n.twenty.prob(N)`, what can you do to be confident that your answer is accurate to two decimal places?

    To simulate twenty coin tosses you can use the command `round(runif(20))`, then interpret a 1 as a head and a 0 as a tail. One way to structure `four.n.twenty()` is to first generate a sequence of twenty coin tosses, then for $i = 1, \ldots, 17$ check to see if tosses $(i, i+1, i+2, i+3)$ are all heads. Suppose that you use 1 for a head and 0 for a tail, and that `coins` is a vector of 0's and 1's of length n, then `coins` is a sequence of n heads if and only if `prod(coins) == 1`.

CHAPTER 14

Random variables

In this chapter we introduce the concept of a random variable, which quantifies the outcome of a random experiment. We define discrete and continuous random variables and consider various ways of describing their distributions, including the distribution function, probability mass function, and probability density function. We then define some important aspects of their distributions (expectation and variance) and their relationships with other random variables (independence and covariance).

We also consider transformations of random variables, which are needed for later simulation results, and the Weak Law of Large Numbers, which describes the change in behaviour of an average as the sample size increases. The Weak Law of Large Numbers also justifies using frequencies to estimate probabilities.

In the following two chapters we look in more detail at specific examples of discrete and continuous random variables.

14.1 Definition and distribution function

Suppose that we have a sample space Ω and a probability measure $\mathbb{P}$ that maps events (subsets of Ω) to the interval $[0, 1]$. A *random variable* (or rv) X is a function from Ω to $\mathbb{R}$, the real line. In other words, a random variable is a value that we associate with each outcome in Ω.

We define $\mathbb{P}(X = x)$ to be the probability of the event $\{\omega \in \Omega : X(\omega) = x\}$. More generally $\mathbb{P}(X \in A) = \mathbb{P}(\{\omega \in \Omega : X(\omega) \in A\})$. In what follows we will use the shorthand $\{X \in A\}$ for $\{\omega \in \Omega : X(\omega) \in A\}$.

For example, suppose we toss a fair coin three times and let X be the number of heads until the first tail and let Y be the total number of heads. X and Y are both random variables. Let ω be an element of Ω, then we have

ω	HHH	HHT	HTH	HTT	THH	THT	TTH	TTT
$\mathbb{P}(\{\omega\})$	1/8	1/8	1/8	1/8	1/8	1/8	1/8	1/8
$X(\omega)$	3	2	1	1	0	0	0	0
$Y(\omega)$	3	2	2	1	2	1	1	0

Thus $\mathbb{P}(X = 0) = \mathbb{P}(\{THH, THT, TTH, TTT\}) = 1/2$ and $\mathbb{P}(Y = 0) = \mathbb{P}(\{TTT\}) = 1/8$.

A note on notation: we usually use capital letters for random variables and lower case letters for their possible values.

An important concept is that we can describe a random variable X without having to describe Ω. We do this using the *(cumulative) distribution function* (cdf or df) of the random variable:

$$F(x) = \mathbb{P}(X \leq x).$$

When we are dealing with more than one rv we will write F_X for the df of X. The distribution function is also called the *law* of the random variable or just 'the distribution'. If you know F then for any interval $(a, b]$ we have $\mathbb{P}(a < X \leq b) = F(b) - F(a)$, and using combinations of such intervals we can recover all possible probabilities involving X. One consequence of this formula is that if $a < b$ then $F(a) \leq F(b)$ so F is a non-decreasing function. It is also true that as $x \to -\infty$, $F(x) \to 0$, and as $x \to \infty$, $F(x) \to 1$.

Random variables are a generalising concept with broad applicability; the same distribution function can appear in many different contexts, allowing us to use results from one context in another.

14.2 Discrete and continuous random variables

We identify two particular types of random variable based on their distribution functions. If the distribution function is a step function then the random variable is called *discrete*, and if the distribution function can be written as the integral of some function (called the density) then the random variable is called *continuous*. It is also possible to have random variables that are mixtures of discrete and continuous, but we will not consider these. Examples of discrete and continuous random variables are provided in Figures 14.1 and 14.2.

Discrete random variables If the df F of X has a jump of size p at the point a, then

$$\begin{aligned}
\mathbb{P}(X = a) &= \mathbb{P}(X \leq a) - \mathbb{P}(X < a) \\
&= \mathbb{P}(X \leq a) - \lim_{\epsilon \downarrow 0} \mathbb{P}(X \leq a - \epsilon) \\
&= F(a) - \lim_{\epsilon \downarrow 0} F(a - \epsilon) \\
&= p.
\end{aligned}$$

We describe a discrete random variable X using its *probability mass function* (pmf), often written p or p_X, where for all x

$$p(x) = \mathbb{P}(X = x).$$

We prefer to use p to describe a random variable, rather than F, as it can be

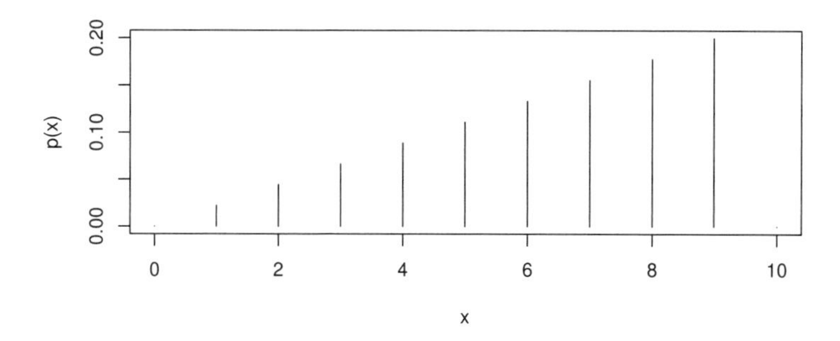

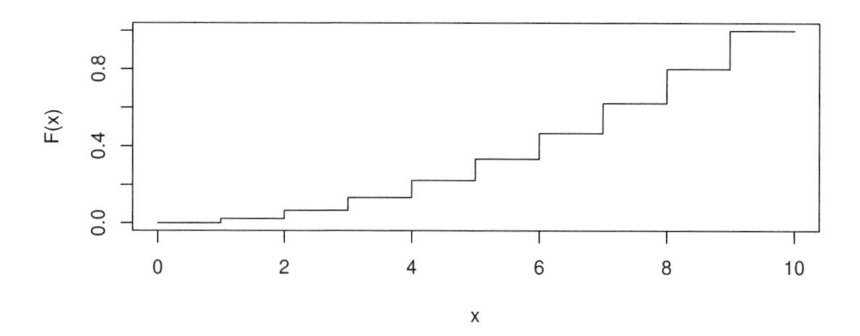

Figure 14.1 *A discrete pmf and the corresponding cdf.*

easily interpreted as 'probability mass'. Given p we can easily recover F and thus the probability of events of interest:

$$\mathbb{P}(X \in (a,b]) = F(b) - F(a) = \sum_{x \in (a,b]} p(x).$$

Of course if we sum the probability mass function over all possible values of the discrete random variable, we get 1.

Continuous random variables For a continuous rv Y, we assume that the cumulative distribution function (cdf) can be written as

$$F(y) = \mathbb{P}(Y \leq y) = \int_{-\infty}^{y} f(u)\, du,$$

where $f = f_Y$ is the probability density function. It follows that F is continuous everywhere and, at any point y for which $f(y)$ is continuous, $F'(y)$ exists and equals $f(y)$ (by the fundamental theorem of calculus). Since F is continuous, for any a we have

$$\mathbb{P}(Y = a) \quad = \quad \mathbb{P}(Y \leq a) - \mathbb{P}(Y < a)$$

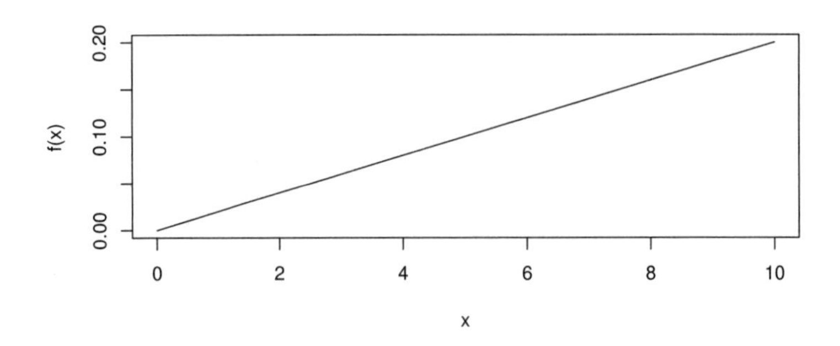

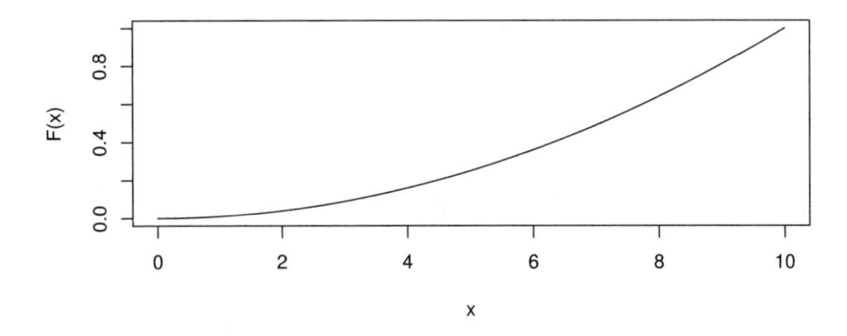

Figure 14.2 *A continuous pdf and the corresponding cdf.*

$$\begin{aligned} &= \mathbb{P}(Y \le a) - \lim_{\epsilon \downarrow 0} \mathbb{P}(Y \le a - \epsilon) \\ &= F(a) - \lim_{\epsilon \downarrow 0} F(a - \epsilon) \\ &= 0. \end{aligned}$$

We can think of $f(y)$ as the density of probability at y:

$$\mathbb{P}(y < Y < y + dy) = f(y)\, dy.$$

In practice a plot of the pdf f is easier to interpret than a plot of the cdf F, but they are equivalent, because F can be obtained by integrating f. The area under the pdf between a and b is $\mathbb{P}(a < Y < b)$, and the total area under the pdf is 1. Note that it does not matter if f is undefined at some points, since the integral of f is not changed if we change f at a single point.

We remark that it is theoretically possible for a random variable to have a continuous cdf, but no density. However such random variables are of limited practical interest, and we do not describe them further.

14.3 Empirical cdf's and histograms

In this section we further develop the idea of probability as a long-term frequency. To do this we need to make precise the concept of a 'sequence of independent trials'.

We start by extending the concept of independence, defined for events in Section 13.3, to random variables. Random variables X and Y are said to be *independent* if any event defined using X is independent of any event defined using Y. That is, for any sets A and B, the events $\{X \in A\}$ and $\{Y \in B\}$ are independent. Independence is usually an assumption that we make, based on a physical understanding of the rv's in question, rather than something we try to prove. For example performing an experiment 100 times in the same conditions, with fresh equipment each time, we might assume that the measurement errors are independent. Informally, X and Y are independent if knowing the value of X tells you nothing new about Y.

Within this book we will say that a *random sample* from the distribution F is a sequence of mutually independent random variables, $X_1, X_2, \ldots, X_n$, with the same distribution function F. Such a sequence is also called an independent and identically distributed (iid) sequence. Given a random sample, for any x we can approximate $F(x)$ using the *empirical distribution function*

$$\hat{F}(x) = \frac{|\{X_i \leq x\}|}{n}. \tag{14.1}$$

If F is from a discrete distribution, then we estimate the probability mass function p using

$$\hat{p}(x) = \frac{|\{X_i = x\}|}{n}. \tag{14.2}$$

$\hat{F}(x)$ and $\hat{p}(x)$ are just the observed frequencies of the events $\{X \leq x\}$ and $\{X = x\}$. Thus our concept of probability as a long-term frequency (Equation 13.1) is equivalent to saying $\hat{F}(x) \to F(x)$ and $\hat{p}(x) \to p(x)$ as the sample size $n \to \infty$. We will in fact prove this rigorously later, thus showing that the idea of probability as long-term frequency is consistent with the probability axioms. The hat notation is used to indicate that $\hat{F}$ and $\hat{p}$ are estimates of F and p, respectively.

For continuous random variables, we estimate the density using a scaled histogram. For small δ we have $F'(x) \approx (F(x+\delta) - F(x))/\delta$, so we put

$$\begin{aligned} \hat{f}_k = \frac{|\{i : k\delta < x_i \leq (k+1)\delta\}|}{n\delta} &= \frac{\hat{F}((k+1)\delta) - \hat{F}(k\delta)}{\delta} \\ &\approx \frac{F((k+1)\delta) - F(k\delta)}{\delta} \approx f(k\delta). \end{aligned}$$

We use $\hat{f}_k$ as an approximation to $f(x)$ for $x \in (k\delta, (k+1)\delta]$.

You should ask yourself what happens to $\hat{f}$ as the sample size $n \to \infty$? And as $\delta \to 0$?

In the context of histograms it is common to call an interval $(k\delta, (k+1)\delta]$ a *bin*. For a continuous random variable X, $\mathbb{P}(X \in (k\delta, (k+1)\delta]) = \mathbb{P}(X \in [k\delta, (k+1)\delta))$, so in theory it doesn't matter whether the bin edges are attached to the bin on the left, or the bin on the right. In practice we have to make a choice, but this is quite arbitrary and different authors make different choices.

Also note that the choice of bin width and the bin start points can affect the apparent distribution of the data. For this reason many analysts prefer to use smooth density estimates, which we do not cover here.

14.3.1 Example: Cavendish's experiments

Here are 29 measurements of the density of the Earth made by Henry Cavendish in 1798, each presented as a multiple of the density of water.

```
> cavendish <- c(5.5, 5.57, 5.42, 5.61, 5.53, 5.47, 4.88,
+      5.62, 5.63, 4.07, 5.29, 5.34, 5.26, 5.44, 5.46, 5.55,
+      5.34, 5.3, 5.36, 5.79, 5.75, 5.29, 5.1, 5.86, 5.58,
+      5.27, 5.85, 5.65, 5.39)
```

Clearly there was some error in the measurement process, which we think of as random. R provides built-in functions for plotting the empirical cdf and scaled histogram.

```
> opar <- par(mfrow = c(2, 1), las = 1, mar = c(4.2, 4, 1, 1))
> plot(ecdf(cavendish),
+       xlab="Density of the Earth", ylab="Cumulative Freq", main="")
> hist(cavendish, freq=TRUE, breaks=20,
+       xlab="Density of the Earth", ylab="Scaled Hist", main="")
> par(opar)
```

The output is given in Figure 14.3. Use `help` for more detail on the `ecdf` and `hist` functions used to generate the figure.

14.4 Expectation and finite approximations

The expectation $\mathbb{E}X$ of a random variable X is akin to its centre of mass, or its 'centre of probability'.

$$\mathbb{E}X = \begin{cases} \sum_x x\, p(x), & X \text{ discrete;} \\ \int x\, f(x)dx, & X \text{ continuous.} \end{cases}$$

Often we will write μ or μ_X for $\mathbb{E}X$. The expectation is also called the expected value or *mean* of the random variable.

The mean is the theoretical analogue of the average. To see this we go back to our concept of probability as a long-term frequency (Equations 13.1 and 14.1).

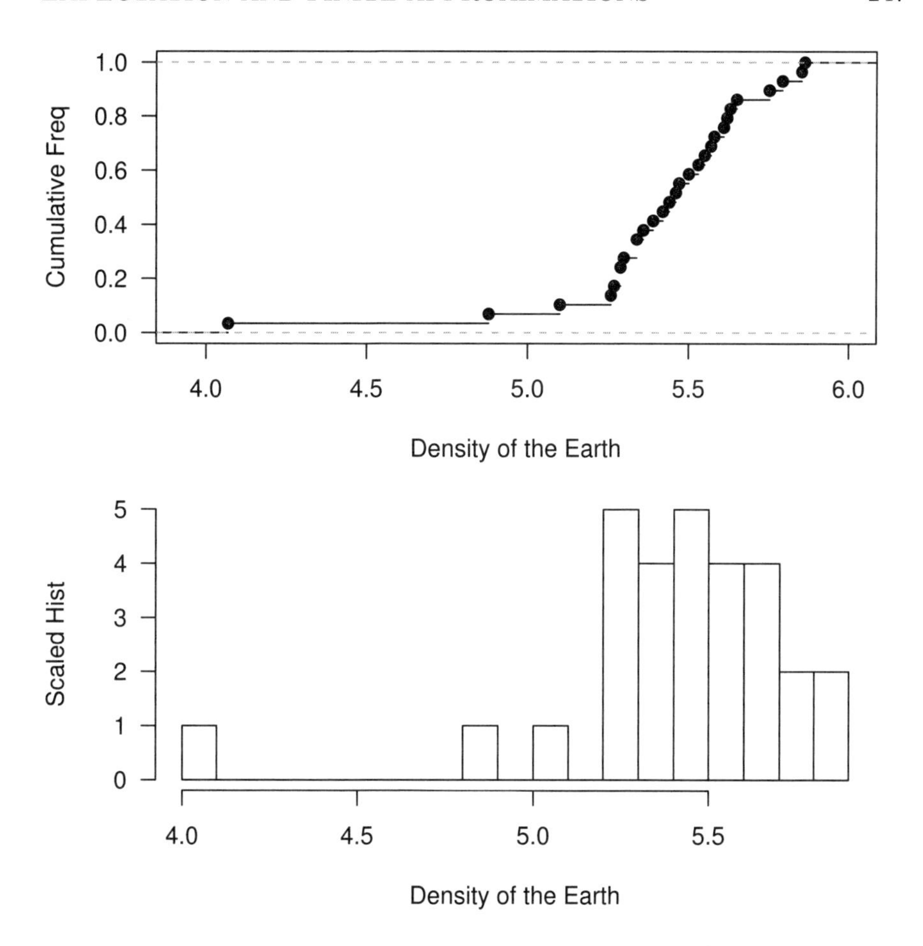

Figure 14.3 *Density of the Earth: empirical cdf and pdf of Cavendish's measurements. See Example 14.3.1.*

Suppose that X is a discrete rv with pmf p and let $X_1, \ldots, X_n$ be an iid sample also with pmf p. Then we have

$$\mathbb{E}X = \sum_x x\, p(x) \approx \sum_x x \frac{|\{X_i = x\}|}{n} = \frac{1}{n} \sum_{i=1}^{n} X_i = \overline{X}.$$

That is, the expectation is approximately the sample average. We will show later that the right-hand side converges to the left-hand side as $n \to \infty$, however we have to be careful what convergence means in this context, as the right-hand side is random.

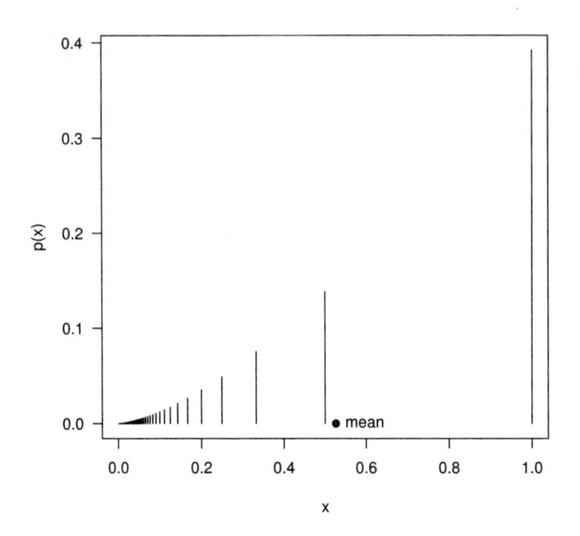

Figure 14.4 *The pmf of a random variable and its mean (Example 14.4.1).*

14.4.1 Example: numerical calculation of the mean `expex.r`

If we know the pmf of a discrete rv X then we can calculate its mean numerically. For example, suppose that $p(x) \propto x^{3/2}$ for x $= 1, 1/2, 1/3, \ldots, 1/1000$. That is, $p(x) = c\,x^{3/2}$ where c is such that $\sum_x p(x) = c\sum_{k=1}^{1000}(1/k)^{3/2} = 1$. Here is some R code for calculating $\mathbb{E}X$. The graphical output is given in Figure 14.4.

```
# program: spuRs/resources/scripts/expex.r
#
# calculating the mean of a discrete rv X

x <- 1/(1000:1)              # possible values for X
pX <- x^1.5                  # probability mass ftn
pX <- pX/sum(pX)             # must have sum(pX) == 1
muX <- sum(x*pX)             # mean

# plot the pmf and mean
par(las=1)
plot(c(0, 1), c(0, max(pX)), type="n", xlab="x", ylab="p(x)")
lines(x, pX, type="h")
points(muX, 0, pch=19)
text(muX, 0, "mean", pos=4)
```

14.4.2 Example: truncated normal

Suppose that X has a normal density truncated to $(0, 1)$. That is, for some constant c,

$$f(x) = \begin{cases} c\exp(-x^2/2) & \text{for } x \in (0,1) \\ 0 & \text{otherwise.} \end{cases}$$

What is $\mathbb{E}X$?

We answer this numerically, using the function `simpson` (see Section 11.3). Note that to find c we use the fact that $\int_0^1 f(x)dx = 1$.

```
> source('../scripts/simpson.r')
> f <- function(x) exp(-x^2/2)
> c <- 1/simpson(f, 0, 1)
> xf <- function(x) x*f(x)
> mu <- simpson(xf, 0, 1)*c
> cat('mean of X is ', mu, '\n')

mean of X is  0.4598622
```

14.4.3 Infinite range

The *range* of a random variable X is the set of values it can take, that is $\{x : X(\omega) = x \text{ for some } \omega \in \Omega\}$. If the range of X is bounded then $\mathbb{E}X$ exists and is finite, and there should be no problem calculating it numerically. However, it is a very different story if the range of X is unbounded, as $\mathbb{E}X$ may be infinite, or just not exist.

For example, suppose X is discrete with the pmf

$$p_X(x) = \frac{6}{\pi^2 x^2} \text{ for } x = 1, 2, \ldots,$$

then

$$\mathbb{E}X = \sum_{x=1}^{\infty} x p_X(x) = \frac{6}{\pi^2} \sum_{x=1}^{\infty} \frac{1}{x} = \infty.$$

For a continuous example, suppose that Y has a Cauchy distribution

$$f_Y(y) = \frac{1}{\pi(1+y^2)} \text{ for } y \in \mathbb{R},$$

then $\mathbb{E}Y$ does not exist, since

$$\int_{-\infty}^{0} \frac{y}{\pi(1+y^2)} dy = -\infty \text{ and } \int_{0}^{\infty} \frac{y}{\pi(1+y^2)} dy = \infty,$$

and $-\infty + \infty$ is not well defined.

Now consider calculating $\mathbb{E}X$ numerically, when X has an infinite range. The problem is that we cannot calculate an infinite sum or integral numerically, so we must use a finite approximation.

For example, let X be a discrete random variable with range $\mathbb{N} = \{0, 1, \ldots\}$ and pmf p, then we use the approximation

$$\begin{aligned} \mathbb{E}X = \sum_{x=0}^{\infty} xp(x) &> \sum_{x=0}^{n-1} xp(x) + n\left(1 - \sum_{x=0}^{n-1} p(x)\right) \\ &= n - \sum_{x=0}^{n-1}(n-x)p(x) = E_n. \end{aligned}$$

If $\mathbb{E}X$ is finite then $E_n \uparrow \mathbb{E}X$ as $n \to \infty$. Given $\delta > 0$ we would like to choose n so that the error $\mathcal{E}_n = \mathbb{E}X - E_n \leq \delta$. But $\mathcal{E}_n$ depends on $p(x)$ for all $x \geq n$, so we cannot find $\mathcal{E}_n$ numerically, and thus we do not know when to stop.

Unless we have some theoretical bound on $\mathcal{E}_n$, the best we can do is stop when, for some predefined tolerance, $\epsilon > 0$,

$$E_n - E_{n-1} = 1 - \sum_{x=0}^{n-1} p(x) \leq \epsilon.$$

We must be aware however, that $E_n - E_{n-1} \leq \epsilon \not\Rightarrow \mathbb{E}X - E_n \leq \epsilon$. Moreover, it is possible that $E_n - E_{n-1} \leq \epsilon$ but $\mathbb{E}X = \infty$. Consider again the discrete example with pmf $p_X(x) = 6/(\pi^2 x^2)$ for $x = 1, 2, \ldots$. We have $E_n - E_{n-1} \leq \epsilon$ when $\sum_{x=1}^{n-1} p_X(x) \geq 1 - \epsilon$, which we can easily check incrementally:

```
> p_X <- function(x) 6/pi^2/x^2
> E_n <- function(n) n - sum( (n - 1:(n-1)) * p_X(1:(n-1)) )
> eps <- 1e-6
> n <- 2
> S <- p_X(1)
> while (S < 1 - eps) {
+   n <- n + 1
+   S <- S + 6/pi^2/(n-1)^2
+ }
> n

[1] 607928

> E_n(n)

[1] 9.05509

> E_n(2*n)

[1] 9.476474
```

The take-home message is, when using a truncation to approximate a sum or integral over an infinite range, to be confident that our approximation is good, we need a theoretical bound on the error.

14.4.4 Example: gamma function

The gamma function is defined by

$$\Gamma(z) = \int_0^\infty x^{z-1} e^{-x} dx, \text{ for } z > 0.$$

(It appears later in Sections 16.3.3 and 16.4.3.) For integers z you can show using integration by parts that $\Gamma(z) = (z-1)!$, and using contour integration you can show that $\Gamma(1/2) = \sqrt{\pi}$ and $\Gamma(z+1/2) = \sqrt{\pi} \prod_{n=1}^{z} (z - n + 1/2)$ for $z > 0$, but for any other z numerical techniques are required.

Let $G_T(z) = \int_0^T x^{z-1} e^{-x} dx$, then $G_T(z) \uparrow \Gamma(z)$ as $T \to \infty$, but how large does T need to be to get a good approximation? Noting that $x^{z-1} e^{-x/2} \to 0$ as $x \to \infty$, we have $x^{z-1} e^{-x} \leq e^{-x/2}$ for all x large enough, so for T large enough

$$\Gamma(z) - G_T(z) = \int_T^\infty x^{z-1} e^{-x} dx \leq \int_T^\infty e^{-x/2} dx = 2e^{-T/2}.$$

So in this case we can choose T to make the truncation error smaller that any specified margin $\delta > 0$.

For example, if $z = 1.1$ then $x^{0.1} \leq e^{x/2}$ for all x, so any T is large enough. If we specify an error of 10^{-16} then it suffices to take T such that $2e^{T/2} \leq 10^{-16}$, that is $T \geq 75.07$ (to 2 decimal places).

14.5 Transformations

A random variable X is a function from Ω to $\mathbb{R}$, so if h is a function from $\mathbb{R}$ to $\mathbb{R}$, then $Y = h(X)$ must also be a random variable.

For any function $h : \mathbb{R} \to \mathbb{R}$ and set $A \subset \mathbb{R}$, we define the *set-valued inverse* as

$$h^{-1}(A) = \{x : h(x) \in A\}.$$

For random variables X and $Y = h(X)$, the cdf of Y can always be described by using the set-valued inverse:

$$\begin{aligned} F_Y(y) &= \mathbb{P}(Y \leq y) \\ &= \mathbb{P}(h(X) \leq y) \\ &= \mathbb{P}(h(X) \in (-\infty, y]) \\ &= \mathbb{P}(X \in h^{-1}((-\infty, y])). \end{aligned}$$

If h is strictly increasing then the usual inverse is well defined, and for y in the range of h we get

$$F_Y(y) = \mathbb{P}(X \in h^{-1}((-\infty, y])) = \mathbb{P}(X \in (-\infty, h^{-1}(y)]) = F_X(h^{-1}(y)).$$

We will see in Chapter 18 that a very important application of variable transformations is the simulation of random variables.

14.5.1 Transforming a discrete rv

If X is discrete then so is $Y = h(X)$, and we can obtain its pmf p_Y from the pmf p_X of X:

$$\begin{aligned} p_Y(y) &= \mathbb{P}(Y = y) = \mathbb{P}(\{\omega \in \Omega : Y(\omega) = y\}) \\ &= \mathbb{P}(\{\omega \in \Omega : h(X(\omega)) = y\}) \\ &= \mathbb{P}(\{\omega \in \Omega : \text{ there exists } x \text{ such that } X(\omega) = x \text{ and } h(x) = y\}) \\ &= \sum_{x : h(x)=y} \mathbb{P}(\{\omega \in \Omega : X(\omega) = x\}) \text{ (disjoint sets)} \\ &= \sum_{x : h(x)=y} p_X(x). \end{aligned}$$

For example suppose X has pmf

x	-2	-1	0	1	2
$p_X(x)$	0.2	0.2	0.2	0.2	0.2

Let $Y = X^2$ then Y has pmf

y	0	1	4
$p_Y(y)$	0.2	0.4	0.4

14.5.2 Example: transforming a continuous rv

Suppose that X has density $f_X(x) = (x+1)/2$ for $-1 < x < 1$, and that $Y = \exp(X)$ and $Z = X^2$. What are the densities of Y and Z?

For $-1 < x < 1$, the cdf of X is given by

$$F_X(x) = \int_{-1}^{x} f_X(u)du = \left.\frac{(u+1)^2}{4}\right|_{-1}^{x} = \frac{(x+1)^2}{4}.$$

Let $h(x) = \exp(x)$ then h is strictly increasing and $h^{-1}(y) = \log(y)$, so Y has cdf

$$\begin{aligned} F_Y(y) &= \mathbb{P}(\exp(X) \le y) \\ &= \mathbb{P}(X \le \log(y)) \\ &= \frac{(\log(y)+1)^2}{4}. \end{aligned}$$

Note that since $-1 < X < 1$ we must have $1/e < Y < e$, so we can assume $1/e < y < e$ above. Differentiating we get, for $1/e < y < e$,

$$f_Y(y) = F_Y'(y) = \frac{\log(y)+1}{2y}.$$

Let $g(x) = x^2$ then g is not one to one for $x \in (-1, 1)$, so we need to take

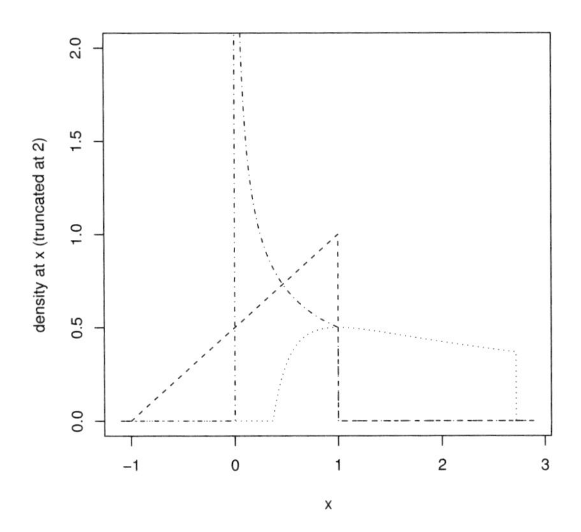

Figure 14.5 *Densities of X, $Y = \exp(X)$ and $Z = X^2$ (Example 14.5.2.)*

care when calculating the cdf of Z. For $x \in (-1,1)$ we get $z \in [0,1)$, so for $z \in [0,1)$ we have

$$\begin{aligned} F_Z(z) &= \mathbb{P}(X^2 \le z) \\ &= \mathbb{P}(-\sqrt{z} \le X \le \sqrt{z}) \\ &= F_X(\sqrt{z}) - F_X(-\sqrt{z}) \\ &= \frac{(\sqrt{z}+1)^2 - (1-\sqrt{z})^2}{4} \\ &= \sqrt{z}. \end{aligned}$$

Differentiating we get, for $z \in [0,1)$,

$$f_Z(z) = F'_Z(z) = \frac{1}{2\sqrt{z}}.$$

See Figure 14.5 for sketches of the transformed densities.

14.5.3 Expectation of a transformed random variable

When calculating the expectation of a transformed rv, we use the following rule: for any function $h : \mathbb{R} \to \mathbb{R}$

$$\mathbb{E}h(X) = \begin{cases} \sum_x h(x)\,p(x), & X \text{ discrete;} \\ \int h(x)\,f(x)dx, & X \text{ continuous.} \end{cases} \tag{14.3}$$

The proof for the discrete case is straightforward. Let $Y = h(X)$ then by definition

$$\begin{aligned}\mathbb{E}Y &= \sum_y y\,\mathbb{P}(Y=y) \\ &= \sum_y y \sum_{x:y=h(x)} p(x) \\ &= \sum_y \sum_{x:y=h(x)} h(x)\,p(x) \\ &= \sum_x h(x)\,p(x).\end{aligned}$$

For example suppose X has pmf

x	-2	-1	0	1	2
$p_X(x)$	0.2	0.2	0.2	0.2	0.2

Let $Y = X^2$ then, using Equation 14.3 we get

$$\begin{aligned}\mathbb{E}Y &= (-2)^2 \times 0.2 + (-1)^2 \times 0.2 + 0^2 \times 0.2 + 1^2 \times 0.2 + 2^2 \times 0.2 \\ &= 2.\end{aligned}$$

Alternatively using the pmf of Y we get

$$\mathbb{E}Y = 0 \times 0.2 + 1 \times 0.4 + 4 \times 0.4 = 2.$$

Equation 14.3 simplifies for linear functions, that is, functions of the form $h(x) = ax + b$. Taking the discrete case we have

$$\mathbb{E}(aX+b) = \sum_x (ax+b)p(x) = a\sum_x x\,p(x) + b\sum_x p(x) = a\mathbb{E}X + b.$$

This result is true for any random variable, not just discrete ones, provided the expectation exists.

It is important to note that in general $\mathbb{E}h(X) \neq h(\mathbb{E}X)$.

14.5.4 Sums of random variables

We prove that for any two discrete random variables X and Y

$$\mathbb{E}(X+Y) = \mathbb{E}X + \mathbb{E}Y.$$

The *joint probability mass function* $p(x,y)$ is defined by

$$p(x,y) = \mathbb{P}(X = x \text{ and } Y = y).$$

If we sum over all possible values of x, we get back the pmf of Y, and conversely if we sum over all possible values of y:

$$\begin{aligned} \sum_x p(x,y) &= \mathbb{P}(Y=y) = p_Y(y) \\ \sum_y p(x,y) &= \mathbb{P}(X=x) = p_X(x) \\ \sum_x \sum_y p(x,y) &= 1 \end{aligned}$$

Now, for any function $h : \mathbb{R} \times \mathbb{R} \to \mathbb{R}$, we have for $Z = h(X,Y)$

$$\begin{aligned} \mu_Z &= \mathbb{E}Z \\ &= \sum_z z\mathbb{P}(Z=z) \\ &= \sum_z z \sum_{(x,y):h(x,y)=z} p(x,y) \\ &= \sum_z \sum_{(x,y):h(x,y)=z} h(x,y)\,p(x,y) \\ &= \sum_x \sum_y h(x,y)\,p(x,y). \end{aligned}$$

In particular, putting $h(x,y) = x + y$ we get

$$\begin{aligned} \mathbb{E}(X+Y) &= \sum_x \sum_y (x+y)\,p(x,y) \\ &= \sum_x \sum_y x\,p(x,y) + \sum_x \sum_y y\,p(x,y) \\ &= \sum_x x \sum_y p(x,y) + \sum_y y \sum_x p(x,y) \\ &= \sum_x x\,p_X(x) + \sum_y y\,p_Y(y) \\ &= \mathbb{E}(X) + \mathbb{E}(Y). \end{aligned}$$

This result, $\mathbb{E}(X+Y) = E(X) + E(Y)$, is true for *any* random variables, discrete or continuous, provided the expected values exist. Since we already know that $\mathbb{E}(aX) = a\mathbb{E}(X)$, we have that for any random variables $X_1, \ldots, X_n$ and scalars (constants) $a_1, \ldots, a_n$,

$$\mathbb{E}(a_1 X_1 + \cdots + a_n X_n) = a_1\mathbb{E}(X_1) + \cdots + a_n\mathbb{E}(X_n). \tag{14.4}$$

14.6 Variance and standard deviation

We define the *variance* of a random variable X to be $\mathbb{E}(X - \mathbb{E}X)^2$. We write $\text{Var}\, X$ or sometimes σ_X^2. The *standard deviation* of X is the square root of the variance.

Both the variance and standard deviation are measures of spread (Figure 14.6). The variance is easier to calculate, but the standard deviation is easier to interpret physically, as it has the same unit of measurement as the random variable. Together the mean and standard deviation of a random variable give a simple *summary* of its distribution.

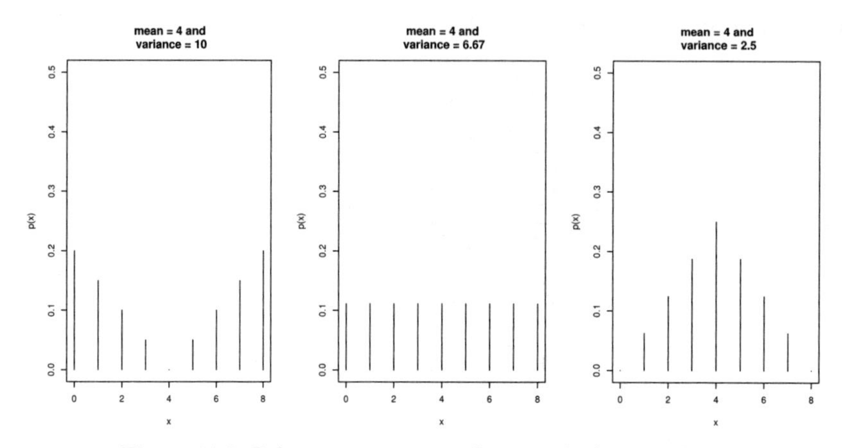

Figure 14.6 *Relating variance to the spread of a distribution.*

Here are some useful results on the variance. All are straightforward to prove for discrete random variables (and are true for all random variables).

- $\text{Var}\, X \geq 0$.
- $\text{Var}\, X = \mathbb{E}X^2 - (\mathbb{E}X)^2$.
- $\text{Var}\, X = 0$ if and only if $X = \mu$ (constant).
- $\text{Var}\,(aX + b) = a^2 \text{Var}\, X$ for any constants a and b.

To obtain a general expression for the variance of the sum of two random variables, we start with the basic definition of $\text{Var}\,(X + Y)$:

$$\begin{aligned}
\text{Var}\,(X+Y) &= \mathbb{E}(X + Y - (\mu_X + \mu_Y))^2 \\
&= \mathbb{E}\big((X - \mu_X) + (Y - \mu_Y)\big)^2 \\
&= \mathbb{E}(X - \mu_X)^2 + 2\mathbb{E}(X - \mu_X)(Y - \mu_Y) + \mathbb{E}(Y - \mu_Y)^2 \\
&= \text{Var}\,(X) + 2\text{Cov}(X, Y) + \text{Var}\,(Y)
\end{aligned}$$

where we define $\text{Cov}(X, Y) = \mathbb{E}(X - \mu_X)(Y - \mu_Y)$.

$\text{Cov}(X, Y)$ is called the *covariance* of X and Y and describes how X and Y

'co-vary' together. For example if Y tends to be above its mean when X is (and vice versa) then $\text{Cov}(X, Y)$ will be positive and the variance of the sum will be increased. In this case we call X and Y *positively correlated*. Similarly, for *negatively correlated* random variables, the variance of the sum is reduced. If $\text{Cov}(X, Y) = 0$ we say that X and Y are *uncorrelated*.

We show below that independent random variables are uncorrelated. To prove this we first show that if X and Y are independent random variables, then $\mathbb{E}(XY) = (\mathbb{E}X)(\mathbb{E}Y)$. Let $p(x, y)$ be the joint pmf of X and Y, then since they are independent, $p(x, y) = p_X(x)p_Y(y)$. Thus

$$\begin{aligned} \mathbb{E}(XY) &= \sum_x \sum_y xy\, p(x, y) \\ &= \sum_x x\, p(x) \sum_y y\, p(y) \\ &= (\mathbb{E}X)(\mathbb{E}Y). \end{aligned}$$

By multiplying out the terms in the definition of $\text{Cov}(X, Y)$, we also find:

$$\text{Cov}(X, Y) = \mathbb{E}(X - \mu_X)(Y - \mu_Y) = \mathbb{E}(XY) - (\mathbb{E}X)(\mathbb{E}Y).$$

So if X and Y are independent they are also uncorrelated and hence

$$\text{Var}\,(X + Y) = \text{Var}\, X + \text{Var}\, Y.$$

This is a very important and frequently used result for independent random variables.

Since we already know that $\text{Var}\,(aX) = a^2 \text{Var}\,(X)$, we thus have that for any *independent* random variables $X_1, \ldots, X_n$ and scalars $a_1, \ldots, a_n$,

$$\text{Var}\,(a_1 X_1 + \cdots + a_n X_n) = a_1^2 \text{Var}\,(X_1) + \cdots + a_n^2 \text{Var}\,(X_n). \tag{14.5}$$

We note that $\text{Cov}(X, Y)$ will change if the units of measurement change for the random variables involved. By standardising the covariance appropriately we obtain a scale invariant measure of how X and Y co-vary, which is called the *correlation coefficient*:

$$\rho(X, Y) = \frac{\text{Cov}(X, Y)}{\sqrt{\text{Var}\,(X)\text{Var}\,(Y)}}$$

The correlation coefficient is an important quantity in statistics and in particular in linear regression.

14.7 The Weak Law of Large Numbers

Let $X_1, \ldots, X_n$ be an iid random sample with mean μ. In this section we make precise our earlier statement that $\overline{X} \approx \mu$.

In statistics $\overline{X}$ is called an *estimator* of μ, and as such is sometimes written $\hat{\mu}$. If x_i is the *observed value* of X_i, then the observed value of $\overline{X}$ is just $\overline{x} = (\sum_{i=1}^n x_i)/n$, which is called a *point estimate* of μ. From a statistical point of view, we think of $\overline{X}$ as describing the potential values of the sample mean, and $\overline{x}$ as a particular realisation. If we were to collect another n observations of X, then we could take their average to get a second observation of $\overline{X}$ (or we could combine the two samples to get a single more accurate estimate).

Let $\mu = \mathbb{E}X_i$ and $\sigma^2 = \operatorname{Var} X_i$, then from (14.4) and (14.5) we have that

$$\begin{aligned}
\mathbb{E}\overline{X} &= \mathbb{E}(X_1/n + \cdots + X_n/n) \\
&= (\mathbb{E}X_1)/n + \cdots (\mathbb{E}X_n)/n \\
&= \mu \\
\operatorname{Var}\overline{X} &= \operatorname{Var}((X_1 + \cdots + X_n)/n) \\
&= \operatorname{Var}(X_1 + \cdots + X_n)/n^2 \\
&= (\operatorname{Var} X_1 + \cdots + \operatorname{Var} X_n)/n^2 \\
&= \sigma^2/n.
\end{aligned}$$

Thus as $n \to \infty$ we have $\mathbb{E}\overline{X} = \mu$ and $\operatorname{Var}\overline{X} \to 0$. That is, $\overline{X}$ starts to look like the constant μ as $n \to \infty$.

Since $\mathbb{E}\overline{X} = \mu$ we say it is an *unbiased* estimator of μ.

Markov's Inequality If X is a random variable that takes on only non-negative values, then for any $a > 0$

$$\mathbb{P}(X \geq a) \leq \frac{\mathbb{E}X}{a}.$$

Proof. Suppose X is discrete with pmf $p_X(x)$, then

$$\mathbb{E}X = \sum_x x\, p_X(x) \geq \sum_{x \geq a} x\, p_X(x) \geq \sum_{x \geq a} a\, p_X(x) = a\mathbb{P}(X \geq a).$$

The continuous case is proved similarly.

Chebyshev's Inequality If X is a random variable with mean μ and variance σ^2, then for any $c > 0$

$$\mathbb{P}(|X - \mu| \geq c\sigma) \leq \frac{1}{c^2}.$$

Proof. The non-negative random variable $(X - \mu)^2$ has the expected value $\mathbb{E}(X - \mu)^2 = \sigma^2$ and so by Markov's inequality with $a = c^2\sigma^2$, we have

$$\mathbb{P}((X - \mu)^2 \geq c^2\sigma^2) \leq \frac{\mathbb{E}(X - \mu)^2}{c^2\sigma^2} = \frac{1}{c^2}.$$

Since $(X - \mu)^2 \geq c^2\sigma^2$ if and only if $|X - \mu| \geq c\sigma$, the result follows.

The Weak Law of Large Numbers Let $X_1, \ldots, X_n$ be an iid random sample each with mean μ and finite variance, then for any $\varepsilon > 0$,

$$\mathbb{P}(|\overline{X} - \mu| > \varepsilon) \to 0 \text{ as } n \to \infty.$$

That is, given a tolerance ε, the probability that $\overline{X}$ is within ε of μ gets as close to 1 as you like, as the sample size increases. We say $\overline{X}$ *converges in probability* to μ as $n \to \infty$, and write

$$\overline{X} \xrightarrow{\mathbb{P}} \mu \text{ as } n \to \infty.$$

Proof. From Chebyshev's inequality, for any random variable with finite mean and variance, $\mathbb{P}(|X - \mu| \geq k\sigma_X) \leq 1/k^2$, so

$$\begin{aligned} \mathbb{P}(|\overline{X} - \mu| \geq \varepsilon) &\leq \frac{\sigma_{\overline{X}}^2}{\varepsilon^2} = \frac{\sigma_X^2}{n\varepsilon^2} \\ \mathbb{P}(|\overline{X} - \mu| \leq \varepsilon) &\geq 1 - \frac{\sigma_X^2}{n\varepsilon^2} \to 1 \text{ as } n \to \infty. \end{aligned}$$

More generally any estimator that is unbiased for a parameter θ and whose standard deviation goes to 0 as $n \to \infty$, will converge in probability to θ.

We note that one can prove a stronger version of this theorem—called the Strong Law of Large Numbers—under weaker conditions. Namely, you can show what is called almost sure convergence, assuming only that you have an iid sequence with finite mean. The proof, however, requires more probability theory than we have covered.

14.7.1 Sample proportion

The estimators $\hat{F}$ and $\hat{p}$ given in (14.1) and (14.2) are in fact special cases of $\hat{\mu}$. Let $1_A(x)$ be the *indicator function* for the set A. That is $1_A(x) = 1$ if $x \in A$ and 0 otherwise. Then, in the case where X is discrete,

$$\mathbb{E}1_A(X) = \sum_x 1_A(x)p(x) = \sum_{x \in A} p(x) = \mathbb{P}(X \in A).$$

Put $A = (-\infty, x]$ to get $F(x)$ or $A = \{x\}$ to get $p(x)$.

In general, suppose that the random variables X_i are *indicator variables*,

$$X_i = \begin{cases} 1, & \text{if item } i \text{ has some property of interest,} \\ 0, & \text{otherwise,} \end{cases}$$

where $p = \mathbb{P}(X_i = 1) = \mathbb{E}X_i$. Then the sample mean is just the sample proportion

$$\overline{X} = \hat{p} = \frac{1}{n}\sum_{i=1}^{n} X_i.$$

We have

$$\mathbb{E}\hat{p} = \mathbb{E}\left(\frac{1}{n}\sum_{i=1}^{n} X_i\right) = \frac{1}{n}\sum_{i=1}^{n} \mathbb{E}X_i = p,$$

and, noting that $\operatorname{Var} X_i = \mathbb{E}X_i^2 - (\mathbb{E}X_i)^2 = p - p^2 = p(1-p)$,

$$\operatorname{Var}\hat{p} = \operatorname{Var}\left(\frac{1}{n}\sum_{i=1}^{n} X_i\right) = \frac{1}{n^2}\sum_{i=1}^{n} \operatorname{Var} X_i = \frac{p(1-p)}{n}.$$

14.7.2 Sample variance

Let $X_1, \ldots, X_n$ be an iid sample with the same distribution as X, with $\mathbb{E}X = \mu$ and $\operatorname{Var} X = \sigma^2$. Since $\sigma^2 = \mathbb{E}(X-\mu)^2$, the Weak Law of Large Numbers suggests that we estimate σ^2 using

$$\frac{1}{n}\sum_{i=1}^{n}(X_i - \mu)^2.$$

Of course, the problem with this is that if we do not know σ^2, then we probably do not know μ either. The way around this is to use an estimate of μ, which leads to the estimator

$$S^2 = \frac{1}{n-1}\sum_{i=1}^{n}(X_i - \overline{X})^2 = \frac{1}{n-1}\left[\sum_{i=1}^{n} X_i^2 - n\overline{X}^2\right].$$

S^2 is called the *sample variance.* As it is an estimator for σ^2 it is often written $\hat{\sigma}^2$. If $x_1, \ldots, x_n$ are the observed sample values, then the observed value of S^2 is $s^2 = \sum_{i=1}^{n}(x_i - \overline{x})^2/(n-1)$, which is the point estimate of σ^2.

Note that in S^2 we divide by $n-1$ rather than n. This choice makes S^2 *unbiased*, that is, $\mathbb{E}S^2 = \sigma^2$.

$$\begin{aligned}
\mathbb{E}S^2 &= \mathbb{E}\frac{1}{n-1}\left[\sum_{i=1}^{n} X_i^2 - n\overline{X}^2\right] \\
&= \frac{1}{n-1}\left[\sum_{i=1}^{n} \mathbb{E}X_i^2 - n\mathbb{E}\overline{X}^2\right] \\
&= \frac{1}{n-1}\left[\sum_{i=1}^{n}(\operatorname{Var} X_i + (\mathbb{E}X_i)^2) - n(\operatorname{Var}\overline{X} + (\mathbb{E}\overline{X})^2)\right] \\
&= \frac{1}{n-1}\left[\sum_{i=1}^{n}(\sigma^2 + \mu^2) - n(\sigma^2/n + \mu^2)\right] \\
&= \sigma^2.
\end{aligned}$$

The *sample standard deviation* $S = \sqrt{S^2}$ is an estimator of the standard deviation σ. However, S is *not* unbiased, because $\mathbb{E}S = \mathbb{E}\sqrt{S^2} \neq \sqrt{\mathbb{E}S^2}$. Nonetheless, the bias is usually small.

If the X_i are indicator variables with $\mu = p$, then $\sigma^2 = p(1-p)$, and it is common to estimate σ^2 using $\hat{p}(1-\hat{p}) = \overline{X}(1-\overline{X})$. This is in fact very close to S^2, but not quite the same. Because $X_i \in \{0,1\}$ we have $X_i = X_i^2$, so

$$\begin{aligned}
\overline{X}(1-\overline{X}) &= \overline{X} - \overline{X}^2 \\
&= \frac{1}{n}\sum_{i=1}^{n} X_i - \overline{X}^2 \\
&= \frac{1}{n}\left[\sum_{i=1}^{n} X_i^2 - n\overline{X}^2\right] \\
&= \frac{n-1}{n} S^2.
\end{aligned}$$

14.8 Exercises

1. Suppose you throw two dice. What values can the following random variables take?

   (a). The minimum face value showing;

   (b). The absolute difference between the face values showing;

   (c). The ratio: minimum face value/other face value.

   Assuming all outcomes in the sample space are equally likely, what are the probability mass functions for these random variables? Give these in a table format and also do a rough sketch.

   Calculate the mean of each random variable.

2. The following is the probability mass function of a discrete random variable X

x	1	2	3	4	5
$\mathbb{P}(X = x)$	$2c$	$3c$	c	$4c$	$5c$

   (a). What is the value of c?

   (b). Find $\mathbb{P}(X \leq 4)$ and $\mathbb{P}(1 < X < 5)$.

   (c). Calculate $\mathbb{E}X$ and $\mathrm{Var}\, X$.

3. A game consists of first tossing an unbiased coin and then rolling a six-sided die. Let the random variable X be the score that is obtained by adding the face value of the die and the number of heads obtained (0 or 1). List the possible values of X and calculate its pmf.

4. This question concerns an experiment with sample space

$$\Omega = \{a, b, c, d\}.$$

(a) List all possible events for this experiment.

(b) Suppose that $\mathbb{P}(\{a\}) = \mathbb{P}(\{b\}) = \mathbb{P}(\{c\}) = \mathbb{P}(\{d\}) = 1/4$.
Find two independent events and two dependent events.

(c) Define random variables X, Y, and Z as follows

ω	a	b	c	d
$X(\omega)$	1	1	0	0
$Y(\omega)$	0	1	0	1
$Z(\omega)$	1	0	0	0

Show that X and Y are independent and that X and Z are dependent.

(d) Let $W = X + Y + Z$. What is the probability mass function (pmf) of W? What is its mean and variance?

5. A discrete random variable has pmf $f(x) = k(1/2)^x$ for $x = 1, 2, 3$; $f(x) = 0$ for all other values of x. Find the value of k and then the mean and variance of the random variable.

6. Consider the discrete random variable X that takes the values $0, 1, 2, \ldots, 9$ each with probability $1/10$. Let Y be the remainder obtained after dividing X^2 by 10 (e.g., if $X = 9$ then $Y = 1$). Y is a function of X and so is also a random variable. Find the pmf of Y.

7. Consider the discrete probability distribution defined by

$$p(x) = \mathbb{P}(X = x) = \frac{1}{x(x+1)} \quad \text{for } x = 1, 2, 3, \ldots$$

(a) Let $S(n) = \mathbb{P}(X \leq n) = \sum_{x=1}^{n} p(x)$. Using the fact that $\frac{1}{x(x+1)} = \frac{1}{x} - \frac{1}{x+1}$, find a formula for $S(n)$ and thus show that p is indeed a pmf.

(b) Write down the formula for the mean of this distribution. What is the value of this sum?

8. For some fixed integer k, the random variable Y has probability mass function (pmf)

$$p(y) = \mathbb{P}(Y = y) = \begin{cases} c(k-y)^2 & \text{for } y = 0, 1, 2, \ldots, k-1 \\ 0 & \text{otherwise.} \end{cases}$$

(a) What is the value of c? (Your answer will depend on k.)
Hint: $\sum_{i=1}^{n} i^2 = n(n+1)(2n+1)/6$.

(b) Give a formula for the cumulative distribution function (cdf) $F(y) = \mathbb{P}(Y \leq y)$ for $y = 0, 1, 2, \ldots, k-1$.

(c) Write a function in R to calculate $F(y)$. Your function should take y and k as inputs and return $F(y)$. You may assume that k is an integer greater than 0 and that $y \in \{0, 1, 2, \ldots, k-1\}$.

9. Toss a coin 20 times and let X be the length of the longest sequence of heads. We wish to estimate the probability function p of X. That is, for $x = 1, 2, \ldots, 20$ we wish to estimate

$$p(x) = \mathbb{P}(X = x).$$

Here is a function `maxheads(n.toss)` that simulates X (using `n.toss = 20`).

```
maxheads <- function(n.toss) {
  # returns the length of the longest sequence of heads
  # in a sequence of n.toss coin tosses
  n_heads = 0    # length of current head sequence
  max_heads = 0 # length of longest head sequence so far
  for (i in 1:n.toss) {
    # toss a coin and work out length of current head sequence
    if (runif(1) < 0.5) { # a head, sequence of heads increases by 1
      n_heads <- n_heads + 1
    } else { # a tail, sequence of heads goes back to 0
      n_heads <- 0
    };
    # see if current sequence of heads is the longest
    if (n_heads > max_heads) {
      max_heads <- n_heads
    }
  }
  return(max_heads)
}
```

Use `maxheads(20)` to generate an iid sample $X_1, \ldots, X_N$ then estimate p using

$$\hat{p}(x) = \frac{|\{X_i = x\}|}{N}.$$

Print out your estimate as a table like this

```
  x  p_hat(x)
--------------
  0    0.0010
  1    0.0500
  .         .
  .         .
 20    0.0000
```

As a supplementary exercise try rewriting the function `maxheads` using the R function `rle`.

10. Suppose the rv X has continuous pdf $f(x) = 2/x^2$, $1 \le x \le 2$. Determine the mean and variance of X and find the probability that X exceeds 1.5.

11. Which of the following functions are probability density functions for a continuous random variable X?

(a).

$$f(x) = \begin{cases} 5x^4 & 0 \le x \le 1 \\ 0 & \text{otherwise} \end{cases}$$

(b).

$$f(x) = \begin{cases} 2x & -1 \le x \le 2 \\ 0 & \text{otherwise} \end{cases}$$

(c).

$$f(x) = \begin{cases} 1/2 & -1 \le x \le 1 \\ 0 & \text{otherwise} \end{cases}$$

(d).

$$f(x) = \begin{cases} 2x/9 & 0 \le x \le 3 \\ 0 & \text{otherwise} \end{cases}$$

For those that are pdfs, calculate $\mathbb{P}(X \le 1/2)$.

12. Suppose a continuous random variable Y has pdf

$$f(y) = \begin{cases} 3y^2 & 0 \le y \le 1 \\ 0 & \text{otherwise} \end{cases}$$

(a). Sketch this pdf and find $\mathbb{P}(0 \le Y \le 1/2)$ and $\mathbb{P}(1/2 \le Y \le 1)$.
(b). Find the cdf $F_Y(y)$ of Y.

13. Suppose a continuous random variable Z has pdf

$$f_Z(z) = \begin{cases} z-1 & 1 \le z \le 2 \\ 3-z & 2 \le z \le 3 \\ 0 & \text{otherwise} \end{cases}$$

(a). Sketch this pdf and find $\mathbb{P}(Z \le 3/2)$ and $\mathbb{P}(3/2 \le Z \le 5/2)$.
(b). Find the cdf $F_Z(z)$ of Z.

14. A random variable X has cdf

$$F_X(x) = \begin{cases} 0 & x \le 0 \\ 1 - e^{-x} & 0 < x < \infty \end{cases}$$

(a). Sketch this cdf.
(b). Is X continuous or discrete? What are the possible values of X?
(c). Find $\mathbb{P}(X \ge 2)$, $\mathbb{P}(X \le 2)$, and $\mathbb{P}(X = 0)$.

15. A random variable X has cdf

$$F_X(x) = \begin{cases} x/2 & 0 < x \le 1 \\ x - 1/2 & 1 < x < 3/2 \end{cases}$$

(a). Sketch this cdf.
(b). Is X continuous or discrete? What are the possible values of X?

(c). Find $\mathbb{P}(X \leq 1/2)$ and $\mathbb{P}(X \geq 1/2)$.

(d). Find a number m such that $\mathbb{P}(X \leq m) = \mathbb{P}(X \geq m) = 1/2$ (the *median*).

16. For Exercises 12–15 above, try to guess the mean by judging the 'centre of gravity' of the pdf. Then check your guess by evaluating the mean analytically.

17. Consider two continuous random variables X and Y with pdfs

$$f_X(x) = \begin{cases} 4x^3 & 0 \leq x \leq 1 \\ 0 & \text{otherwise} \end{cases}$$

$$f_Y(y) = \begin{cases} 1 & 0 \leq y \leq 1 \\ 0 & \text{otherwise} \end{cases}$$

(a). Sketch both these pdfs and try to guess the means of X and Y. Check your guesses by actually calculating the means.

(b). From the sketches, which random variable do you think would be more variable, X or Y? Check your guess by actually calculating the variances.

18. It is known that a good model for the variation, from item to item, of the quality of a certain product is the random variable X with probability density function $f(x) = 2x/\lambda^2$ for $0 \leq x \leq \lambda$. Here λ is a parameter that depends on the manufacturing process, and can be altered.

During manufacture, each item is tested. Items for which $X > 1$ are passed, and the rest are rejected. The cost of a rejected item is $c = a\lambda + b$ and the profit on a passed item is $d - c$, for constants a, b, and d.

Find λ such that the expected profit is maximised.

19. The variable X has pdf $f(x) = \frac{1}{8}(6 - x)$ for $2 \leq x \leq 6$. A sample of two values of X is taken. Denoting the lesser of the two values by Y, use the cdf of X to write down the cdf of Y. Hence obtain the pdf and mean of Y. Show that its median is approximately 2.64. (The median is the point m for which $\mathbb{P}(Y \leq m) = 0.5$.)

20. Discrete and continuous are not the only possible types of random variable. For example, what sort of distribution is the time spent waiting in a bank queue? If we suppose that there is a strictly positive probability of waiting no time at all, then the cumulative distribution function will have a jump at 0. However, if there are people ahead of you, the time you wait could be any value in $(0, \infty)$, so that this part of the cumulative distribution will be a continuous function. Thus this distribution is a mixture of a discrete and continuous part.

Let X be the length of time that a customer is in the queue, and suppose that

$$F(x) = 1 - pe^{-\lambda x} \text{ for } x \geq 0, \lambda > 0 \text{ and } 0 < p < 1.$$

Find $\mathbb{P}(X = 0)$ and the cdf for $X \mid X > 0$. Hence, find the mean queuing time, noting that (from the Law of Total Probability)

$$\begin{aligned} \mathbb{E}X &= \mathbb{E}(X \mid X = 0)\mathbb{P}(X = 0) + \mathbb{E}(X \mid X > 0)\mathbb{P}(X > 0) \\ &= 0 + \mathbb{E}(X \mid X > 0)\mathbb{P}(X > 0). \end{aligned}$$

21. Let $X_1, \ldots, X_n$ be an iid sample with mean μ and variance σ^2. Show that you can write

$$(n-1)S^2 = \sum_{i=1}^{n}(X_i - \overline{X})^2 = \sum_{i=1}^{n}(X_i - \mu)^2 - n(\mu - \overline{X})^2.$$

Now suppose that $\mathbb{E}(X_i - \mu)^4 < \infty$, and use the Weak Law of Large Numbers to show that

$$S^2 \xrightarrow{\mathbb{P}} \sigma^2 \text{ as } n \to \infty.$$

CHAPTER 15

Discrete random variables

This chapter builds on the general framework for random variables provided in the previous chapter. We study some of the most common and important discrete random variables, and summarise the R functions relating to them. In particular we cover the Bernoulli distribution, binomial distribution, geometric distribution, negative binomial distribution, and the Poisson distribution.

In the next chapter we cover continuous random variables.

15.1 Discrete random variables in R

R has built-in functions for handling the most commonly encountered probability distributions. Suppose that the random variable X is of type `dist` with parameters `p1, p2, ...`, then

`ddist(x, p1, p2, ...)` equals $\mathbb{P}(X = x)$ for X discrete, or the density of X at x for X continuous;

`pdist(q, p1, p2, ...)` equals $\mathbb{P}(X \leq q)$;

`qdist(p, p1, p2, ...)` equals the smallest q for which $\mathbb{P}(X \leq q) \geq p$ (the $100p$ %-point);

`rdist(n, p1, p2, ...)` is a vector of `n` pseudo-random numbers from distribution type `dist`.

The inputs `x`, `q`, and `p` can all be vector valued, in which case the output is vector valued.

Here are some of the discrete distributions provided by R, together with the names of their parameter inputs.

Distribution	R name (dist)	Parameter names
Binomial	`binom`	`size, prob`
Geometric	`geom`	`prob`
Negative binomial	`nbinom`	`size, prob`
Poisson	`pois`	`lambda`

15.2 Bernoulli distribution

The Bernoulli, binomial, geometric, and negative binomial distributions all arise from the context of a sequence of independent trials. Each of these random variables describes a different aspect of such an experiment.

In any random trial we can always partition the sample space by arbitrarily characterising some outcomes as 'successes' and the complementary outcomes as 'failures'. We suppose that each trial is successful with probability p and unsuccessful otherwise.

A Bernoulli random variable B is based on a single trial and takes on the value 1 if the trial is a success or 0 otherwise. We use the notation $B \sim \text{Bernoulli}(p)$ to indicate that B has a Bernoulli distribution with parameter p.

$$\begin{aligned}
\mathbb{P}(B = x) &= \begin{cases} p & \text{for } x = 1; \\ 1 - p & \text{for } x = 0; \end{cases} \\
\mathbb{E}B &= 1 \cdot p + 0 \cdot (1 - p) = p; \\
\operatorname{Var} B &= \mathbb{E}(B - p)^2 = (1 - p)^2 \cdot p + (0 - p)^2 \cdot (1 - p) = p(1 - p).
\end{aligned}$$

The Bernoulli random variable is also referred to as an *indicator variable*, as it indicates or signals the occurrence of a success.

15.3 Binomial distribution

Let X be the number of successes in n independent trials, with probability of success p, then X is said to have a binomial distribution with parameters n and p. We write $X \sim \text{binom}(n, p)$.

Let $B_1, \ldots, B_n$ be independent $\text{Bernoulli}(p)$ random variables, then

$$X = B_1 + \cdots + B_n \sim \text{binom}(n, p).$$

For $x = 0, 1, \ldots, n$ we have

$$\begin{aligned}
\mathbb{P}(X = x) &= \binom{n}{x} p^x (1 - p)^{n-x}; \\
\mathbb{E}X &= \mathbb{E}(B_1 + \cdots + B_n) = \mathbb{E}B_1 + \cdots + \mathbb{E}B_n = np; \\
\operatorname{Var} X &= \operatorname{Var}(B_1 + \cdots + B_n) = \operatorname{Var} B_1 + \cdots + \operatorname{Var} B_n = np(1 - p).
\end{aligned}$$

The variance result uses the fact that the B_i are independent. To prove the formula for $\mathbb{P}(X = x)$, we use the fact that the number of ways you can choose x trials (the successful ones) from a set of n is $\binom{n}{x}$.

Clearly the Bernoulli distribution is the same as a $\text{binom}(1, p)$ distribution. You should check that in the case $n = 1$, the formulae for the distribution, mean, and variance are the same as those for the Bernoulli.

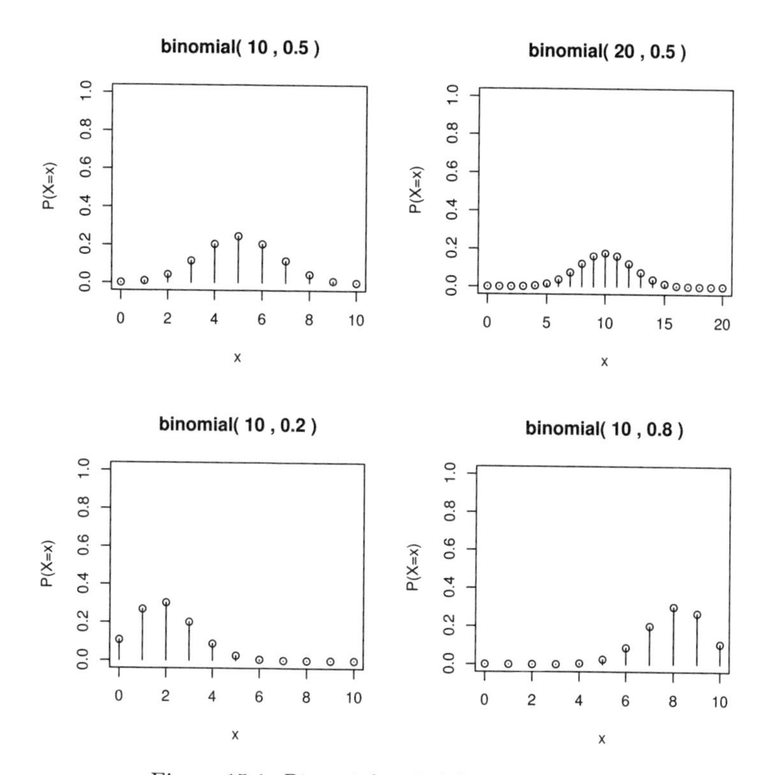

Figure 15.1 *Binomial probability mass functions.*

The binomial random variable gets its name from the *binomial expansion*: for any a and b we have

$$(a+b)^n = \sum_{x=0}^{n} \binom{n}{x} a^x b^{n-x}.$$

You can use this identity to show that $\sum_{x=0}^{n} \mathbb{P}(X = x) = 1$. Note that $0!$ is defined to be 1.

Figure 15.1 shows the probability mass function of several binomial distributions.

15.3.1 Example: sampling a manufacturing line

Suppose that items on a manufacturing line each have a probability 0.01 of being faulty. If you test a randomly selected sample of n items, how large does n have to be to have a 95% chance of having a faulty item in the sample?

Let X be the number of faulty items in the sample, then we want to know how large n has to be to get $\mathbb{P}(X \geq 1) \geq 0.95$. Assuming that the sample items are faulty independently, we have $X \sim \text{binom}(n, 0.01)$, so

$$\begin{aligned} \mathbb{P}(X \geq 1) &= 1 - \mathbb{P}(X = 0) \\ &= 1 - \binom{n}{0} 0.01^0 0.99^n = 1 - 0.99^n \\ &\geq 0.95. \end{aligned}$$

Solving the inequality for n we get $n \geq 299$ (rounding up to the nearest integer).

Alternatively, we might want to know what is the probability that a thousand randomly selected items will have less than 20 failures. We can resolve this in R by using the cdf function for the binomial distribution.

```
> pbinom(19, size = 1000, prob = 0.01)

[1] 0.9967116
```

15.4 Geometric distribution

Let $B_1, B_2, \ldots$ be an infinite sequence of independent Bernoulli(p) random variables and let Y be such that $B_1 = \cdots = B_Y = 0$ and $B_{Y+1} = 1$, then Y is said to have a geometric distribution with parameter p. That is, Y is the number of trials up to (but not including) the first success. We write $Y \sim \text{geom}(p)$, and we have, for $y = 0, 1, \ldots,$

$$\begin{aligned} \mathbb{P}(Y = y) &= \mathbb{P}(B_1 = 0, \ldots, B_y = 0, B_{y+1} = 1) \\ &= \mathbb{P}(B_1 = 0) \cdots \mathbb{P}(B_y = 0)\mathbb{P}(B_{y+1} = 1) = (1-p)^y p; \\ \mathbb{E}Y &= \sum_{y=0}^{\infty} y(1-p)^y p = \frac{1-p}{p}; \\ \operatorname{Var} Y &= \mathbb{E}Y^2 - (\mathbb{E}Y)^2 = \mathbb{E}Y(Y-1) + \mathbb{E}Y - (\mathbb{E}Y)^2 = \frac{1-p}{p^2}. \end{aligned}$$

The formula for the mean and variance require some algebra (omitted here).

The geometric random variable gets its name from the formula for the sum of a *geometric progression*. For any $\alpha \in (-1, 1)$ we have

$$\sum_{n=0}^{\infty} \alpha^n = \frac{1}{1-\alpha}.$$

Use this to show that $\sum_{y=0}^{\infty} \mathbb{P}(Y = y) = 1$. Similarly we can use the identity $\sum_{n=0}^{\infty} n\alpha^n = \alpha/(1-\alpha)^2$ to find $\mathbb{E}Y$.

Figure 15.2 shows the probability mass function of several geometric distributions.

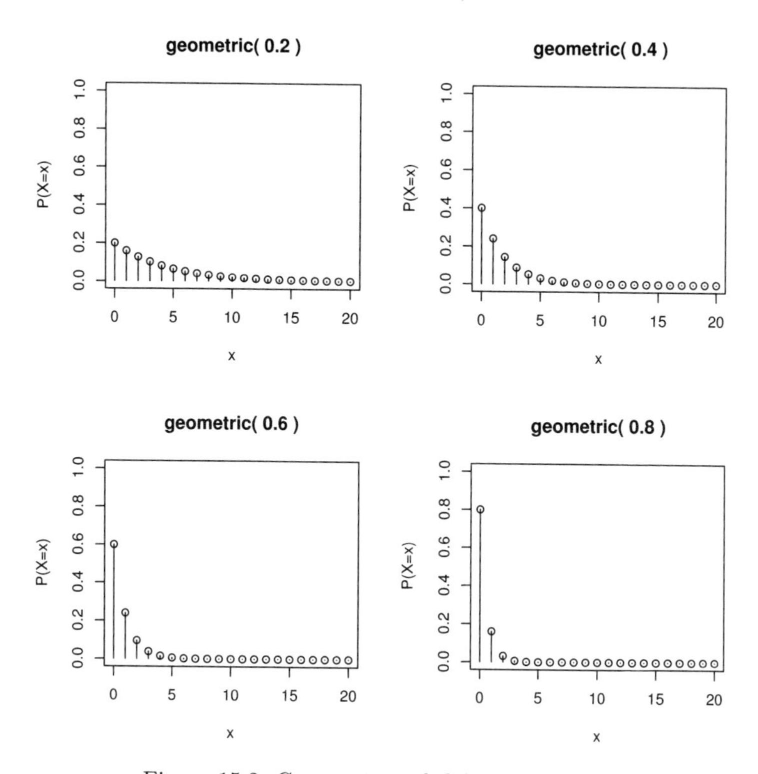

Figure 15.2 *Geometric probability mass functions*

Warning: some authors define a geometric rv to be the number of trials up to *and including* the first success. That is $Y+1$ rather than Y.

15.4.1 Example: lighting a Barbeque

You are trying to light a barbeque with matches on a windy day. Each match has a chance $p = 0.1$ of lighting the barbeque and you only have four matches. What is the probability you get the barbeque lit before you run out of matches?

Imagine initially that we have an infinite supply of matches, and let Y be the number of failed attempts before you light the barbeque. Then $Y \sim \text{geom}(0.1)$ and the required probability is

$$\begin{aligned}
\mathbb{P}(Y \leq 3) &= \sum_{y=0}^{3} p(1-p)^y \\
&= 1-(1-p)^4 \text{ (geometric sum)} \\
&= 0.3439 \text{ (four decimal places).}
\end{aligned}$$

In R,

```
> pgeom(3, 0.1)
[1] 0.3439
```

Now suppose that using matches two at a time, the probability of successfully lighting the barbeque increases to 0.3 each time. Is it a good idea to use the matches two at a time?

Let W be the number of failed attempts to light the barbeque using matches two at a time, then $W \sim \text{geom}(0.3)$ and we have

$$\mathbb{P}(W \le 1) = 0.3 + 0.7 \times 0.3 = 0.51.$$

```
> pgeom(1, 0.3)
[1] 0.51
```

We conclude that you should use the matches two at a time.

15.4.2 Example: two-up

Two-up is a simple gambling game that was popular with Australian servicemen in the first and second World Wars, and can now be played legally in Australian casinos and also throughout Australia on ANZAC day. Two coins are tossed and players bet on whether or not the coins show two heads or two tails. If there is one of each then the coins are tossed again. In casinos the house takes all bets if the number of tosses exceeds five. What is the probability of this occurring?

Let X be the number of tosses with no result. Assuming that the coins are fair and that tosses are independent (generally not the case when the game is played outside casinos), we have that $X \sim \text{geom}(0.5)$. The required probability is then

$$\begin{aligned}\mathbb{P}(X \ge 5) &= \sum_{x=5}^{\infty} p(1-p)^x \\ &= (1-p)^5 \text{ (geometric sum)} \\ &= (0.5)^5 \;=\; 1/32.\end{aligned}$$

In R,

```
> 1 - pgeom(4, 0.5)
[1] 0.03125
```

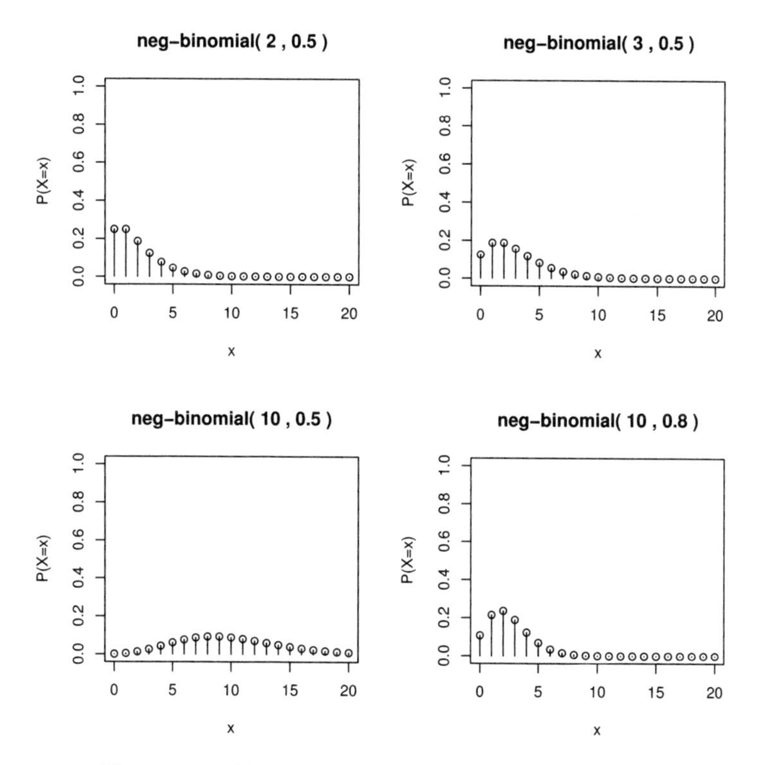

Figure 15.3 *Negative binomial probability mass functions.*

15.5 Negative binomial distribution

Let Z be the number of failures before the r-th success, in a sequence of iid Bernoulli(p) trials, then Z is said to have a *negative binomial* distribution. We write $Z \sim \text{nbinom}(r, p)$. Let $Y_1, \ldots, Y_r$ be iid geom(p) rv's, then

$$Z = Y_1 + \cdots + Y_r \sim \text{nbinom}(r, p).$$

It follows immediately that

$$\begin{aligned} \mathbb{E}Z &= r(1-p)/p; \\ \text{Var}\, Z &= r(1-p)/p^2. \end{aligned}$$

If the r-th success is on trial x, then the previous $r-1$ successes can occur anywhere in the previous $x-1$ trials. Thus there are $\binom{x-1}{r-1}$ ways we can get the r-th success on the x-th trial. Each of these ways has probability $p^r(1-p)^{x-r}$, so putting $z = x - r$ we have for $z = 0, 1, \ldots$,

$$\mathbb{P}(Z = z) = \binom{r+z-1}{r-1} p^r (1-p)^z.$$

You should check that this formula agrees with the geometric in the case $r = 1$.

Figure 15.3 shows the probability mass function of several negative binomial distributions.

Like the geometric, some authors define the negative binomial to be the number of trials (successes and failures) up to and including the r-th success, rather than just counting the failures.

15.5.1 Example: quality control

A manufacturer tests the production quality of its product by randomly selecting 100 from each batch. If there are more than two faulty items, then they stop production and try to fix the problem.

Suppose that each item is faulty independently of the others, with probability p. Let X be the number of faults in a sample of size 100, then $X \sim \text{binom}(100, p)$ and

$$\mathbb{P}(\text{stopping production}) = \mathbb{P}(X \geq 3).$$

If $p = 0.01$ then the probability of stopping production is

```
> 1 - pbinom(2, 100, 0.01)
[1] 0.0793732
```

In practice, rather than test every sample item, we test sequentially and stop when we get three faults. Let Z be the number of working items we check before we find three faults, then $Z \sim \text{nbinom}(3, p)$ and

$$\mathbb{P}(\text{stopping production}) = \mathbb{P}(Z + 3 \leq 100).$$

Note that $Z + 3$ is the total number of checks up to and including finding the third fault. In R we have

```
> pnbinom(97, 3, 0.01)
[1] 0.0793732
```

15.6 Poisson distribution

We say X has a Poisson distribution with parameter λ, and write $X \sim \text{pois}(\lambda)$, if X has pmf

$$\mathbb{P}(X = x) = \frac{e^{-\lambda}\lambda^x}{x!} \text{ for } x = 0, 1, \ldots.$$

Note that 0! is defined to be 1.

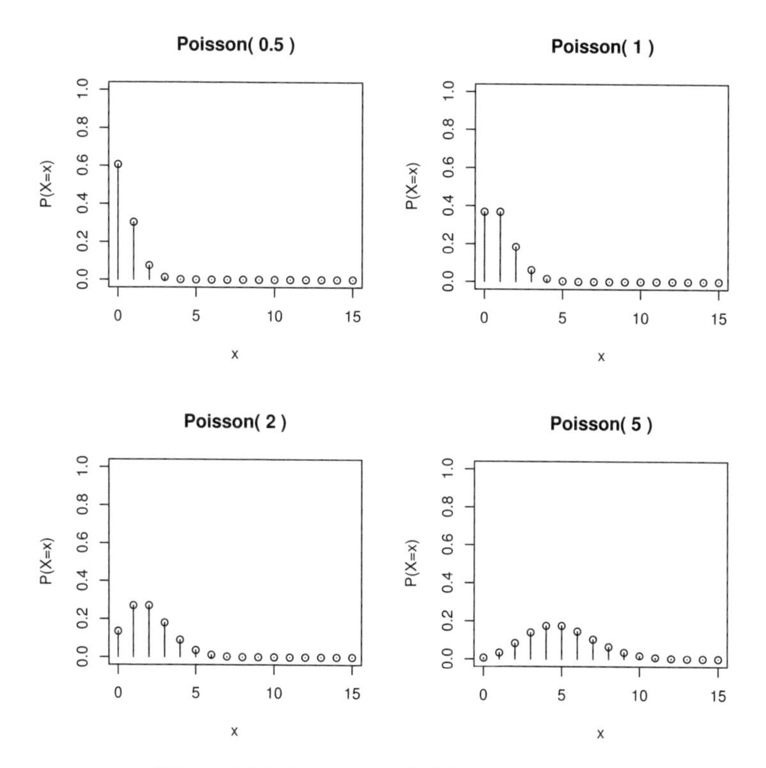

Figure 15.4 *Poisson probability mass functions.*

The Poisson distribution is used as a model for rare events and events occurring at random over time or space. Examples are the number of accidents in a year; the number of misprints on a page; the number of gamma particles released in a second; the number of phone calls arriving at an exchange in an hour; the number of companies going bankrupt in a year; the number of deaths due to horse-kick in the Prussian army in a year (a famous initial application); etc.

Figure 15.4 shows the probability mass function of several Poisson distributions.

An infinite Taylor's series expansion of e^λ about $\lambda = 0$ gives

$$e^\lambda = \sum_{n=0}^{\infty} \frac{\lambda^n}{n!}.$$

From this we see that $\sum_{x=0}^{\infty} \mathbb{P}(X = x) = 1$, as required. We also use this fact to calculate the mean and variance

$$\mathbb{E}X = \sum_{x=0}^{\infty} x\mathbb{P}(X = x)$$

$$\begin{aligned}
&= \sum_{x=0}^{\infty} x \frac{e^{-\lambda}\lambda^x}{x!} \\
&= \sum_{x=1}^{\infty} \frac{e^{-\lambda}\lambda^x}{(x-1)!} \\
&= \lambda \sum_{x=1}^{\infty} \frac{e^{-\lambda}\lambda^{(x-1)}}{(x-1)!} \\
&= \lambda \sum_{y=0}^{\infty} \frac{e^{-\lambda}\lambda^y}{y!} \qquad (y = x-1) \\
&= \lambda.
\end{aligned}$$

For the variance we use $\mathrm{Var}\, X = \mathbb{E}X^2 - (\mathbb{E}X)^2 = \mathbb{E}(X(X-1)) + \mathbb{E}X - (\mathbb{E}X)^2 = \mathbb{E}X(X-1) + \lambda - \lambda^2$. Using the same method that we used to calculate $\mathbb{E}X$ we can show that $\mathbb{E}(X(X-1)) = \lambda^2$, so that $\mathrm{Var}\, X = \lambda$.

15.6.1 Example: the dreaded lurgy

The dreaded lurgy is a disease introduced to humans by the Goons. Suppose that deaths due to the dreaded lurgy over the last seven years were 2, 3, 3, 2, 2, 1, 1. Now suppose that this year we get four deaths due to the dreaded lurgy. Deaths have increased four-fold, should we panic?

Let X_i be the number of deaths in year i and assume that $X_1, \ldots, X_7$ are an iid sample with a pois(λ) distribution, for some unknown λ. From the Weak Law of Large Numbers we have $\overline{X} = 2 \approx \mathbb{E}X_i = \lambda$, so we take $\hat{\lambda} = 2$ as an estimate of λ. Given this estimate, we have

$$\mathbb{P}(X_8 \geq 4) = 1 - \mathbb{P}(X_8 < 4) = 1 - 0.135 - 0.271 - 0.271 - 0.180 = 0.143.$$

From R,

```
> 1 - ppois(3, 2)
[1] 0.1428765
```

That is, we estimate the chance of getting four or more deaths in a year at roughly 14%. Perhaps not cause enough to panic yet.

15.6.2 Poisson as a binomial limit

Suppose that $X \sim \mathrm{binom}(n, p)$ and that for some fixed λ, we have $p = \lambda/n$, then

$$\lim_{n\to\infty} \mathbb{P}(X = x) \to \mathbb{P}(\Lambda = x) \text{ where } \Lambda \sim \mathrm{pois}(\lambda).$$

That is, for n large the binom(n,p) distribution is approximately pois(np).

We have

$$\begin{aligned}
\mathbb{P}(X=x) &= \binom{n}{x} p^x (1-p)^{n-x} \\
&= n\cdot(n-1)\cdots(n-x+1)\frac{1}{x!}\left(\frac{\lambda}{n}\right)^x \left(1-\frac{\lambda}{n}\right)^{n-x} \\
&= 1\cdot\frac{n-1}{n}\cdots\frac{n-x+1}{n}\frac{\lambda^x}{x!}\left(1-\frac{\lambda}{n}\right)^n \left(1-\frac{\lambda}{n}\right)^{-x} \\
&\to 1\cdot 1\cdots 1\frac{\lambda^x}{x!}e^{-\lambda}1 \text{ as } n\to\infty \\
&= \mathbb{P}(\Lambda = x).
\end{aligned}$$

Here we used the following fundamental result on the exponential:

$$\lim_{n\to\infty}\left(1+\frac{x}{n}\right)^n = e^x.$$

This result shows us why the Poisson distribution is a good model for events occurring randomly in time, at a constant average rate. Suppose that a certain event occurs on average λ times per unit time. Split the time interval $(0,1]$ into n intervals, $(0,1/n],(1/n,2/n],\ldots,((n-1)/n,1]$, then suppose that at most one event can occur in each interval. This assumption is reasonable for large n. If the probability of an event occurring in any given interval is p then, writing X_n for the total number of events occurring, we have $X_n \sim$ binom(n,p) and $\mathbb{E}X_n = np$. As we assumed $\mathbb{E}X_n = \lambda$ we must have $p=\lambda/n$, so sending $n\to\infty$ we find that in the limit the number of events has a pois(λ) distribution.

Another consequence of this result is that, if X and Y are independent Poisson rv's with parameters λ and μ, then $X+Y\sim$ pois$(\lambda+\mu)$.

15.7 Exercises

1. The probability of recovery from a certain disease is 0.15. Nine people have contracted the disease. What is the probability that at most 2 of them recover? What is the expected number that will recover?
2. On a multiple choice exam with five possible answers for each of the ten questions, what is the probability that a student would get three or more correct answers just by guessing (choosing an answer at random)? What is the expected number of correct answers the student would get just by guessing?
3. An airline knows that on average 10% of people making reservations on a certain flight will not show up. So they sell 20 tickets for a flight that can only hold 18 passengers.

(a). Assuming individual reservations are independent, what is the probability that there will be a seat available for every passenger that shows up?

(b). Now assume there are 15 flights of this type in one evening. Let N_0 be the number of these flights on which everyone who shows up gets a seat and N_1 be the number of these flights that leave just one disgruntled person behind. What are the distributions of N_0 and N_1? What are their means and variances?

(c). The independence assumption in (a) is not really very realistic. Why? Try to describe what might be a more realistic model for this situation.

4. In the board game Monopoly you can get out of jail by throwing a double (on each turn you throw two dice). Let N be the number of throws required to get out of jail this way. What is the distribution of N, $\mathbb{E}(N)$, and $\text{Var}(N)$?

5. A couple decides to keep having children until they have a daughter. That is, they stop when they get a daughter even if she is their first child. Let N be the number of children they have. Assume that they are equally likely to have a boy or a girl and that the sexes of their children are independent.

(a). What is the distribution of N? $\mathbb{E}(N)$? $\text{Var}(N)$?

(b). Write down the probabilities that N is 1, 2, or 3.

Another couple decides to do the same thing but they don't want an only child. That is they have two children and then only keep going if they haven't yet had a daughter. Let M be the number of children they have.

(c). Calculate $\mathbb{P}(M = 1)$, $\mathbb{P}(M = 2)$, and $\mathbb{P}(M = 3)$.

(d). Explain why we must have $\mathbb{P}(N = i) = \mathbb{P}(M = i)$ for any $i \geq 3$.

(e). Using the above information calculate $\mathbb{E}(M)$.

Hint: use the known value of $\mathbb{E}(N)$ and consider the difference $\mathbb{E}(N) - \mathbb{E}(M)$.

6. A random variable $Y \sim \text{pois}(\lambda)$ and you are told that λ is an integer.

(a). Calculate $\mathbb{P}(Y = y)/\mathbb{P}(Y = y + 1)$ for $y = 0, 1, \ldots$

(b). What is the most likely value of Y?

Hint: what does it mean if the ratio in (a) is less than one?

7. If X has a Poisson distribution and $\mathbb{P}(X = 0) = 0.2$, find $\mathbb{P}(X \geq 2)$.

8. Suppose $X \sim \text{pois}(\lambda)$.

(a). Find $\mathbb{E}X(X - 1)$ and thus show that $\text{Var}\, X = \lambda$.

(b). Using the fact that $\text{binom}(n, \lambda/n)$ probabilities converge to $\text{pois}(\lambda)$ probabilities, as $n \to \infty$, again show that $\text{Var}\, X = \lambda$.

9. Large batches of components are delivered to two factories, A and B. Each batch is subjected to an acceptance sampling scheme as follows:

   Factory A: Accept the batch if a random sample of 10 components contains less than two defectives. Otherwise reject the batch.

   Factory B: Take a random sample of five components. Accept the batch if this sample contains no defectives. Reject the batch if this sample contains two or more defectives. If the sample contains one defective, take a further sample of five and accept the batch if this sample contains no defectives.

   If the fraction defective in the batch is p, find the probabilities of accepting a batch under each scheme.

   Write down an expression for the average number sampled in factory B and find its maximum value.

10. A new car of a certain model may be assumed to have X minor faults where X has a Poisson distribution with mean μ. A report is sent to the manufacturer listing the faults for each car that has at least one fault. Write down the probability function of Y, the number of faults listed on a randomly chosen report card and find $\mathbb{E}(Y)$. Given $\mathbb{E}(Y) = 2.5$, find μ correct to two decimal places.

11. A contractor rents out a piece of heavy equipment for t hours and is paid \$50 per hour. The equipment tends to overheat and if it overheats x times during the hiring period, the contractor will have to pay a repair cost \$$x^2$. The number of times the equipment overheats in t hours can be assumed to have a Poisson distribution with mean $2t$. What value of t will maximise the expected profit of the contractor?

12. Calculating binomial probabilities using a recursive function.

    Let $X \sim \text{binom}(k, p)$ and let $f(x, k, p) = \mathbb{P}(X = x) = \binom{k}{x} p^x (1-p)^{k-x}$ for $0 \leq x \leq k$ and $0 \leq p \leq 1$. It is easy to show that

$$
\begin{aligned}
f(0, k, p) &= (1-p)^k; \\
f(x, k, p) &= \frac{(k-x+1)p}{x(1-p)} f(x-1, k, p) \text{ for } x \geq 1.
\end{aligned}
$$

    Use this to write a recursive function `binom.pmf(x, k, p)` that returns $f(x, k, p)$.

    You can check that your function works by comparing it with the built-in function `dbinom(x, k, p)`.

13. An airline is selling tickets on a particular flight. There are 50 seats to be sold, but they sell $50 + k$ as there are usually a number of cancellations.

    Suppose that the probability a customer cancels is $p = 0.1$ and assume that individual reservations are independent. Suppose also that the airline makes a profit of \$500 for each passenger who travels (does not cancel and does get a seat), but loses \$100 for each empty seat on the plane and loses \$500 if a customer does not get a seat because of overbooking. The loss because of an empty seat is due to the fixed costs of flying a plane, irrespective of how many passengers it has. The loss if a customer does not get a seat represents both an immediate cost—for example they may get bumped up to first class—as well as a the cost of lost business in the future.

What value of k maximises the airline's expected profit?

14. Write a program to calculate $\mathbb{P}(X + Y + Z = k)$ for arbitrary discrete non-negative rv's X, Y, and Z.

CHAPTER 16

Continuous random variables

In this chapter we further enrich our knowledge of random variables by introducing a number of important continuous random variables. These models supplement the general theory introduced in Chapter 14 and the discrete random variables introduced in the last chapter. We consider the theory, application and implementation in R of the uniform, exponential, Weibull, gamma, normal, χ^2, and t distributions.

16.1 Continuous random variables in R

R has built-in functions for handling the most commonly encountered probability distributions. Suppose that the random variable X is of type `dist` with parameters `p1, p2, ...`, then

`ddist(x, p1, p2, ...)` equals $\mathbb{P}(X = x)$ for X discrete, or the density of X at x for X continuous;

`pdist(q, p1, p2, ...)` equals $\mathbb{P}(X \leq q)$;

`qdist(p, p1, p2, ...)` equals the smallest q for which $\mathbb{P}(X \leq q) \geq p$ (the $100p$ %-point);

`rdist(n, p1, p2, ...)` is a vector of `n` pseudo-random numbers from distribution type `dist`.

The inputs `x`, `q`, and `p` can all be vector valued, in which case the output is vector valued.

Here are some of the continuous distributions provided by R, together with the names of their parameter inputs. Default values are indicated using `=`.

Distribution	R name (dist)	Parameter names
Uniform	`unif`	`min = 0, max = 1`
Exponential	`exp`	`rate = 1`
χ^2	`chisq`	`df`
Gamma	`gamma`	`shape, rate = 1`
Normal	`norm`	`mean = 0, sd = 1`
t	`t`	`df`
Weibull	`weibull`	`shape, scale = 1`

The parameter `rate` that appears in the exponential and gamma distributions will be called λ below; the parameter `shape` used for the gamma is called m below. For the normal distribution R uses as parameters the mean μ and standard deviation σ, rather than the variance σ^2, which we will use. The parameters of the Weibull distribution are explained in Section 16.3.3.

16.2 Uniform distribution

If the probability that X lies in a given subinterval of $[a, b]$ depends only on the length of the subinterval and not on its location, then X is said to have a uniform (or rectangular) distribution on $[a, b]$. Write $X \sim U(a, b)$. The pdf, mean, and variance are

$$\begin{aligned} f(x) &= \frac{1}{b-a} \text{ for } a \leq x \leq b \\ \mu &= \frac{a+b}{2} \\ \sigma^2 &= \frac{(b-a)^2}{12}. \end{aligned}$$

More generally, if S is a bounded subset of $\mathbb{R}^d$ then we say X is uniformly distributed over S if for any $A \subset S$, $\mathbb{P}(X \in A) = |A|/|S|$. Here $|A|$ indicates the size of A, which could be length, area, volume, etc., depending on d.

A trivial example from R is:

```
> punif(0.5, 0, 1)
[1] 0.5
```

16.3 Lifetime models: exponential and Weibull

Let $X \geq 0$ be the time until some event occurs, such as the the breakdown of some mechanical component, in which case X is called the *lifetime* of that component. Let f and F be the pdf and cdf of X, then we define the *survivor function* as $G(x) = \mathbb{P}(X > x) = 1 - F(x)$. That is, $G(x)$ is the probability that the component will survive until time x.

The (age specific) failure rate is called the *hazard function* $\lambda(x)$. $\lambda(x)$ is the rate at which failure occurs at time x, that is

$$\begin{aligned} \lambda(x)\, dx &= \mathbb{P}(\text{lifetime of } X \text{ between } x \text{ and } x + dx \mid \text{lifetime of } X > x) \\ &= \mathbb{P}(\text{component fails between } x \text{ and } x + dx \mid \text{still working at } x) \\ &= \frac{f(x)\, dx}{G(x)}, \\ \lambda(x) &= \frac{f(x)}{G(x)}. \end{aligned}$$

We can find the density f from λ as follows:

$$\begin{aligned} f(x) &= \frac{dF(x)}{dx} = \frac{d}{dx}(1 - G(x)) = -\frac{dG(x)}{dx} \\ \lambda(x) &= \frac{f(x)}{G(x)} = \frac{-G'(x)}{G(x)} = -\frac{d}{dx}\log G(x) \\ G(x) &= \exp\left(-\int_0^x \lambda(u)\,du\right) \\ f(x) &= \lambda(x)\exp\left(-\int_0^x \lambda(u)\,du\right) \end{aligned}$$

16.3.1 Exponential distribution

If $\lambda(x) = \lambda$, that is a constant rate of failure, then we say X has an *exponential* distribution and write $X \sim \exp(\lambda)$. In this case

$$\begin{aligned} f(x) &= \lambda e^{-\lambda x} \\ \mu &= 1/\lambda \\ \sigma^2 &= 1/\lambda^2 \end{aligned}$$

To say $\lambda(x)$ is constant is to say that ageing has no effect, that is, the component fails at random. This property of the exponential is called the *memoryless property.* It is usually expressed as follows, for $s, t \geq 0$,

$$\begin{aligned} \mathbb{P}(X > s + t \,|\, X > s) &= \mathbb{P}(X > s + t \text{ and } X > s)/\mathbb{P}(X > s) \\ &= \mathbb{P}(X > s + t)/\mathbb{P}(X > s) \\ &= e^{-\lambda(s+t)}/e^{-\lambda s} \\ &= e^{-\lambda t} = \mathbb{P}(X > t). \end{aligned}$$

In other words, given that you have survived until age s, the probability of surviving an additional time t is the same as if you had just been born.

Figure 16.1 shows the probability density function of several exponential distributions.

16.3.2 Example: radioactive decay

Uranium-238 decays into thorium-234 at some rate λ per year (releasing an alpha particle in the process), constant over time. The half life of uranium-238 is 4.47×10^9 years, and is defined as the (expected) time it takes for half of some lump of uranium-238 to decay into thorium-234. That is, if X is the time to decay of a single atom, then $X \sim \exp(\lambda)$ and

$$\mathbb{P}(X > 4.47 \times 10^9) = 0.5.$$

But $\mathbb{P}(X > x) = e^{-\lambda x}$ so we have $\lambda = \log 2/(4.47 \times 10^9) = 1.55 \times 10^{-8}$.

A gram of uranium-238 contains approximately 2.53×10^{21} atoms. What is

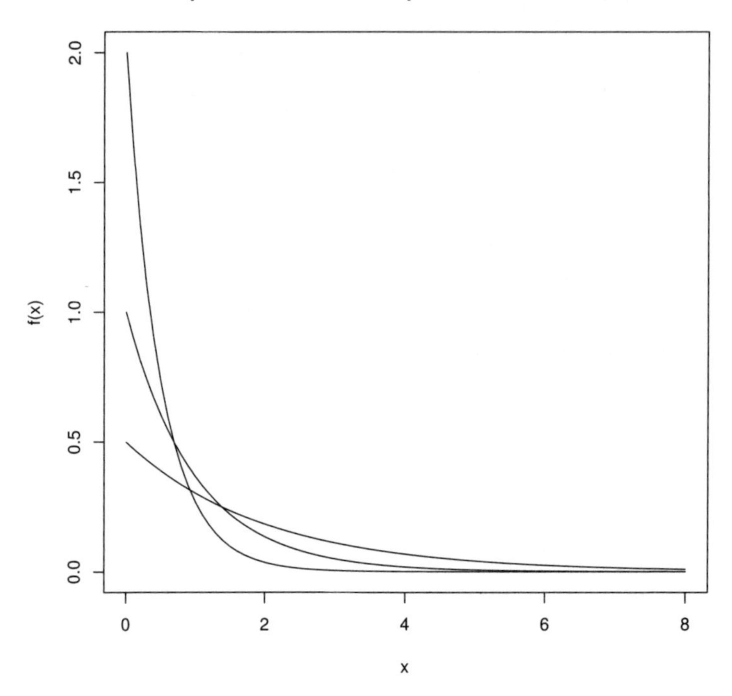

Figure 16.1 *Some exponential densities.*

the expected time until the first release of an alpha particle? Until the first decay we have 2.53×10^{21} atoms each decaying at rate 1.55×10^{-8} per year, so the total rate of decay is roughly 3.9×10^{13} per year. That is, the time to the first release of an alpha particle has an $\exp(3.9 \times 10^{13})$ distribution, with mean of 2.6×10^{-14} years (less than one millionth of a second).

We have implicitly used here the fact that the minimum of n independent exponential random variables is also exponential, with rate given by the sum of the original n rates (see Exercise 2).

16.3.3 Weibull distribution

X has a Weibull distribution with parameters λ and m if it has hazard function $\lambda(x) = m\lambda x^{m-1}$, for m and $\lambda > 0$. We write $X \sim \text{Weibull}(\lambda, m)$.

Clearly a Weibull$(\lambda, 1)$ rv is the same as an $\exp(\lambda)$ rv. More generally, we have

$$\begin{aligned} G(x) &= \exp\left(-\textstyle\int_0^x \lambda(u)\,du\right) = \exp(-\lambda x^m), \\ f(x) &= \lambda(x)\exp\left(-\textstyle\int_0^x \lambda(u)\,du\right) = m\lambda x^{m-1}e^{-\lambda x^m} \text{ for } x \geq 0. \end{aligned}$$

Using these we can show that

$$\begin{aligned}\mu &= \lambda^{-1/m}\Gamma(1+1/m)\\ \sigma^2 &= \lambda^{-2/m}(\Gamma(1+2/m)-\Gamma(1+1/m)^2)\end{aligned}$$

where Γ is the gamma function:

$$\begin{aligned}\Gamma(p) &= \int_0^\infty x^{p-1}e^{-x}\,dx \text{ for } p>0;\\ \Gamma(p) &= (p-1)\Gamma(p-1) \text{ for all } p>1;\ \Gamma(1)=1;\ \Gamma(1/2)=\sqrt{\pi};\\ \Gamma(n) &= (n-1)! \text{ for integer valued } n.\end{aligned}$$

For p not equal to an integer or an integer plus $^1/_2$, we need to use numerical integration to calculate $\Gamma(p)$.

Figure 16.2 shows the hazard functions and probability density functions of several Weibull distributions.

Note that the R parameterisation of the Weibull distribution differs from that presented here. To evaluate Weibull probabilities in R, use m for the `shape` argument and $\lambda^{-1/m}$ for the `scale` argument. Thus, to reproduce the three lower panels in Figure 16.2, use the following functions.

```
curve(dweibull(x, shape = 0.5, scale = 2^(-1/0.5)), from = 0, to = 4)
curve(dweibull(x, shape = 1.5, scale = 2^(-1/1.5)), from = 0, to = 4)
curve(dweibull(x, shape = 3, scale = 2^(-1/3)), from = 0, to = 4)
```

16.3.4 Example: time to the next disaster

Suppose that the chance of a nuclear power station having a major accident in any given year is proportional to its age. Also suppose that we keep building nuclear power stations at a rate of one per year, until we have a major accident. Let T be the time until the first major accident. What (approximately) is the distribution of T?

Let αt be the chance that a single power station age t has an accident in the next year. This is essentially equivalent to saying that at age t it has accidents at a rate of αt per year. After t years there are t power stations operating, so the total rate of accidents is αt^2. Thus (approximately) $T \sim$ Weibull$(\alpha/3, 3)$. T is only approximately Weibull because in reality we can only have a whole number of power stations, and here we have allowed a fractional number.

For example, let α be one in one million. Then the probability that the first major accident is within the next 50 years would be

```
> pweibull(50, 3, (1e-06/3)^(-1/3))
[1] 0.04081054
```

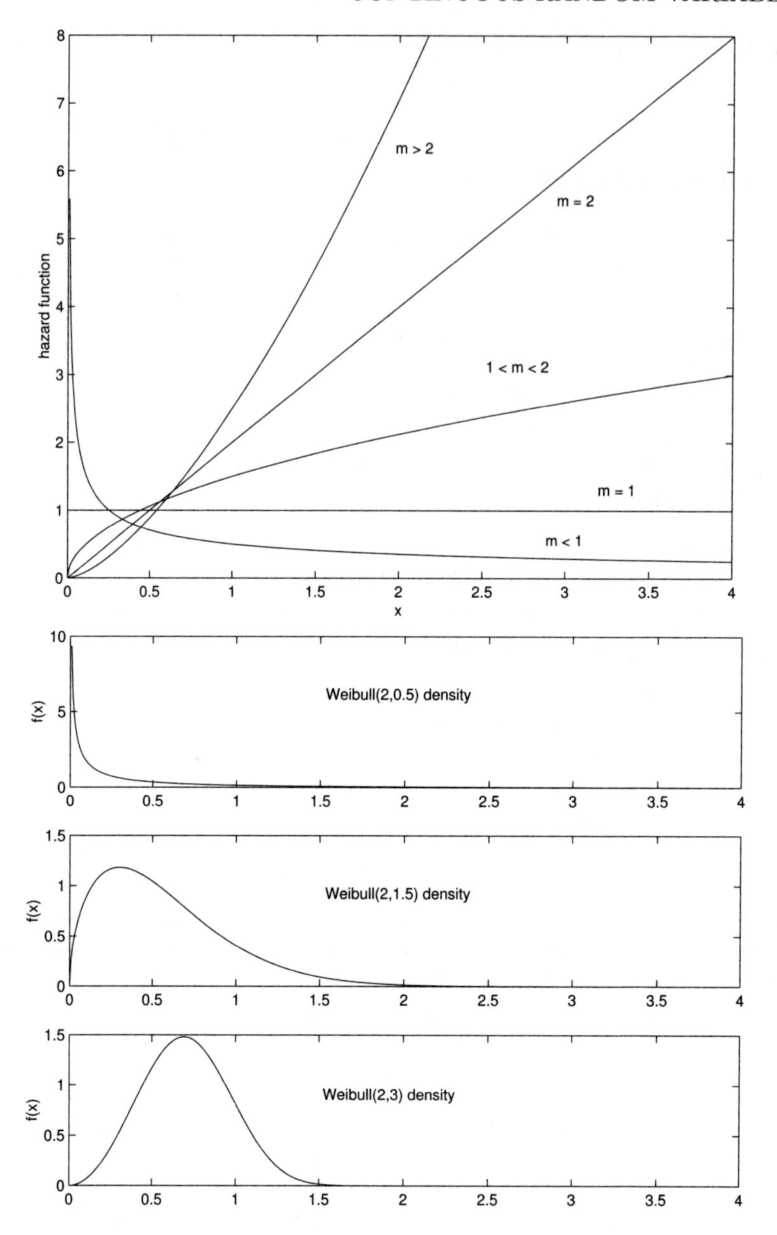

Figure 16.2 *Hazard functions and densities of some Weibull random variables.*

16.4 The Poisson process and the gamma distribution

A Poisson process is the continuous-time analogue of a sequence of independent trials.

We suppose that we have a sequence of events, occurring at some rate λ per unit time. That is, the *expected* number of events occurring in the time interval (s,t) is $\lambda(t-s)$, and the *infinitesimal probability* that an event occurs in the time interval $(t, t+dt)$ is $\lambda\, dt$.

Let T_k be the time between the $k-1$ and k-th events, and let $N(s,t)$ be the number of events that have occurred during the interval (s,t). It can be shown that the $\{T_k\}_{k=1}^{\infty}$ are iid $\exp(\lambda)$ random variables and that $N(s,t) \sim \text{pois}(\lambda(t-s))$. Moreover, if the intervals (a,b) and (s,t) are disjoint, then $N(a,b)$ and $N(s,t)$ are independent.

To understand the Poisson process it is useful to consider a discrete approximation. Take the time interval $[0,t]$ and split it into n subintervals of length t/n. The probability of an event occurring in the i-th interval is approximately $\lambda t/n$, independently of all the others. The total number of events occurring in $[0,t]$ thus has a $\text{binom}(n, \lambda t/n)$ distribution, which converges to a $\text{pois}(\lambda t)$ distribution as $n \to \infty$ (see Section 15.6). The number of intervals between any two events, X say, has a $\text{geom}(\lambda t/n)$ distribution. Thus the time between any two events is given by $Y = (t/n)X$ and we have

$$\begin{aligned}
\mathbb{P}(Y > y) &= \mathbb{P}(X > ny/t) \\
&= \sum_{x=\lceil ny/t \rceil}^{\infty} (\lambda t/n)(1-\lambda t/n)^x \\
&= \left(1 - \frac{\lambda t}{n}\right)^{\lceil ny/t \rceil} \\
&\to (e^{-\lambda t})^{y/t} = e^{-\lambda y} \text{ as } n \to \infty.
\end{aligned}$$

But this is just the probability that an $\exp(\lambda)$ random variable is larger than y, as required.

Figure 16.3 shows a realisation of a Poisson process.

16.4.1 A paradox?

Suppose we have a Poisson process of rate λ, and we turn up at some random time t to observe it. On average, we will arrive halfway between two arrivals. Thus the expected time until the next arrival will be half the expected time between any two arrivals, that is $1/(2\lambda)$. But the memoryless property of the exponential tells us that the time from our appearance to the next arrival should still be exponential(λ), with mean $1/\lambda$, a contradiction!

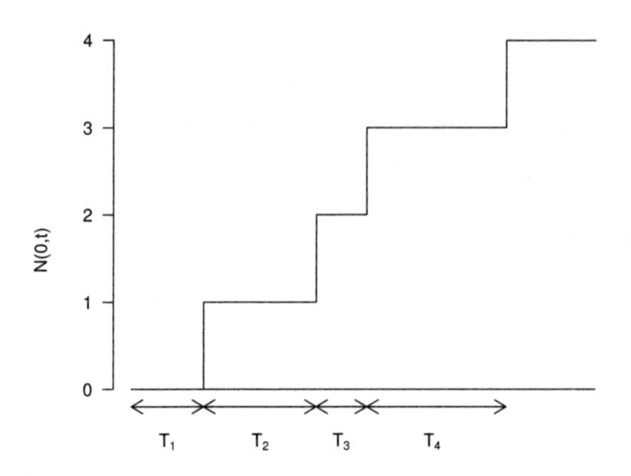

Figure 16.3 *A Poisson process.*

This seeming paradox is not in fact real, as there is a flaw in the above argument. If we turn up at a random time, then we are more likely to turn up between two widely spaced arrivals than between two closely spaced arrivals. Thus the interarrival period in which we turn up will be larger on average than the norm, and so its expected length will be larger than the norm (in fact, exactly twice the norm).

16.4.2 Merging and Thinning

The Poisson process has many useful properties. Two of these concern merging and thinning, which are illustrated in Figure 16.4.

If we merge a Poisson process rate λ_1 with an independent Poisson process rate λ_2, then the result is a Poisson process rate $\lambda_1 + \lambda_2$. By merging we mean that we add all of the events together.

We thin a process by tossing a (biased) coin for each event: heads we keep it; tails it is discarded. If we start with a Poisson process rate λ, and the probability of keeping an event is p, then the result is a Poisson process rate $p\lambda$.

Both of these results are intuitively clear from the discrete approximation.

16.4.3 Gamma distribution

The exponential distribution is the continuous analogue of the geometric distribution. If we sum independent geometric distributions we get a negative

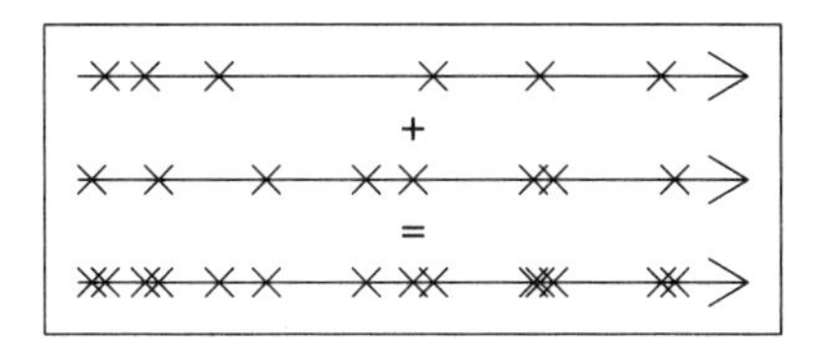

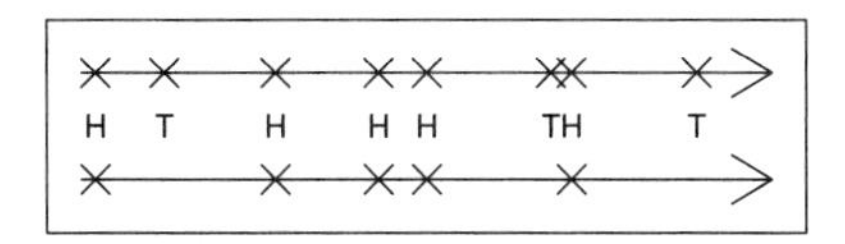

Figure 16.4 *Merging and thinning of Poisson processes.*

binomial. The continuous analogue of the negative binomial is the *gamma* distribution (see Table 16.1).

Let X be the sum of m independent $\exp(\lambda)$ random variables, then it can be shown that X has the following pdf, mean, and variance

$$\begin{aligned} f(x) &= \frac{1}{\Gamma(m)}\lambda^m x^{m-1} e^{-\lambda x} \text{ for } x \geq 0 \text{ and } m, \lambda > 0 \\ \mu &= m/\lambda \\ \sigma^2 &= m/\lambda^2 \end{aligned}$$

We write $X \sim \Gamma(\lambda, m)$. Note that this definition actually holds for all $m > 0$, not just integer values. In the special case where m is integer valued, the gamma distribution is more properly known as the *Erlang* distribution. $F(x) = \int_0^x f(u)du$ must be calculated numerically when m is not an integer.

Examples of gamma densities are presented in Figure 16.5. To produce these density plots use

```
> curve(dgamma(x, shape = 0.5, rate = 2), from = 0, to = 4)
> curve(dgamma(x, shape = 1.5, rate = 2), from = 0, to = 4)
> curve(dgamma(x, shape = 3, rate = 2), from = 0, to = 4)
```

Note that in R, the default order for the parameters of the gamma distribution is (m, λ) rather than (λ, m).

16.4.4 Example: discrete simulation of a queue

Consider a store where customers queue to pay for their goods at the check-out.

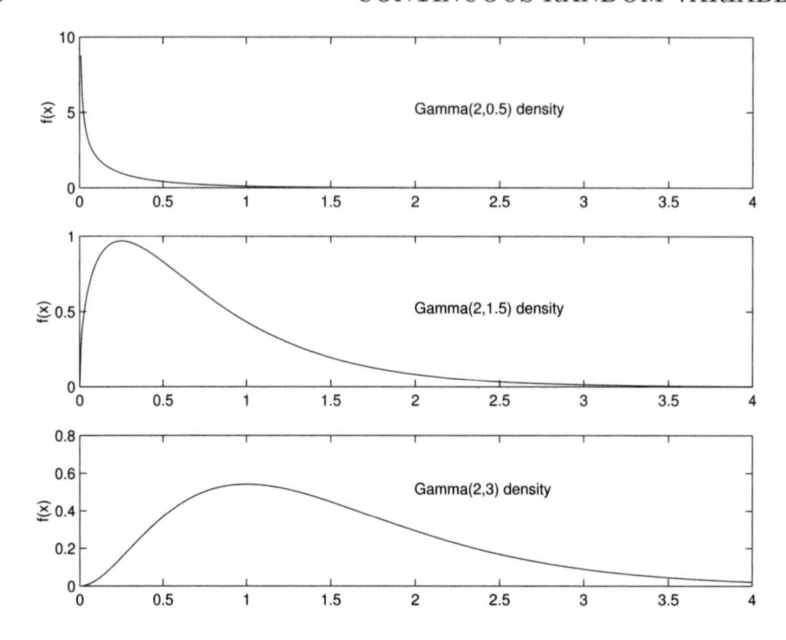

Figure 16.5 *Some gamma densities.*

Table 16.1 *Some correspondences between discrete and continuous distributions*

Context	Discrete case	Continuous case
Random Process	Sequence of independent trials	Poisson process
Number of events in an interval	Binomial	Poisson
Time between two events	Geometric	Exponential
Time between multiple events	Negative binomial	Gamma

We will use a Poisson process to model the arrival of customers at the checkout. That is, customers arrive randomly at some constant rate λ. This assumption is reasonably realistic.

We will also suppose that the time taken to serve a single customer has an exponential distribution, with parameter μ. This assumption is less realistic than the arrival assumption, but we make it because it simplifies our analysis enormously. The reason for the simplification is that the exponential distribution has the memoryless property.

When customer i arrives at the head of the queue, we associate with him a random variable S_i, which is how long he will take to be served. If he arrives at time t then he will depart when his service finishes at time $t + S_i$.

If $S_i \sim \exp(\mu)$ then this is equivalent to the following: if at time $t+s$ customer i is still being served, the probability that service finishes in the next small interval of time $(t + s, t + s + dt)$ is $\mu\, dt$. That is, the points at which services finish are just like the events in a Poisson process with rate μ. Thus, with exponential service times, we can determine the times when people depart the queue using a 'service process', which is a Poisson process of rate μ.

The only thing we have to worry about is if a service event occurs when the queue is empty, but we can in fact just discard these.

Here is a program for simulating a queuing system. We use a discrete approximation to the arrival and service processes. That is, we split time into small intervals of length δ, then the chance of an arrival in any interval is $\lambda\delta$ and the chance of a departure is $\mu\delta$ (provided the queue is not empty). Note that we do not allow an arrival and a departure to occur in the same interval. (This has probability $\lambda\mu\delta^2$, which will be very small if δ is small.) Also note that we use the command `set.seed(rand.seed)`. The function `set.seed` will be explained in Section 18.1.2, but can be ignored at this point.

The output is given in Figure 16.6. What would happen if $\mu < \lambda$? Try it and see.

```
# program: spuRs/resources/scripts/discrete_queue.r
# Discrete Queue Simulation

# inputs
lambda <- 1       # arrival rate
mu <- 1.1         # service rate
t.end <- 100      # duration of simulation
t.step <- 0.05    # time step
rand.seed <- 99 # seed for random number generator

# simulation
set.seed(rand.seed)
queue <- rep(0, t.end/t.step + 1)
for (i in 2:length(queue)) {
  if (runif(1) < lambda*t.step) { # arrival
    queue[i] <- queue[i-1] + 1
  } else if (runif(1) < mu*t.step) { # potential departure
    queue[i] <- max(0, queue[i-1] - 1)
  } else { # nothing happens
    queue[i] <- queue[i-1]
  }
}

# output
```

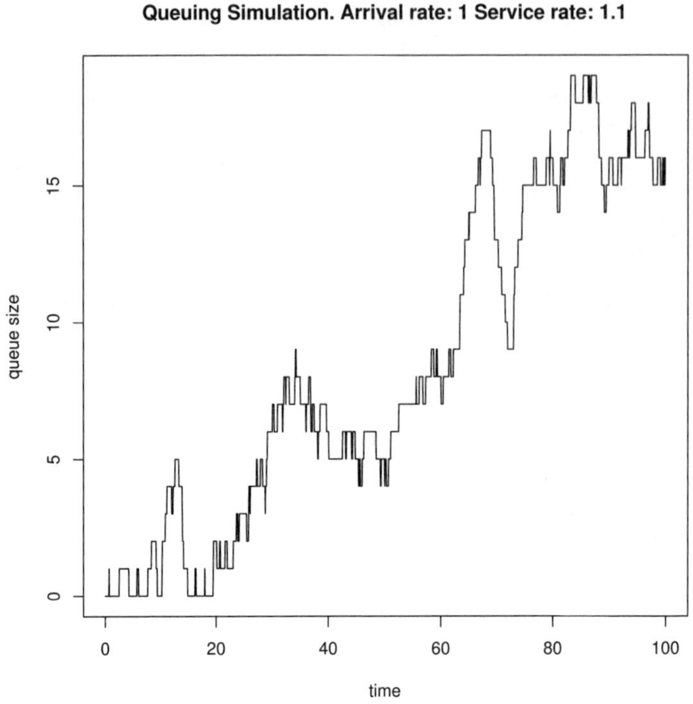

Queuing Simulation. Arrival rate: 1 Service rate: 1.1

Figure 16.6 *Output from* `discrete_queue.r`*: a simulation of a single server queue.*

```
plot(seq(from=0, to=t.end, by=t.step), queue, type='l',
    xlab='time', ylab='queue size')
title(paste('Queuing Simulation. Arrival rate:', lambda,
    'Service rate:', mu))
```

16.5 Sampling distributions: normal, χ^2, and t

The following types of distribution are important in Statistics, because they appear naturally when dealing with random samples.

16.5.1 Normal or Gaussian distribution

The importance of the normal (or Gaussian) distribution comes from the Central Limit Theorem (see Chapter 17), which tells us that when you take the average of a sufficiently large iid sample, the distribution of the sample averages looks like that of a normal random variable. The normal distribution is also commonly used to model measurement errors, as well as many natural phenomena. We write $X \sim N(\mu, \sigma^2)$, where $\mu = \mathbb{E}X$ and $\sigma^2 = \text{Var}\, X$. The normal density is

$$f(x) = \frac{1}{\sqrt{2\pi\sigma^2}} e^{-(x-\mu)^2/(2\sigma^2)} \text{ for } -\infty < x < \infty.$$

The case $\mu = 0$, $\sigma^2 = 1$ is called the standard normal. If $Z \sim N(0, 1)$ then $\sigma Z + \mu \sim N(\mu, \sigma^2)$.

The distribution function F of X cannot be obtained analytically, instead we must use numerical integration. The density of the standard normal is denoted ϕ and the distribution function is denoted Φ. F can be obtained from Φ via $F(x) = \Phi((x - \mu)/\sigma)$.

Figure 16.7 presents several normal densities.

16.5.2 Example: normal percentage points

In Statistics much use is made of Φ^{-1}, and it is common for textbooks to give tables of Φ^{-1}, called percentage points or quantiles. One important use of quantiles is in the calculation of confidence intervals: see Section 17.3. For example, if $Z \sim N(0, 1)$, then $\mathbb{P}(Z > 1.6449) = 5\%$ and $\mathbb{P}(Z > 1.9600) = 2.5\%$. That is, $\Phi^{-1}(0.95) = 1.6449$ and $\Phi^{-1}(0.975) = 1.9600$. So 1.6449 is the 95th percentage point or the 0.95 quantile of $N(0, 1)$.

Let $z_\alpha = \Phi^{-1}(\alpha)$, then z_α is the unique root of the function $\Phi(z) - \alpha$. Thus, if we can calculate Φ then we can find z_α using a root-finding algorithm, which is what the following code does.

```
# program: spuRs/resources/scripts/ppoint.r

phi <- function(x) return(exp(-x^2/2)/sqrt(2*pi))

ppoint <- function(p, pdf = phi, z.min = -10, tol = 1e-9) {
  # calculate a percentage point
  #
  # p is assumed to be between 0 and 1
  # pdf is assumed to be a probability density function
  #
  # let F(x) be the integral of pdf from -infinity to x
  # we apply the Newton-Raphson algorithm to find z_p such that F(z_p) = p
  # that is, to find z_p such that F(z_p) - p = 0
  # note that the derivative of F(z) - p is just pdf(z)
  #
  # we approximate -infinity by z.min (that is we assume that the integral
  # of pdf from -infinity to z.min is negligible)

  # do first iteration
  x <- 0
```

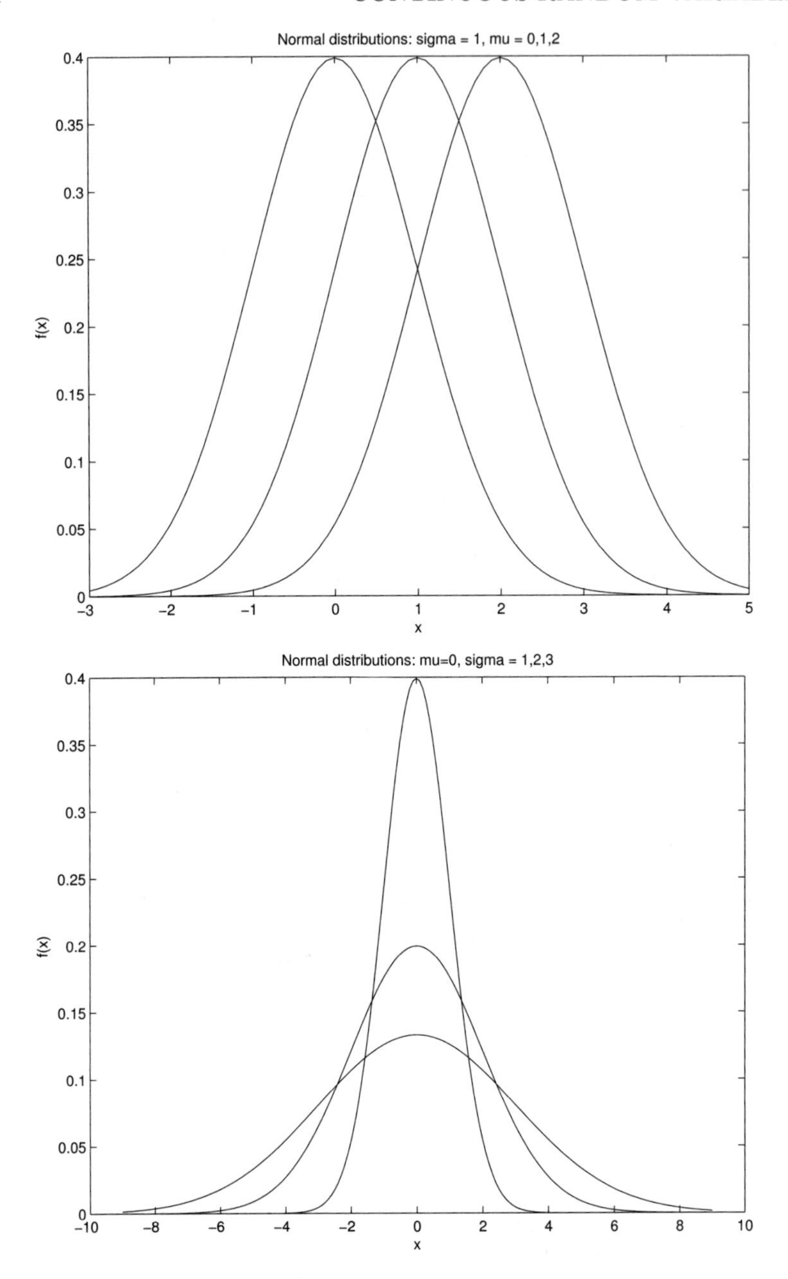

Figure 16.7 *Normal density: the effect of μ and σ^2.*

```
  f.x <- simpson_n(pdf, z.min, x) - p
  # continue iterating until stopping conditions are met
  while (abs(f.x) > tol) {
    x <- x - f.x/pdf(x)
    f.x <- simpson_n(pdf, z.min, x) - p
  }
  return(x)
}
```

```
> source("../scripts/simpson_n.r")
> source("../scripts/ppoint.r")
> ppoint(0.95)
[1] 1.644853
> ppoint(0.975)
[1] 1.959966
```

We note that R provides convenient built-in functions for determining percentage points for a variety of distributions: see Section 16.1.

16.5.3 The sum of independent normals

A remarkable result, which we will not be proving, is that if $X \sim N(\mu_1, \sigma_1^2)$ and $Y \sim N(\mu_2, \sigma_2^2)$ independent of X, then $X + Y \sim N(\mu_1 + \mu_2, \sigma_1^2 + \sigma_2^2)$.

Even if we cannot prove that the sum of two independent normals is normal, we can verify the theorem experimentally. R provides the function `rnorm` for simulating normal random variables. We can check that `rnorm` actually works by simulating an iid sample of $N(0, 1)$ rv's, and checking that their histogram looks like a normal density.

```
> z <- rnorm(10000)
> par(las = 1)
> hist(z, breaks = seq(-5, 5, 0.2), freq = F)
> phi <- function(x) exp(-x^2/2)/sqrt(2 * pi)
> x <- seq(-5, 5, 0.1)
> lines(x, phi(x))
```

The output is given in Figure 16.8. Happy that `rnorm` does what is says on the box, we can now check our theorem on the sum of independent normals.

```
> z1 <- rnorm(10000, mean=1, sd=1)
> z2 <- rnorm(10000, mean=1, sd=2)
> z <- z1 + z2  # mean = 2, var = 1^2 + 2^2 = 5
> par(las=1)
> hist(z, breaks=seq(-10, 14, .2), freq=F)
> f <- function(x) exp(-(x-2)^2/10)/sqrt(10*pi)  # N(2, 5) density
> x <- seq(-10, 14, .1)
> lines(x, f(x))
```

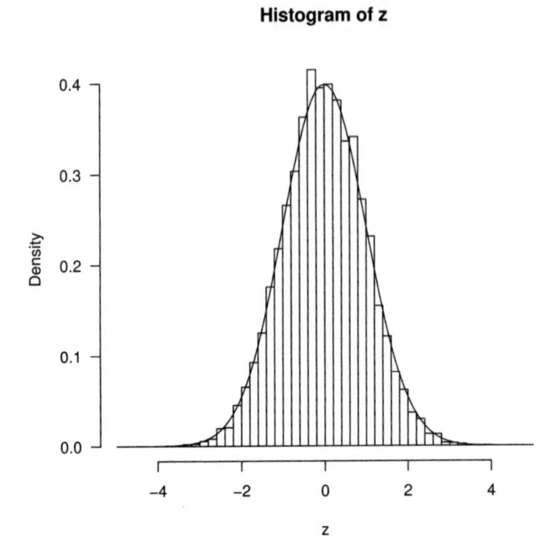

Figure 16.8 *Validating* `rnorm`.

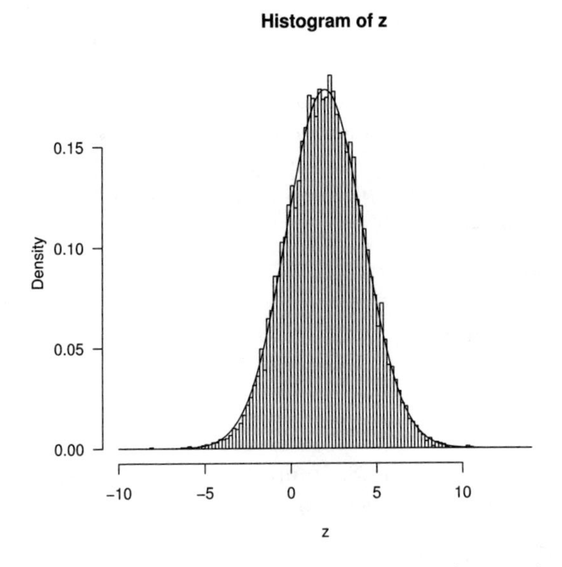

Figure 16.9 *Checking that the distribution of the sum of two independent normals is normal.*

The output is given in Figure 16.9. We see that the scaled histogram is very close to the theoretical density, which supports the theory.

16.5.4 χ^2 distribution

Suppose $Z_1, \ldots, Z_\nu$ are iid $N(0,1)$, then $X = Z_1^2 + \cdots + Z_\nu^2$ is said to have a χ_ν^2 distribution. We say X has a chi-squared distribution with ν degrees of freedom, and write $X \sim \chi_\nu^2$. It can be shown that a χ_ν^2 has the same distribution as a $\Gamma(1/2, \nu/2)$. Thus $\mathbb{E}X = \nu$ and $\operatorname{Var} X = 2\nu$.

16.5.5 Student's t distribution

If $X \sim N(0,1)$ and $Y \sim \chi_\nu^2$ independently of X, then

$$T = \frac{X}{\sqrt{Y/\nu}}$$

is said to have a t distribution with ν degrees of freedom, and written $T \sim t_\nu$. T has density

$$f(x) = \frac{\Gamma((\nu+1)/2)}{\sqrt{\nu\pi}\,\Gamma(\nu/2)} \left(1 + \frac{x^2}{\nu}\right)^{-(\nu+1)/2} \quad \text{for } -\infty < x < \infty.$$

The t_ν distribution is symmetric, and similar in shape to the $N(0,1)$, but with fatter tails. As $\nu \to \infty$, the t_ν density converges to the standard normal density.

The t distribution is also called Student's t distribution, after the pseudonym 'Student' of William Sealy Gosset, who first described it. Gosset used the pseudonym because his then employer, the Guinness brewery in Dublin, prohibited the publication of any papers by its employees. The t distribution is used to construct confidence intervals for the mean when the population variance is unknown: see Section 17.3.3.

Several t densities are presented in Figure 16.10.

16.6 Exercises

1. A random variable U has a $U(a,b)$ distribution if $\mathbb{P}(U \in (u,v)) = (v-u)/(b-a)$ for all $a \le u \le v \le b$.

   Show that if $U \sim U(a,b)$ then so is $a + b - U$.

2. Show that if $X \sim \exp(\lambda)$ and $Y \sim \exp(\mu)$, independently of X, then $Z = \min\{X, Y\} \sim \exp(\lambda + \mu)$.

   Hint: $\min\{X,Y\} > z \iff X > z$ and $Y > z$.

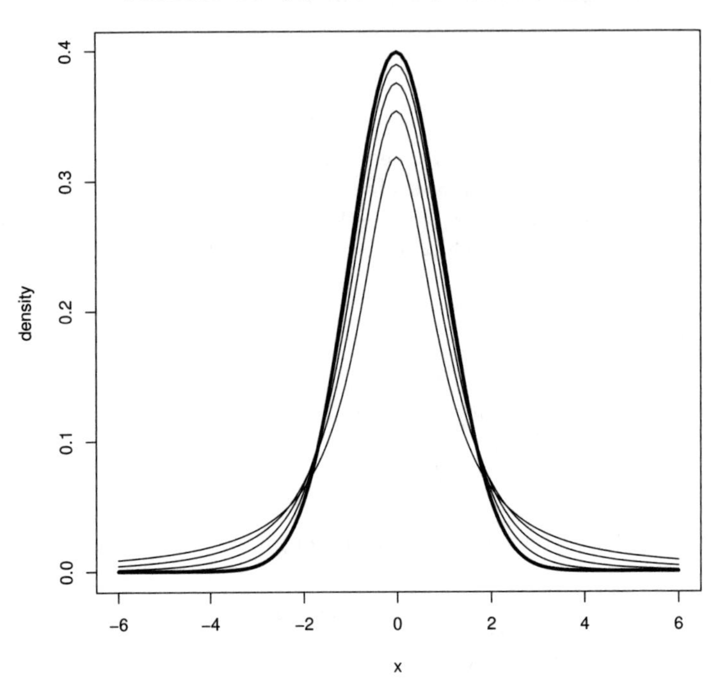

Figure 16.10 *t densities for* $\nu = 1, 2, 4, 10,$ *and* ∞.

3. The time to failure of a new type of light bulb is thought to have an exponential distribution.

   Reliability is defined as the probability that an article will not have failed by a specified time. If the reliability of this type of light bulb at 10.5 weeks is 0.9, find the reliability at 10 weeks.

   One hundred bulbs of this type are put in a new shop. All the bulbs that have failed are replaced at 20-week intervals and none are replaced at other times. If R is the number of bulbs that have to be replaced at the end of the first interval, find the mean and variance of R.

   Explain why this result will hold for *any* such interval and not just the first.

4. The length of a certain type of battery is normally distributed with mean 5.0 cm and standard deviation 0.05 cm. Find the probability that such a battery has a length between 4.92 and 5.08 cm.

   Tubes are manufactured to contain four such batteries. 95% of the tubes have lengths greater than 20.9, and 10% have lengths greater than 21.6 cm.

Assuming that the lengths are also normally distributed, find the mean and standard deviation, correct to two decimal places.

If tubes and batteries are chosen independently, find the probability that a tube will contain four batteries with at least 0.75 cm to spare.

5. A man travels to work by train and bus. His train is due to arrive at 08:45 and the bus he hopes to catch is due to leave at 08:48. The time of arrival of the train has a normal distribution with mean 08:44 and standard deviation three mins; the departure time of the bus is independently normally distributed with mean 08:50 and standard deviation one minute. Calculate the probabilities that:

   - The train is late;
   - The bus departs before the train arrives;
   - In a period of five days there are at least three days on which the bus departs before the train arrives.

6. Suppose $X \sim U(0,1)$ and $Y = X^2$.

   Use the cdf of X to show that $P(Y \leq y) = \sqrt{y}$ for $0 < y < 1$, and thus obtain the pdf of Y. Hence or otherwise evaluate $\mathbb{E}(Y)$ and $\text{Var}\,(Y)$.

7. A mechanical component is only usable if its length is between 3.8 cm and 4.2 cm. It is observed that on average 7% are rejected as undersized, and 7% are rejected oversized. Assuming the lengths are normally distributed, find the mean and standard deviation of the distribution.

8. Telephone calls arrive at a switchboard in accordance with a Poisson process of rate $\lambda = 5$ per hour.

   (a). What is the distribution of N_1 = the number of calls that arrive in any one hour period?

   (b). What is the distribution of N_2 = the number of calls that arrive in any half hour period?

   (c). Find the probability that the operator is idle for the next half hour.

9. Glass sheets have faults called 'seeds', which occur in accordance with a Poisson process at a rate of 0.4 per square metre. Find the probability that rectangular sheets of glass of dimensions 2.5 metres by 1 metre will contain:

   (a). No seeds.

   (b). More than one seed.

   If sheets with more than one seed are rejected, find the probability that in a batch of 10 sheets, at most one is rejected.

10. Cars pass through an intersection in accordance with a Poisson process with rate $\lambda = 3$ per minute. A pedestrian takes s seconds to cross at the intersection and chooses to start to cross irrespective of the traffic conditions. Assume that if he is on the intersection when a car passes by, then he is injured. Find the probability that the pedestrian crosses safely for s = 5, 10, and 20.

11. We examine blood under a microscope for red blood cell deficiency, using a small fixed volume that will contain on the average five red cells for a normal person. What is the probability that a specimen from a normal person will contain only two red cells or fewer (assume that cells are independently and uniformly distributed throughout the volume)?

12. Defects occur in an optical fibre in accordance with a Poisson process with rate $\lambda = 4.2$ per kilometre. Let N_1 be the number of defects in the first kilometre of fibre and N_2 be the number of defects in the second and third kilometres of fibre.

    (a). What are the distributions of N_1 and N_2?

    (b). Are N_1 and N_2 dependent or independent?

    (c). Let $N = N_1 + N_2$. What is the distribution of N?

13. The time (in hours) until failure of a transistor is a random variable $T \sim \exp(1/100)$.

    (a). Find $\mathbb{P}(T > 10)$.

    (b). Find $\mathbb{P}(T > 100)$.

    (c). It is observed that after 90 hours the transistor is still working. Find the conditional probability that $T > 100$, that is, $\mathbb{P}(T > 100 \,|\, T > 90)$. How does this compare with part (a)? Explain this result.

14. Jobs submitted to a computer system have been found to require a CPU time T, which is exponentially distributed with mean 150 milliseconds. If a job doesn't complete within 90 milliseconds is suspended and put back at the end of the queue. Find the probability that an arriving job will be forced to wait for a second quantum.

15. An insurance company has received notification of five pending claims. Claim settlement will not be complete for at least one year. An actuary working for the company has been asked to determine the size of the reserve fund that should be set up to cover these claims. Claims are independent and exponentially distributed with mean \$2,000. The actuary recommends setting up a claim reserve of \$12,000. What is the probability that the total claims will exceed the reserve fund?

16. Suppose that $X \sim U(0,1)$.

    (a). Put $Y = h(X)$ where $h(x) = 1 + x^2$. Find the cdf F_Y and the pdf f_Y of Y.

    (b). Calculate $\mathbb{E}Y$ using $\int y f_Y(y)\,dy$ and $\int h(x) f_X(x)\,dx$.

    (c). The function `runif(n)` simulates n iid $U(0,1)$ random variables, thus `1 + runif(n)^2` simulates n iid copies of Y.

    Estimate and plot the pdf of Y using a simulated random sample. Experiment with the bin width to get a good-looking plot: it should be reasonably detailed but also reasonably smooth. How large does your sample have to be to get a decent approximation?

17. Let $N(t)$ be the number of arrivals up to and including time t in a Poisson process of rate λ, with $N(0) = 0$. In this exercise we will verify that $N(t)$ has a pois(λt) distribution.

    We define the Poisson process in terms of the times between arrivals, which are independent with an exp(λ) distribution. The first part of the task is to simulate $N(t)$ by simulating all the arrival times up until time t. Let $T(k)$ be the time of the first arrival, then

    $$T(1) \sim \exp(\lambda) \text{ and } T(k) - T(k-1) \sim \exp(\lambda).$$

    Given the arrival times we get $N(t) = k$ where k is such that

    $$T(k) \leq t < T(k+1).$$

    Thus to simulate $N(t)$ we simulate $T(1), T(2), \ldots$, and keep going until we get $T(n) > t$, then put $N(t) = n - 1$.

    Once you have code that can simulate $N(t)$, use it to generate a sample with $\lambda = 0.5$ and $t = 10$. Now check the distribution of $N(t)$ by using the sample to estimate the probability function of $N(t)$. That is, for $x \in \{0, 1, 2, \ldots\}$ (stop at around 20), we calculate $\hat{p}(x)$ = proportion of sample with value x, and compare the estimates with the theoretical Poisson probabilities

    $$p(x) = e^{-\lambda t} \frac{(\lambda t)^x}{x!}.$$

    An easy way to compare the two is to plot $\hat{p}(x)$ for each x and then on the same graph plot the true probability function using vertical lines with heights $p(x)$,

    You might also like to try plotting the *sample path* of a Poisson process. That is, plot $N(t)$ as a function of t.

CHAPTER 17

Parameter Estimation

An important practical challenge in model fitting is as follows. Imagine that we have a set of data that we believe comes from some distribution or other. How can we find values of the relevant parameters so that the distribution represents the data?

In parametric model fitting (also called parametric inference), we specify a priori what type of distribution we will fit, for example a normal distribution, then choose the parameters that best fit the data (μ and σ^2 in the case of the normal distribution).

In this chapter we cover a range of approaches to finding the single best estimate of a parameter, given some data and a model, as well as approaches to determining a range of possible values that a parameter could take.

Another important practical task is to choose the distribution that best fits an observed sample; we do not cover this challenge here.

17.1 Point Estimation

We start with the problem of finding values for the parameters that provide the best fit between the model and the data, called *point estimates*. First, we need to define what we mean by 'best fit'. There are two commonly used criteria:

Method of moments chooses the parameters so that the sample moments (for example the sample mean and variance) match the theoretical moments of our chosen distribution.

Maximum likelihood chooses the parameters to maximise a function of the data called the likelihood, which measures how likely it is to observe our given sample.

We will demonstrate both approaches through two examples.

17.1.1 Example: Kew rainfall

The rainfall at Kew Gardens in London has been systematically measured since 1697. Figure 17.1 gives a histogram of total July rainfall in millimetres, over the years 1697 to 1999.[1] The gamma distribution is often a good fit to aggregated rainfall data, and will be our candidate distribution in this case.

Method of moments We read the Kew rainfall data from the file `kew.txt` and calculate the sample mean and variance. The data are in units of 0.1 mm, so we first divide by 10 to get millimetres.

```
> kew <- read.table("../data/kew.txt", col.names = c("year",
+     "jan", "feb", "mar", "apr", "may", "jun", "jul", "aug",
+     "sep", "oct", "nov", "dec"))
> kew[, 2:13] <- kew[, 2:13]/10
> kew.mean <- apply(kew[-1], 2, mean)
> kew.var <- apply(kew[-1], 2, var)
```

Here the command `apply(kew[-1], 2, mean)` applies the `mean` function to the columns of `kew[-1]`, that is, to all columns apart from the first. If $X \sim \Gamma(\lambda, m)$ then it has mean m/λ and variance m/λ^2. Let $X_1, \ldots, X_n$ be an iid sample from X. Using the method of moments, we choose m and λ so that the sample and theoretical mean and variance match, giving us a (non-linear) system of equations for m and λ:

$$\begin{aligned} \hat{\mu} &= \overline{X} = m/\lambda \\ \hat{\sigma}^2 &= S^2 = m/\lambda^2. \end{aligned}$$

In this case the system has an easy solution: $\lambda = \overline{X}/S^2$ and $m = \overline{X}^2/S^2$. Using these equations we can estimate λ and m for each month. To judge how well our chosen distribution fits the data, we plot a histogram of the July figures (scaled to integrate to 1) and superimpose the density of our fitted distribution (Figure 17.1).

```
> lambda.mm <- kew.mean/kew.var
> m.mm <- kew.mean^2/kew.var
> hist(kew$jul, breaks = 20, freq = FALSE, xlab = "rainfall (mm)",
+     ylab = "density", main = "July rainfall at Kew, 1697 to 1999")
> t <- seq(0, 200, 0.5)
> lines(t, dgamma(t, m.mm[7], lambda.mm[7]), lty = 2)
```

The distribution seems like a reasonable candidate based on this figure.

Maximum likelihood Maximum likelihood fitting is usually more work than the method of moments, but it is preferred as the resulting estimator is known to have good theoretical properties. We will restrict ourselves to the mechanics of maximum likelihood fitting, for a theoretical justification please read up on statistical inference.

[1] Data obtained from the U.S. National Climatic Data Center, Global Historical Climatology Network data base (GHCN-Monthly Version 2) `http://www.ncdc.noaa.gov/oa/climate/ghcn-monthly/`.

Suppose $X_1, \ldots, X_n$ are iid continuous random variables with density function f, then for scalars $x_1, \ldots, x_n$, we have

$$\begin{aligned} &\mathbb{P}(x_1 < X_1 \le x_1 + dx, \ldots, x_n < X_n \le x_n + dx) \\ &= \prod_{i=1}^{n} \mathbb{P}(x_i < X_i \le x_i + dx) = \prod_{i=1}^{n} f(x_i)dx. \end{aligned}$$

Thus $\prod_{i=1}^{n} f(x_i)$ gives us a measure of how likely it is to observe values $x_1, \ldots, x_n$. Maximum likelihood fitting consists of choosing f to maximise $\prod_{i=1}^{n} f(x_i)$, for a given set of observations. In practice it is usually easier to solve the equivalent problem of maximising $\log(\prod_{i=1}^{n} f(x_i)) = \sum_{i=1}^{n} \log f(x_i)$, which is called the *log likelihood.*

Let x_i be the observed July rainfall in year i. We suppose that the x_i are iid observations from a $\Gamma(\lambda, m)$ distribution, so the log likelihood is

$$\begin{aligned} l(\lambda, m) &= \sum_{i=1697}^{1999} \log\left(\lambda^m x_i^{m-1} e^{-\lambda x_i} / \Gamma(m)\right) \\ &= n\left(m \log \lambda + (m-1)\overline{\log x} - \lambda \overline{x} - \log \Gamma(m)\right), \end{aligned}$$

where $n = 1999 - 1696 = 303$ and the bar indicates an average over all i. We choose λ and m to maximise $l(\lambda, m)$.

Note that l is infinite if any $x_i = 0$. This is theoretically impossible if our model is correct, but may happen in practice, invalidating the method. To avoid this problem we increase any observations of 0 to 0.1.

```
> x <- kew$jul
> x[x == 0] <- 0.1
```

The partial derivative with respect to λ is

$$\frac{\partial l(\lambda, m)}{\partial \lambda} = n\left(\frac{m}{\lambda} - \overline{x}\right).$$

Setting this to zero we get $\lambda = m/\overline{x}$. Substituting this back into l, we see that we need to choose m to maximise

$$l(m) = n\left(m \log(m/\overline{x}) + (m-1)\overline{\log x} - m - \log \Gamma(m)\right).$$

Thus, differentiating and setting the derivative to zero, m must satisfy

$$l'(m) = \log m - \log \overline{x} + \overline{\log x} - \frac{\Gamma'(m)}{\Gamma(m)} = 0.$$

We cannot solve this exactly, so instead we use the Newton-Raphson root-finding algorithm.

We think of l as a function of m (and λ), but it also depends on the sample $x_1, \ldots, x_n$. The sample remains fixed as we optimise l, but it is useful to be able to pass it as a parameter. Accordingly we use the following modification of our function `newtonraphson` (Section 10.3).

```
> newtonraphson <- function(ftn, x0, tol = 1e-9, max.iter = 100, ...) {
+   # find a root of ftn(x, ...) near x0 using Newton-Raphson
+   # initialise
+   x <- x0
+   fx <- ftn(x, ...)
+   iter <-  0
+   # continue iterating until stopping conditions are met
+   while ((abs(fx[1]) > tol) && (iter < max.iter)) {
+     x <- x - fx[1]/fx[2]
+     fx <- ftn(x, ...)
+     iter <-  iter + 1
+   }
+   # output depends on success of algorithm
+   if (abs(fx[1]) > tol) {
+     stop("Algorithm failed to converge\n")
+   } else {
+     return(x)
+   }
+ }
```

To apply `newtonraphson` we need a function that returns the vector $(l'(m), l''(m))$. Let $a = \log \overline{x} - \overline{\log x}$, then we have

```
> dl <- function(m, a) {
+     return(c(log(m) - digamma(m) - a, 1/m - trigamma(m)))
+ }
```

Here we have used two built-in functions: `digamma(x)` returns $\Gamma'(x)/\Gamma(x)$ and `trigamma(x)` returns $(\Gamma(x)\Gamma''(x) - \Gamma'(x)^2)/\Gamma(x)^2$. (If we wished we could write our own functions instead, using one of our numerical integration routines.)

We can now find m and thus λ. We do this using the July data and then plot the corresponding density over the scaled histogram (Figure 17.1). As a starting point for the Newton-Raphson algorithm, we use the estimate of m obtained using the method of moments.

```
> m.ml <- newtonraphson(dl, m.mm[7], a = log(mean(x)) -
+     mean(log(x)))
> lambda.ml <- m.ml/mean(x)
> lines(t, dgamma(t, m.ml, lambda.ml))
```

The curve that represents the maximum likelihood fit also seems to provide a reasonable match to the observed data.

17.1.2 Example: truncated normal

The truncated normal distribution appears in a variety of settings, usually as a result of measurement problems or sampling restrictions.

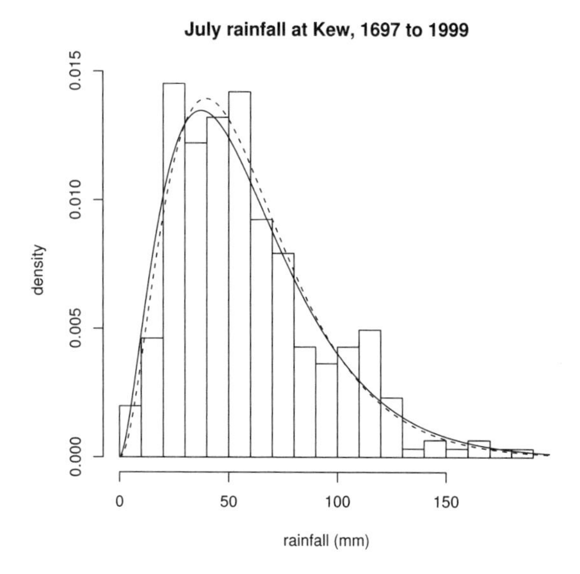

Figure 17.1 *Histogram of July rainfall at Kew with two fitted gamma densities: the dashed line is from the method of moments and the solid line from maximum likelihood. See Example 17.1.1.*

For example, demographers use historical military records to see how the distribution of height has changed over time. Most armies keep good records of their soldiers, including height, but typically they only accept recruits above a certain minimum height, say 150 cm. Given that adult height is normally distributed, the height of men in the army is then *truncated normal*.

For another example, consider trace elements. In many biomedical samples the log-concentration of a given trace element follows a normal distribution. However even our best measuring devices cannot accurately measure very small concentrations, and so we have to discard measurements below a certain level, therefore our observations are truncated below.

The truncated normal is also often seen in the health and economics literatures when observations are censored in some way.

There is a subtle difference between our first two examples. In the first case we do not know how many potential recruits were too short, but in the second case we do know how many measurements were discarded. For this example we will suppose that we are in the first situation. Suppose that we observe Y truncated below at a (where a is known). More specifically, we only observe Y if it is greater than a, which is to say we observe Y conditioned to be greater than a. Thus if X is the distribution of our observation, then

$$\mathbb{P}(X \leq x) = \mathbb{P}(Y \leq x \,|\, Y > a) = \mathbb{P}(a < Y \leq x)/\mathbb{P}(Y > a).$$

Let f_Y and F_Y be the density and cdf of Y, then X has the density

$$f_X(x) = \frac{f_Y(x)}{1 - F_Y(a)} \text{ for } x > a.$$

Suppose $Y \sim N(\mu, \sigma^2)$ and $x_1, \ldots, x_n$ are independent observations of X. We can use maximum likelihood to find μ and σ. That is, we choose μ and σ to maximise the log-likelihood l, given by

$$l(\mu, \sigma) = \sum_{i=1}^{n} \log f_X(x_i).$$

Using R's built-in functions for the density and distribution function, we can encode the log-likelihood easily.

```
> ell <- function(theta, a, x) {
+     mu <- theta[1]
+     si <- theta[2]
+     sum(log(dnorm(x, mu, si)) - log(1 - pnorm(a, mu, si)))
+ }
```

To test how well maximum likelihood performs in this situation, we simulate 10,000 $N(0, 1)$ random variables, conditioned to be greater than $a = -1$, then estimate μ and σ. To maximise the likelihood we use `optim` (with the default Nelder-Mead algorithm), with starting values of 0 and 1 for $\hat{\mu}$ and $\hat{\sigma}$. Note that `optim` minimises rather than maximises, however the argument `control = list(fnscale = -1)` instructs `optim` to multiply `ell` by -1 first.

```
> # inputs
> mu <- 0
> si <- 1
> a <- -1
> # generate sample
> set.seed(890)
> x <- rnorm(10000, mu, si)
> x.small <- (x <= a)
> while (sum(x.small) > 0) {
+     x[x.small] <- rnorm(sum(x.small), mu, si)
+     x.small <- (x <= a)
+ }
> # maximise the likelihood
> ell.optim <- optim(c(mu, si), ell, a = a, x = x,
+                    control = list(fnscale = -1))
> cat("ML estimate of mu", ell.optim$par[1], "and sigma",
+     ell.optim$par[2], "\n")

ML estimate of mu 0.03018642 and sigma 0.9819888
```

In this case our estimates are accurate to one decimal place.

17.2 The Central Limit Theorem

The Central Limit Theorem (CLT) is one of the most important results in probability theory, largely because it provides the theoretical justification for many statistical procedures. We will use it principally to tell us how precise $\overline{X}$ is as an estimate of $\mathbb{E}X$, which we do using confidence intervals.

Suppose that $X_1, X_2, \ldots, X_n$ are independent and identically distributed, with mean μ and finite variance σ^2. Put $\overline{X} = (X_1 + X_2 + \cdots + X_n)/n$, then for all $x \in (-\infty, \infty)$,

$$\mathbb{P}\left(\frac{\overline{X} - \mathbb{E}\overline{X}}{\sqrt{\operatorname{Var}\overline{X}}} \leq x\right) = \mathbb{P}\left(\frac{\overline{X} - \mu}{\sigma/\sqrt{n}} \leq x\right) \to \Phi(x) \text{ as } n \to \infty,$$

where Φ is the cumulative distribution function of a standard normal random variable.

We say that $\sqrt{n}(\overline{X} - \mu)/\sigma$ *converges in distribution* to Z, where $Z \sim N(0, 1)$, and write

$$\frac{\overline{X} - \mu}{\sigma/\sqrt{n}} \xrightarrow{\text{d}} Z \text{ as } n \to \infty.$$

The process of transforming a random variable by subtracting the mean and dividing by the standard deviation is called *standardisation*. A standardised random variable always has mean 0 and variance 1.

The CLT is used loosely in the following ways

$$\begin{aligned} \overline{X} &\approx N(\mu, \sigma^2/n) \text{ for large } n, \\ \sum_i X_i &\approx N(n\mu, n\sigma^2) \text{ for large } n. \end{aligned}$$

Here we interpret $\approx$ as meaning the cdf of the left-hand side is approximately equal to the distribution on the right-hand side.

17.2.1 Proof of the Central Limit Theorem

A rigourous and general proof of the CLT requires a working knowledge of the Fourier transform, which is the complex conjugate of the characteristic function. This is properly beyond the scope of an introductory programming course, but we nonetheless give a brief sketch of the proof here, because it is so important.

For any random variable X we can define the characteristic function

$$\psi_X(t) = \mathbb{E}e^{itX},$$

where $i = \sqrt{-1}$. Let $\{Y_n\}_{n=1}^{\infty}$ be a sequence of random variables, then it turns out that $Y_n \xrightarrow{\text{d}} Z$ if and only if $\psi_{Y_n}(t) \to \psi_Z(t)$ for all real t. It can also be shown that if the random variables U and V are independent, then $\psi_{U+V}(t) = \psi_U(t)\psi_V(t)$ for all real t.

A second-order Taylor series expansion of ψ_X about 0 gives us

$$\psi_X(t) = \psi_X(0) + t\psi'_X(0) + t^2\psi''_X(0)/2 + o(t^2),$$

where the term $o(t^2)$ goes to 0 faster than t^2. To calculate ψ'_X and ψ''_X we do the differentiation *inside* the expectation (a step that requires some mathematical justification) to get $\psi'_X(0) = i\mathbb{E}X$ and $\psi''_X(0) = -\mathbb{E}X^2$, so

$$\psi_X(t) = 1 + it\mu_X - t^2(\sigma_X^2 + \mu_X^2)/2 + o(t^2).$$

Now let $U_i = (X_i - \mu)/\sigma$ and $Y_n = \sum_{i=1}^n U_i/\sqrt{n} = \sqrt{n}(\overline{X} - \mu)/\sigma$. Since $\mu_{U_i} = 0$ and $\sigma_{U_i}^2 = 1$ we have

$$\begin{aligned} \psi_{Y_n}(t) &= \psi_{\sum_i U_i}(t/\sqrt{n}) \\ &= \prod_i \psi_{U_i}(t/\sqrt{n}) \\ &= \left(1 - \frac{t^2}{2n} + o\left(\frac{t^2}{n}\right)\right)^n \\ &\to e^{-t^2/2} \text{ as } n \to \infty. \end{aligned}$$

Thus $Y_n \xrightarrow{\text{d}} Z$, where Z is a random variable with characteristic function $\psi_Z(t) = e^{-t^2/2}$. It can be shown that the random variable with this characteristic function is the standard normal.

17.2.2 Normal approximation to the binomial

Suppose that $X_1, \ldots, X_n$ are iid Bernoulli(p) random variables. Then $Y = \sum_{i=1}^n X_i \sim \text{binom}(n, p)$ and by the CLT, for large n,

$$Y \approx N(\mu_Y, \sigma_Y^2) = N(np, np(1-p)).$$

That is, the binomial distribution can be approximated by the normal distribution provided n is large enough. As a rule of thumb, this approximation is reasonable provided $np > 5$ and $n(1-p) > 5$; see Figure 17.2.

17.2.3 Continuity correction

If we are approximating a discrete random variable X (for example, a binomial random variable) by a continuous random variable Y, how do we make sense of probabilities such as $\mathbb{P}(X = 28)$ or the difference between $\mathbb{P}(X > 32)$ and $\mathbb{P}(X \geq 32)$? In the case where X is integer valued we use the following *continuity correction*:

$$\mathbb{P}(X = x) \approx \mathbb{P}(x - \tfrac{1}{2} < Y < x + \tfrac{1}{2}).$$

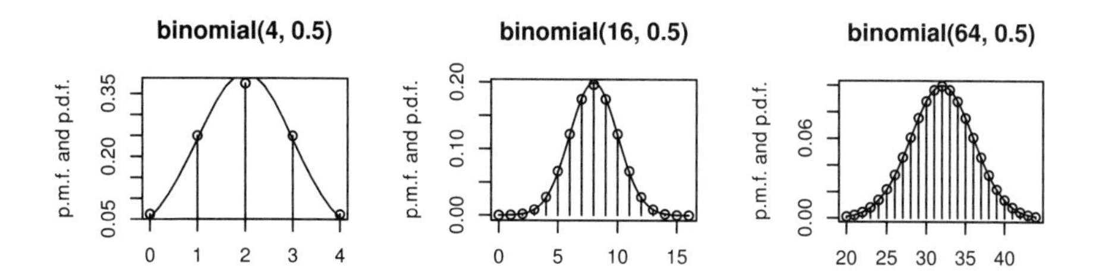

Figure 17.2 *The normal approximation to the binomial distribution. In each plot the vertical lines give the pmf of a binomial distribution and the continuous curve is the pdf of the corresponding normal approximation.*

Thus we approximate $\mathbb{P}(X > 32)$ by $\mathbb{P}(Y > 32.5)$ and approximate $\mathbb{P}(X \geq 32)$ by $\mathbb{P}(Y > 31.5)$. Formally, we have transformed the continuous pdf of Y to a discrete pmf, by concentrating all the mass in the interval $(x - \frac{1}{2}, x + \frac{1}{2})$ onto the point x.

From this discussion it should be clear that the normal approximation for the left-most example given in Figure 17.2 is poor.

17.2.4 Example: insurance risk

A car insurance company is estimating the risk on a block of 250 annual policies. Given that historically 10% of policyholders have at least one claim in a year, what is the probability that more than 12% of the policyholders in this block will have at least one claim?

Let X be the number of policyholders in this block with at least one claim. We want to know $\mathbb{P}(X > 30)$. Assuming the policyholders act independently and in line with history, X is modelled as a binomial random variable:

$$X \sim \text{binom}(250, 0.1)$$

We can calculate $\mathbb{P}(X > 30)$ as $1 - \mathbb{P}(X \leq 30) = 1 - \sum_{k=0}^{30} \mathbb{P}(X = k) = 1 - \sum_{k=0}^{30} \binom{250}{k} 0.1^k 0.9^{250-k}$. However it is easier to use the normal approximation to the binomial, which is justified since here n is large and $np > 5$. We approximate $\mathbb{P}(X > 30)$ by $\mathbb{P}(Y > 30.5)$ where $Y \sim N(25, 22.5)$. Let $Z \sim N(0, 1)$ then we have

$$\begin{aligned}
\mathbb{P}(X > 30) &\approx \mathbb{P}(Y > 30.5) \\
&= \mathbb{P}\left(\frac{Y - 25}{\sqrt{22.5}} > \frac{30.5 - 25}{\sqrt{22.5}}\right) \\
&= \mathbb{P}(Z > 1.1595) = 1 - \Phi(1.1595).
\end{aligned}$$

We can use the function `pnorm(1.1595)` to calculate $\Phi(1.1595)$, or use our own numerical integration function:

```
> source("../scripts/simpson_n.r")
> phi <- function(x) return(exp(-x^2/2)/sqrt(2*pi))
> Phi <- function(z) return(simpson_n(phi, -10, z))
> Phi(1.1595)

[1] 0.8768741
```

Hence $\mathbb{P}(X > 30)$ is approximately 0.123. We can confirm this computation using the built-in function:

```
> 1 - pbinom(30, 250, 0.1)

[1] 0.1246714
```

17.2.5 Normal approximation to the Poisson

Fix λ and choose n and p so that $\lambda = np$. From Section 15.6.2 we know that for n large enough (equivalently p small enough)

$$\text{pois}(\lambda) \approx \text{binom}(n, p).$$

Moreover, from our rule of thumb, if $\lambda = np > 5$ then

$$\text{binom}(n, p) \approx N(np, np(1-p)).$$

Sending $p \to 0$ we get $np(1-p) = \lambda(1-p) \to \lambda$. Thus, for $\lambda > 5$ we have

$$\text{pois}(\lambda) \approx N(\lambda, \lambda).$$

The approximation gets better as $\lambda \to \infty$; see Figure 17.3.

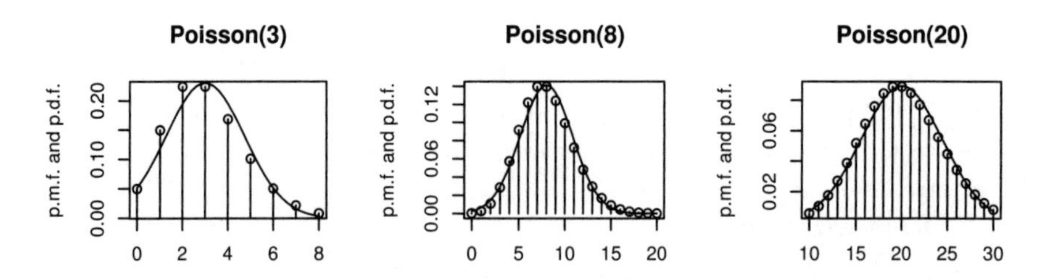

Figure 17.3 *The normal approximation to the Poisson distribution. In each plot the vertical lines give the pmf of a Poisson distribution and the continuous curve is the pdf of the corresponding normal approximation.*

The normal approximation can be used to calculate Poisson probabilities when 'exact' methods fail. Suppose that $X \sim \text{pois}(150)$, and we would like to know

$$\mathbb{P}(X \le 180) = \sum_{k=0}^{180} \mathbb{P}(X = k) = \sum_{k=0}^{180} \frac{150^k e^{-150}}{k!}.$$

We try calculating this numerically as follows:

```
> poispmf <- function(k, lambda) {
+   # returns P(X = k) where X ~ pois(lambda)
+   return(lambda^k*exp(-lambda)/prod(1:k))
+ }
> poiscdf <- function(k, lambda) {
+   # returns P(X <= k) where X ~ pois(lambda)
+   return(sum(sapply(0:k, poispmf, lambda=lambda)))
+ }
> poiscdf(180, 150)

[1] NaN
```

The calculation fails because for large k the values $\mathbb{P}(X = k)$ become impossible to calculate:

```
> sapply(141:180, poispmf, lambda = 150)

 [1] 0.02548978        Inf        Inf        Inf        Inf        Inf
 [7]        Inf        Inf        Inf        Inf        Inf        Inf
[13]        Inf        Inf        Inf        Inf        Inf        Inf
[19]        Inf        Inf        Inf        Inf        Inf        Inf
[25]        Inf        Inf        Inf        Inf        Inf        Inf
[31]        NaN        NaN        NaN        NaN        NaN        NaN
[37]        NaN        NaN        NaN        NaN
```

You can check that for $k \ge 142$, 150^k evaluates to ∞, and for $k \ge 171$, $k!$ evaluates to ∞ (and ∞/∞ is not defined). To some extent the problems with these calculations can be avoided by recursively calculating each $\mathbb{P}(X = k)$ from $\mathbb{P}(X = k - 1)$. However inaccuracies due to the computation of e^λ for large λ remain.

Using a normal approximation we can estimate the probability easily. $X \approx Y \sim N(150, 150)$ so, using a continuity correction,

$$\begin{aligned}\mathbb{P}(X \le 180) &\approx \mathbb{P}(Y < 180.5) \\ &= \mathbb{P}\left(\frac{Y - 150}{\sqrt{150}} < \frac{180.5 - 150}{\sqrt{150}}\right) \\ &= \mathbb{P}(Z \le 2.4903) = \Phi(2.4903) = 0.9936,\end{aligned}$$

where $Z \sim N(0, 1)$ and has cdf Φ.

Of course, R's built-in functions can also handle this computation:

```
> ppois(180, 150)
[1] 0.9923574
```

17.2.6 Normal approximation to the negative binomial and gamma

Let $X = \sum_{i=1}^{r} Y_i$ where $Y_i \sim \text{geom}(p)$, then $X \sim \text{nbinom}(r, p)$. Thus for large r

$$X \approx N(r(1-p)/p, r(1-p)/p^2).$$

Let $X = \sum_{i=1}^{n} Y_i$ where $Y_i \sim \exp(\lambda)$, then $X \sim \text{gamma}(n, \lambda)$. Thus for large n

$$X \approx N(n/\lambda, n/\lambda^2).$$

17.3 Confidence intervals

We know from the Weak Law of Large Numbers that $\overline{X} \xrightarrow{\mathbb{P}} \mathbb{E}X$, but how fast does it converge? For an estimate to be really useful, we need to know how precise it is.

One way to judge how precise an estimate $\overline{X}$ is, is to plot how it changes as the sample size increases. For example, suppose we are given a sample of n iid Poisson(λ) rv's, and we wish to estimate the mean λ using $\overline{X}$. Let $\overline{X}(k) = \sum_{i=1}^{k} X_i/k$ be the sample mean of the first k sample points. By plotting $\overline{X}(k)$ against k we get an idea of whether or not $\overline{X}(k)$ has converged by the time k reaches n. We do this in the code below, using the built-in function `rpois` to simulate Poisson random variables. The output is given in Figure 17.4.

```
set.seed(100)
n <- 2000
la <- 2
x <- rpois(n, la)
xbar <- cumsum(x)/1:n
plot(1:n, xbar, type = "l",
    xlab="sample size k", ylab="k point average", col='blue')
abline(la, 0)
```

Unfortunately this approach is often misleading. In the example above, we see that $\overline{X}(k)$ seems to have settled down around 2.05, which we know is incorrect. If we could increase the sample size ad infinitum, then we would eventually see $\overline{X}(k)$ converge to $\lambda = 2$, but if we did not know the true value of λ, then there is no way we could tell this just by looking at the graph.

A better way of judging how precise $\overline{X}$ is, is to estimate how variable it is, which we can do by repeating the whole experiment a number of times. We do this in the following code, and plot the output in Figure 17.5.

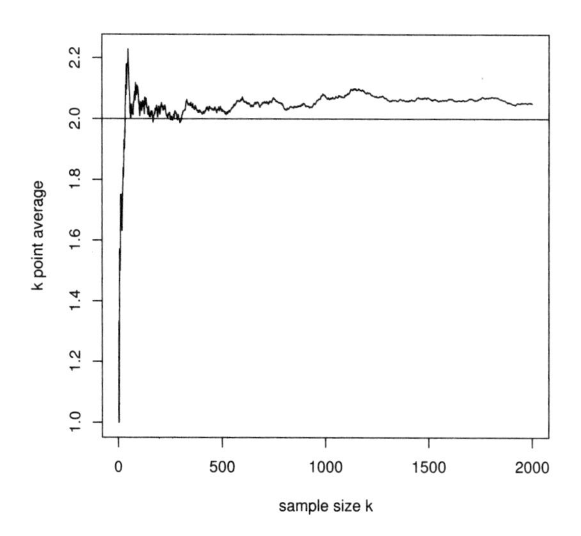

Figure 17.4 *Convergence of $\overline{x}$ to μ as the sample size increases.*

```
set.seed(100)
n <- 2000
la <- 2
plot(c(1, n), c(la-sqrt(la), la+sqrt(la)), type = "n",
    xlab = "sample size k", ylab = "k point average")
for (i in 1:20) {
  x <- rpois(n, la)
  xbar <- cumsum(x)/1:n
  lines(1:n, xbar, type = "l", col='blue')
}
abline(la, 0)
lines(1:n, la + 2*sqrt(la/1:n))
lines(1:n, la - 2*sqrt(la/1:n))
```

Figure 17.5 shows two important things. First, for a sample of size $n = 2000$, it is not unusual to find $\overline{X}$ anywhere between 1.95 and 2.05. Second, as k increases, the range of values displayed by $\overline{X}(k)$ has a width roughly equal to $c/\sqrt{k}$, for some constant c. But what is c? We will answer this question using the Central Limit Theorem.

Up till now we have been content to use one number to estimate the mean (*point estimation*). It would be much more informative if we had an interval telling us where the mean was likely to be. That is, the width of the interval would give us an idea of the margin for error in the point estimate. Such intervals are called *confidence intervals* (CIs) and the process of estimating them is called *interval estimation*.

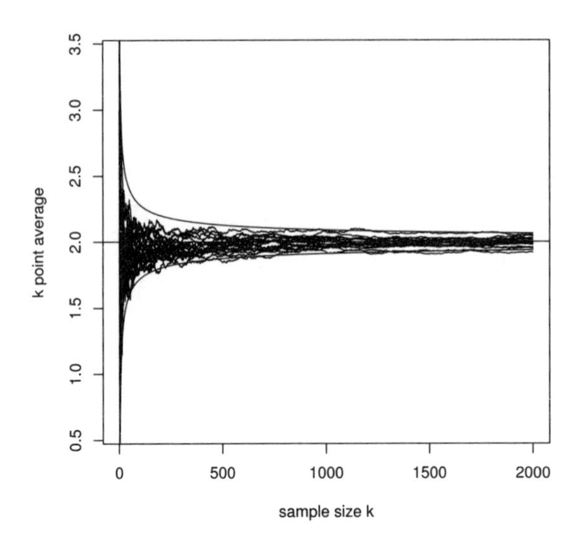

Figure 17.5 *20 plots of the sample mean against sample size, showing how variable $\overline{x}$ is as an estimator of μ, as the sample size increases.*

Suppose $X_1, \ldots, X_n$ are iid with mean μ and variance σ^2, then by the CLT we have $\sqrt{n}(\overline{X} - \mu)/\sigma \xrightarrow{\text{d}} N(0,1)$ and by the Weak Law of Large Numbers $S^2 \xrightarrow{\mathbb{P}} \sigma^2$. Let $Z \sim N(0,1)$, then for large n we have

$$\begin{aligned} 0.95 &= \mathbb{P}(-1.96 < Z < 1.96) \\ &\approx \mathbb{P}\left(-1.96 < \frac{\overline{X} - \mu}{\sigma/\sqrt{n}} < 1.96\right) \\ &\approx \mathbb{P}\left(-1.96 < \frac{\overline{X} - \mu}{S/\sqrt{n}} < 1.96\right) \\ &= \mathbb{P}\left(-1.96\frac{S}{\sqrt{n}} < \overline{X} - \mu < 1.96\frac{S}{\sqrt{n}}\right) \\ &= \mathbb{P}\left(\overline{X} - 1.96\frac{S}{\sqrt{n}} < \mu < \overline{X} + 1.96\frac{S}{\sqrt{n}}\right). \end{aligned}$$

If $X_1, \ldots, X_n$ are an iid sample with mean μ and finite variance, then we say

$$\left(\overline{X} - 1.96\frac{S}{\sqrt{n}},\ \overline{X} + 1.96\frac{S}{\sqrt{n}}\right) \text{ is a 95\% CI for } \mu.$$

The way we interpret this is that in *repeated sampling*, 95% of the time this interval will cover the true value of μ. Our best guess for μ is the point estimate

$\overline{X}$; the size of the CI about $\overline{X}$ gives us an idea of how reliable an estimate it is. Note that sometimes people just use 2 instead of 1.96, to give a slightly more conservative interval estimate.

For the example above we calculate a 95% CI as follows. We use the built-in function `sd` for calculating the sample standard deviation.

```
> set.seed(100)
> n <- 2000
> la <- 2
> x <- rpois(n, la)
> xbar <- mean(x)
> S <- sd(x)
> L <- xbar - 1.96 * S/sqrt(n)
> U <- xbar + 1.96 * S/sqrt(n)
> cat("estimate is", xbar, "\n")

estimate is 2.05

> cat("95% CI is (", L, ", ", U, ")\n", sep = "")

95% CI is (1.986073, 2.113927)
```

Different-sized confidence intervals—90%, 98%, 99%—may be constructed similarly. Let z_α be such that $\mathbb{P}(Z < z_\alpha) = \alpha$, that is $z_\alpha = \Phi^{-1}(\alpha)$. z_α is called the 100α%-point of the standard normal distribution. Then $\mathbb{P}(z_{\alpha/2} < Z < z_{1-\alpha/2}) = 1 - \alpha$, so that a $100(1-\alpha)$% CI for μ is given by

$$\left(\overline{X} - z_{1-\alpha/2}\frac{S}{\sqrt{n}},\ \overline{X} + z_{1-\alpha/2}\frac{S}{\sqrt{n}}\right).$$

Note that because the standard normal density is symmetric about 0, $z_{\alpha/2} = -z_{1-\alpha/2}$ (see Figure 17.6).

CI	:	90%	95%	98%	99%
α	:	0.1	0.05	0.02	0.01
$z_{1-\alpha/2}$	:	1.6449	1.9600	2.3263	2.5758

We see that to be more confident that the interval contains μ, the interval has to be wider. By far the most commonly used confidence interval is the 95%, but this is just convention.

17.3.1 Confidence interval for a proportion

If $X \sim \text{binom}(n, p)$, then using $n\hat{p}(1-\hat{p}) = n(X/n)(1 - X/n)$ as an estimate of $\text{Var}\, X = np(1-p)$, an approximate 95% CI for p is

$$\left(\frac{X}{n} - 1.96\sqrt{\frac{(X/n)(1-X/n)}{n}},\ \frac{X}{n} + 1.96\sqrt{\frac{(X/n)(1-X/n)}{n}}\right).$$

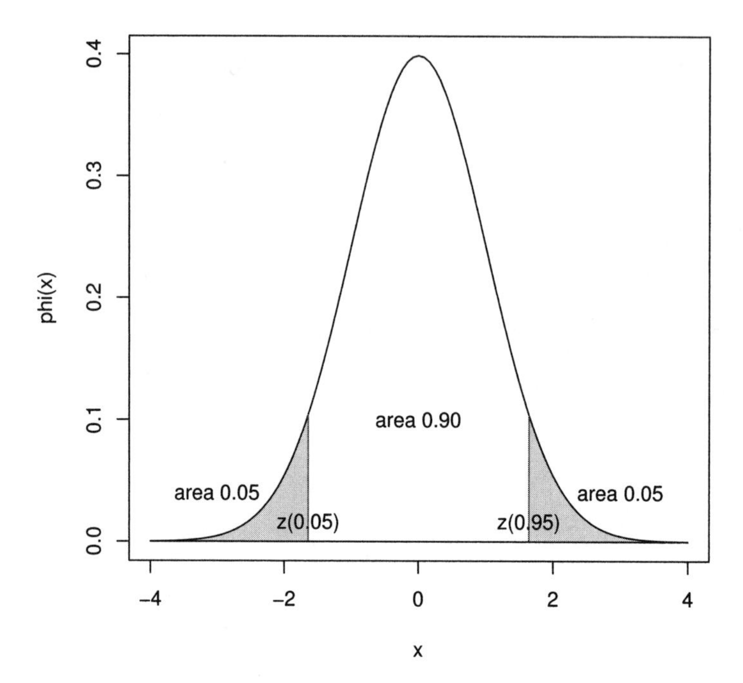

Figure 17.6 *The 5% and 95% percentage points of the standard normal distribution.*

Observe that for $p \in [0, 1]$, the maximum value of $p(1-p)$ is $1/4$ when $p = 1/2$. Thus, $\operatorname{Var} X \leq n/4$. Using this bound to construct a confidence interval we get, for large n,

$$\mathbb{P}\left(p \in \left(\frac{X}{n} - \frac{1.96}{2\sqrt{n}}, \frac{X}{n} + \frac{1.96}{2\sqrt{n}}\right)\right) \geq 0.95.$$

This is a *conservative* confidence interval in the sense that it will contain p at least 95% of the time, but possibly more than that because it is wider than it really needs to be. In particular this CI may significantly overestimate the variability of $\hat{p}$ when p is close to 0 or 1. Its advantage is that you don't have to know $\hat{p}$ to estimate how large the sample should be to achieve some required precision.

We remark that exact confidence intervals for a proportion, called Clopper-Pearson intervals, are available from the `binom` package on CRAN.

17.3.2 Example: accuracy of an opinion poll

In an exit poll of 1000 voters, 443 said they voted for the ALP (Australian Labor Party). A 95% confidence interval for the actual proportion p that voted for the ALP, is

$$0.443 \pm 1.96\sqrt{\frac{0.443 \times 0.557}{1000}} = 0.443 \pm 0.031 = (0.412, 0.474).$$

Notice that the width of the confidence interval is roughly ±3%, called in this context the 'sampling error', which is typical of opinion polls, since in opinion polls p is often near 0.5 and n is usually around a thousand.

How large would n have to be to reduce the sampling error to ±1%? Taking the worst-case, $p = 0.5$, the sampling error is $\pm 1.96/(2\sqrt{n})$, so we require $1.96/(2\sqrt{n}) \leq 0.01$. Thus, $n \geq 9604$.

17.3.3 Small sample confidence intervals

In applying the Central Limit Theorem (CLT) to obtain a confidence interval for μ, we had to assume that the sample size n was large. In practice $n \geq 100$ is usually enough, but the larger n is, the better.

For smaller sample sizes it is still possible to obtain a confidence interval, provided the sample comes from a normal distribution. Suppose that $X_1, \ldots, X_n$ are iid $N(\mu, \sigma^2)$, then it can be shown that for all n

$$T = \frac{\overline{X} - \mu}{S/\sqrt{n}} \sim t_{n-1},$$

where t_ν is the Student-t distribution with ν degrees of freedom. The proof of this result is non-trivial and requires the use of quadratic forms. As $n \to \infty$ the t_{n-1} distribution converges to a $N(0, 1)$ distribution, in accordance with the CLT.

Even though we cannot prove this result, we can test it numerically. We will make use of the function `rnorm` for simulating normal random variables and the function `dt`, which gives the density of a student-t distribution. To test that $T \sim t_{n-1}$, we estimate the probability density function of T using a scaled histogram. This means we need to generate a large sample of T's. Without loss of generality (wlog), we take the case $X_i \sim N(0, 1)$. Suitable code is given below, and the output appears as Figure 17.7.

```
set.seed(99)
n <- 5         # size of X sample
nT <- 10000  # size of T sample
# simulate T sample
Tsample <- rep(0, nT)
for (i in 1:nT) {
```

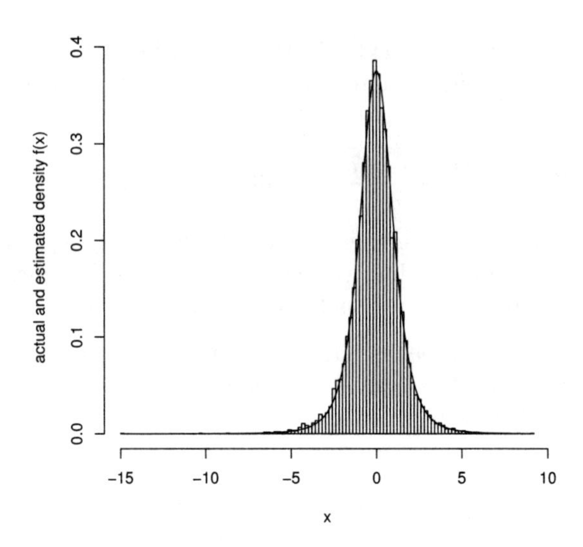

Figure 17.7 *Obtaining the t distribution by standardising the sample mean of an iid normal sample.*

```
 Xsample <- rnorm(n)
 Tsample[i] <- sqrt(n)*mean(Xsample)/sd(Xsample)
}
# plot scaled histogram of T sample
hist(Tsample, breaks=sqrt(nT), freq=F,
    xlab='x', ylab='actual and estimated density f(x)', main='')
# plot target density on top
x <- seq(min(Tsample), max(Tsample), 0.01)
lines(x, sapply(x, dt, df=n-1))
```

Let $t_{\eta,\nu}$ be the $100\eta\%$-point of the t_ν distribution. That is

$$\mathbb{P}(T < t_{\eta,\nu}) = \eta \text{ for } T \sim t_\nu.$$

As the t_ν distribution is symmetric about 0, $t_{\eta,\nu} = -t_{1-\eta,\nu}$.
We form a $100(1-\alpha)\%$ confidence interval for μ as follows:

$$\begin{aligned}
1-\alpha &= \mathbb{P}\left(t_{\alpha/2,n-1} < \frac{\overline{X}-\mu}{S/\sqrt{n}} < t_{1-\alpha/2,n-1}\right) \\
&= \mathbb{P}\left(\mu - t_{1-\alpha/2,n-1}\frac{S}{\sqrt{n}} < \overline{X} < \mu - t_{\alpha/2,n-1}\frac{S}{\sqrt{n}}\right) \\
&= \mathbb{P}\left(\mu \in \left(\overline{X} - t_{1-\alpha/2,n-1}\frac{S}{\sqrt{n}},\ \overline{X} + t_{1-\alpha/2,n-1}\frac{S}{\sqrt{n}}\right)\right).
\end{aligned}$$

$t_{\eta,\nu}$ decreases as n increases, with limiting value z_η, the $100\eta\%$ point of the standard normal. Thus, we see that for small samples, the confidence interval is wider than we would expect from using the Central Limit Theorem. This is because of the extra uncertainty caused by having to estimate σ from the sample. For example, for a 95% CI we have

ν	:	2	5	20	50	100	∞
$t_{0.975,\nu}$	:	4.3027	2.5706	2.0860	2.0086	1.9840	1.9600

17.4 Monte-Carlo confidence intervals

In Section 17.3 we noted that a qualitative method of seeing how precise an estimator is, is to generate several independent realisations of the estimator and observe how variable they are. We can quantify this procedure somewhat.

Suppose that $E_1, \ldots, E_k$ are independent, continuous, and unbiased estimators of μ. That is, for each i, $\mathbb{E}E_i = \mu$. We also assume that μ is the *median* point for each E_i, so that $\mathbb{P}(E_i < \mu) = 0.5$.

For example, suppose $E_i = (X_1^i, \ldots, X_{n(i)}^i)/n(i)$ where the $\{X_j^i\}_{j=1}^{n(i)}$ are an iid sample with mean μ and finite variance. Each E_i is an unbiased estimator of μ, and from the CLT E_i is approximately normal, so μ is also (approximately) the median. Note that we have not assumed that the $n(i)$ are all equal.

Let $E_{(1)}, E_{(2)}, \ldots, E_{(k)}$ be the *ordered sample*, so that $E_{(1)} < E_{(2)} < \cdots < E_{(k)}$ (because they are continuous random variables, the probability of a tie is 0). Then we have

$$\begin{aligned}\mathbb{P}(E_{(1)} \leq \mu \leq E_{(k)}) &= 1 - \mathbb{P}(E_{(1)} > \mu) - \mathbb{P}(E_{(k)} < \mu) \\ &= 1 - \mathbb{P}(\text{ all } E_i > \mu) - \mathbb{P}(\text{ all } E_i < \mu) \\ &= 1 - 0.5^k - 0.5^k \;=\; 1 - 0.5^{k-1}.\end{aligned}$$

Put $k = 6$, then we get $1 - 0.5^5 = 0.96875 \approx 0.97$.

> If $E_1, \ldots, E_6$ are independent, continuous, and unbiased estimators of μ, such that μ is also the median for each E_i, then
>
> $$(E_{(1)},\ E_{(6)}) = \left(\min_i E_i,\ \max_i E_i\right) \text{ is a 97\% CI for } \mu.$$

We do not have to restrict ourselves to the smallest and largest E_i when forming a confidence interval. Suppose $1 \leq a < b \leq k$, then

$$\begin{aligned}&\mathbb{P}(E_{(a)} \leq \mu \leq E_{(b)}) \\ &\quad= \mathbb{P}(\text{at least } a \text{ of the } E_i < \mu \text{ and at most } b-1 \text{ of the } E_i < \mu).\end{aligned}$$

Let $N = |\{E_i \,:\, E_i < \mu\}|$ then $N \sim \text{binom}(k, 0.5)$, and

$$\mathbb{P}(E_{(a)} \leq \mu \leq E_{(b)}) = \mathbb{P}(a \leq N < b) = \sum_{i=a}^{b-1} \binom{k}{i} 0.5^k.$$

This technique provides a quick and simple way to estimate the precision of an estimate. Better tools exist, such as the bootstrap or jackknife, and we recommend them to the interested reader.

17.4.1 Example: meta-analysis of opinion polls

Suppose that eight independent polls report on the proportion of Australians who plan to vote Green in the next federal election, with the following results

$$9.7\%,\ 8.6\%,\ 11.5\%,\ 10.5\%,\ 10.4\%,\ 10.8\%,\ 9.1\%,\ 12.5\%.$$

We will use the second and seventh points in the ordered sample to form a confidence interval. We have

$$\begin{aligned}\sum_{1=2}^{6}\binom{8}{i}0.5^8 &= 1-\left(\binom{8}{0}+\binom{8}{1}+\binom{8}{7}+\binom{8}{8}\right)0.5^8\\ &= 1-18\times 0.5^8 = 0.930 \text{ to 3 significant figures.}\end{aligned}$$

Thus a 93% CI for the true proportion is (0.091, 0.115).

Suppose now that we also know the number of people surveyed in each poll:

$$1000,\ 1000,\ 600,\ 800,\ 1000,\ 500,\ 1000,\ 400.$$

The total number of people surveyed was thus 6300 and the total number who said they planned to vote Green was:

$$\begin{aligned}&1000\times 0.097+1000\times 0.086+600\times 0.115+800\times 0.105\\ &\quad +1000\times 0.104+500\times 0.108+1000\times 0.091+400\times 0.125=635.\end{aligned}$$

A 93% CI for the true proportion p is $\hat{p}\pm z_{0.965}\sqrt{\hat{p}(1-\hat{p})/n}$. Here $\hat{p}=635/6300=0.1008$ (to four significant figures) and $n=6300$, so we get a 93% confidence interval of (0.0970, 0.1046).

Because the second confidence interval uses more information than the first—the information about the sample sizes—we suspect that it is a better interval estimate for p. That is, it gives a better estimate of the variability of $\hat{p}$.

17.5 Exercises

1. Using a normal approximation, find the probability that a Poisson variable with mean 20 takes the value 20. Compare this with the true value; to how many decimal places do they agree?
2. Migrating geese arrive at a certain wetland at a rate of 220 per day during the migration season. Suggest a model for X, the number of geese that arrive per hour (assume the arrival rate remains constant throughout the day).

What is $\mathbb{P}(X > 10)$? Give the answer exactly, based on your model (it is sufficient to express the probability as a finite sum), and approximately, using the Central Limit Theorem.

3. The weights of 20 people are measured, and the resulting sample mean and sample standard deviation are

$$\overline{x} = 71.2 \text{ kg}, \qquad s = 4.9 \text{ kg}.$$

Calculate a 95% CI for the mean μ of the underlying population. Assume that the weights are iid normal.

4. A random sample of size n is taken without replacement from a very large sample of components and r of the sample are found to be defective. Write down an approximate 99% confidence interval for the proportion of the population that are defective stating clearly three reasons why your interval is only approximate.

If $n = 400$ show that the longest the confidence interval can be is about 0.13.

5. Assume a manager is using the sample proportion $\hat{p}$ to estimate the proportion p of a new shipment of computer chips that are defective. He doesn't know p for this shipment, but in previous shipments it has been close to 0.01, that is 1% of chips have been defective.

   (a). If the manager wants the standard deviation of $\hat{p}$ to be about 0.02, how large a sample should she take based on the assumption that the rate of defectives has not changed dramatically?

   (b). Now suppose something went wrong with the production run and the actual proportion of defectives in the shipment is 0.3, that is 30% are defective. Now what would be the actual standard deviation of $\hat{p}$ for the sample size you choose in (a)?

6. A company fills plastic bottles with orange juice. The bottles are supposed to contain 250 ml. In fact, the contents vary according to a normal distribution with mean $\mu = 242$ ml and standard deviation $\sigma = 12$ ml.

   (a). What is the probability that one bottle contains less than 250 ml?

   (b). What is the probability that the mean contents of a carton with 12 bottles is less than 250 ml?

7. The number of accidents per week at a hazardous intersection follows a Poisson distribution with mean 2.2. We observe the intersection for a full year (52 weeks) and calculate $\overline{X}$ the mean number of accidents per week.

   (a). What is the approximate distribution of $\overline{X}$ according to the Central Limit Theorem?

   (b). What is the approximate probability that $\overline{X}$ is less than 2?

   (c). What is the approximate distribution of T, the total number of accidents in the year?

(d). What is the probability that there are fewer than 90 accidents at the intersection during the year?

8. A scientist is observing the radioactive decay of a substance. The waiting time between successive decays has an exponential distribution with a mean of 10 minutes.

   (a). What is the probability that the first waiting time exceeds 12 minutes?

   (b). The scientist observes 50 successive waiting times and calculates the mean. What is the probability that this mean exceeds 12 minutes?

   (c). In another experiment the scientist waits until the 80th decay. What is the probability that he waits longer than 14 hours?

9. An actuary has received notification that 100 claims on an account have been filed but are still in the course of settlement. The actuary has been asked to determine the size of an appropriate reserve fund for these 100 claims. Claim sizes are independent and exponentially distributed with mean \$300. The actuary recommends setting up a claim reserve of \$31,000. What is the probability that the total claims will exceed the reserve fund?

   Hint: use an appropriate approximation.

10. Suppose that 55% of the voting population are Democrat voters. If 200 people are selected at random from the population, what is the probability that more than half of them are Democrat voters?

11. Approximate the probability that the proportion of heads obtained will be between 0.50 and 0.52 when a fair coin is tossed

    (a). 50 times.

    (b). 500 times.

12. A course can cater for 200 new students. Not all offers to students are accepted, so 250 offers are made based on previous rejection rates. Assume that for this current round of offers the actual rejection rate is 35% and that students make their decisions independently.

    (a). State the distribution of N, the number of students who accept, and state its mean and standard deviation.

    (b). Find the approximate probability that less than 180 students accept.

    (c). Find the approximate probability that more than 200 students accept.

13. A survey of 900 people asked whether they play any competitive sport. In fact only 5% of the surveyed population plays a competitive sport.

    (a). Find the mean and standard deviation of the proportion of the sample who play competitive sport.

    (b). What sample size would be required to reduce the standard deviation of the sample proportion to one-half the value you found in (a)?

14. Cards with different shapes printed on them are used to test if a subject has extrasensory perception (ESP). The subject has to guess the shape on the card being viewed by the experimenter without viewing the card itself. Assume we use a large pack containing cards marked with one of four different shapes in equal proportions. That is, we can assume that on each draw, each shape is equally likely, and that successive draws are independent. We test subjects (who are all just guessing at random) on 800 cards each.

    (a). What is the probability that any one subject guesses correctly on any one trial?

    (b). What are the mean and standard deviation of the proportion of successes among the 800 attempts?

    (c). What is the probability that any one subject is successful in at least 26% of the 800 attempts?

    (d). Assume you decide to do further tests on any subject whose proportion of successes is so large that there is only a probability of 0.02 that they could do that well or better simply by guessing. What proportion of successes must a subject have to meet this standard?

    (e). How many subjects will the researcher need to assess so that the probability at least one of them will be tested further is 0.75?

15. You take a random sample of size n from a population which is uniform on the interval $(0, \theta)$, where θ is an unknown parameter.

    (a). Using the Central Limit Theorem, about which point do you think the distribution of the sample mean will become concentrated as the sample size increases? Consequently, what function of the sample mean would you suggest to estimate θ?

    (b). Arguing intuitively, what do you think will happen to the distribution of the sample maximum as the sample size increases?

    (c). Suppose that $X \sim U(0, \theta)$; write down the pdf and cdf of X. Hence find the cdf and pdf of the sample maximum.

    (d). Calculate the expected value of the sample maximum. Use this result to suggest a function of the sample maximum which would give you an unbiased estimate of the unknown parameter θ.

16. Calculating the confidence interval.

    Write a function that takes as input a vector `x`, then returns as output the vector `(m, lb, ub)`, where `m` is the mean and `(lb, ub)` is a 95% confidence interval for `m`. That is

$$
\begin{aligned}
m &= \overline{x}, \\
lb &= \overline{x} - 1.96\sqrt{s^2/n}, \\
ub &= \overline{x} + 1.96\sqrt{s^2/n},
\end{aligned}
$$

where

$$s^2 = \frac{1}{n-1}\sum_{i=1}^{n}(x_i - \overline{x})^2 = \frac{1}{n-1}\left(\sum_{i=1}^{n} x_i^2 - n\overline{x}^2\right).$$

Write a program that applies your subroutine to the following sample

```
11 52 87 45 39 95 42 38 10 03 48 56
```

To four decimal places you should be getting (43.8333, 27.9526, 59.7140).

17. Gaining confidence with confidence intervals.

 We know that the $U(-1, 1)$ rv has mean 0. Use a sample of size 100 to estimate the mean and give a 95% confidence interval. Does the confidence interval contain 0?

 Repeat the above a large number of times. What percentage of time does the confidence interval contain 0? Write your code so that it produces output similar to the following

```
Number of trials: 10

Sample mean  lower bound  upper bound  contains mean?

   -0.0733      -0.1888       0.0422               1
   -0.0267      -0.1335       0.0801               1
   -0.0063      -0.1143       0.1017               1
   -0.0820      -0.1869       0.0230               1
   -0.0354      -0.1478       0.0771               1
   -0.0751      -0.1863       0.0362               1
   -0.0742      -0.1923       0.0440               1
    0.0071      -0.1011       0.1153               1
    0.0772      -0.0322       0.1867               1
   -0.0243      -0.1370       0.0885               1

 100 percent of CI's contained the mean
```

18. Use `rnorm(10, 1, 1)` to generate a sample of 10 independent $N(1, 1)$ random variables. Form a 90% CI for the mean of this sample (using a t distribution). Does this CI include 1?

 Repeat the above 20 times. How many times did the CI include 1? How many times do you expect the CI to include 1?

19. A bottle-washing plant has to discard many bottles because of breakages. Bottles are washed in batches of 144. Let X_i be the number of broken bottles in batch i and let p be the probability that a given bottle is broken.

 (a). Assuming that each bottle breaks independently of the others, what is the distribution of X_1? Also, what is the distribution of $Y = X_1 + X_2 + \cdots + X_{100}$?

 (b). Data are collected from 100 batches of bottles; the total number of broken bottles was 220. Using this data give an estimate and 95% CI for p.

(c). Which would be the more suitable approximation for X_1, a Normal approximation or a Poisson approximation?

20. Consider a normal distribution Y, with mean μ and variance σ^2, truncated so that only observations above some limit a are observed. In Example 17.1.2 we used the method of maximum likelihood to estimate μ and σ; in this exercise we use the method of moments.

    Let $\mu_X = g_1(\mu, \sigma)$ and $\sigma_X^2 = g_2(\mu, \sigma)$ be the mean and variance of the truncated random variable X. That is,

$$\begin{aligned} \mu_X &= \int_a^\infty \frac{x f_Y(x)}{1 - F_Y(a)} dx \text{ and} \\ \sigma_X^2 &= \int_a^\infty \frac{(x - \mu_X)^2 f_Y(x)}{1 - F_Y(a)} dx, \end{aligned}$$

    where the pdf and df of Y (f_Y and F_Y respectively), depend on μ and σ^2. Given μ and σ^2, μ_X and σ_X^2 can be calculated numerically.

    If $X_1, \ldots, \boldsymbol{X}_n$ is a sample from X, then you estimate μ and σ by solving

$$\begin{aligned} \hat{\mu}_X &= \overline{X} = g_1(\mu, \sigma), \\ \hat{\sigma}_X^2 &= S^2 = g_2(\mu, \sigma). \end{aligned}$$

    Put $\theta = (\mu, \sigma)^T$, then this is equivalent to solving $g(\theta) = \theta$, where

$$g(\theta) = \theta + A \begin{pmatrix} \overline{X} - g_1(\mu, \sigma) \\ S^2 - g_2(\mu, \sigma) \end{pmatrix},$$

    for any non-singular 2×2 matrix A.

    One way to solve $g(\theta) = \theta$ is to find an A such that g is a contraction mapping (by trial and error), then use the fixed-point method (see Chapter 10, Exercise 8). Test your method using the same sample used in Example 17.1.2.

PART IV

Simulation

CHAPTER 18

Simulation

Most stochastic simulations have the same basic structure:

1. Identify a random variable of interest X and write a program to simulate it.
2. Generate an iid sample $X_1, \dots, X_n$ with the same distribution as X.
3. Estimate $\mathbb{E}X$ (using $\overline{X}$) and assess the accuracy of the estimate (using a confidence interval).

Step 1 is an example of *model building.* Typically we build up a complex model from simple components, which in this case are independent rv's with known distributions. In other words, random variables are the building blocks of stochastic simulations. As we have seen, R has built-in functions for simulating all the common rv's we encountered in Chapters 15 and 16. The purpose of this chapter is to see how to do this for ourselves, so that we have the tools for simulating the random variables that R does not provide for us. We consider discrete random variables, the inversion and rejection methods for simulating continuous random variables, and then look at particular techniques for simulating normals.

It turns out that all random variables can be generated by manipulating $U(0, 1)$ rv's, so that is where we start.

18.1 Simulating iid uniform samples

We cannot generate truly random numbers on a computer. Instead we generate *pseudo-random numbers*, which have the appearance of random numbers, but are in fact completely deterministic. Pseudo-random numbers can be generated by chaotic dynamical systems, which have the characteristic that the future is very hard to predict given the present.

A very important advantage of using pseudo-random numbers is that, because they are deterministic, any experiment performed using pseudo-random numbers can be repeated exactly.

18.1.1 *Congruential generators*

Congruential generators were the first reasonable class of pseudo-random number generators. At the time of writing R uses a pseudo-random number generator called the *Mersenne-Twister*, which has similar properties to congruential generators, but with a much longer cycle length.

Given an initial number $X_0 \in \{0, 1, \ldots, m-1\}$ and two big numbers A and B we define a sequence of numbers $X_n \in \{0, 1, \ldots, m-1\}, n = 0, 1, \ldots$, by

$$X_{n+1} = (AX_n + B) \bmod m.$$

We get a sequence of numbers $U_n \in [0, 1), n = 0, 1, \ldots$, by putting $U_n = X_n/m$. If m, A, and B are well chosen then the sequence $U_0, U_1, \ldots$, is almost impossible to distinguish from an iid sequence of $U(0, 1)$ random variables.

In practice it is sensible to discard the value 0 when it occurs, as we often divide by U_n. This is justifiable since for a true uniform, the probability of taking on the value 0 is zero. The value 1 can also be a problem, but note that as defined, $U_n < 1$ for all n.

For example if we take $m = 10$, $A = 103$, and $B = 17$, then for $X_0 = 2$, we have

$$\begin{aligned} X_1 &= 223 \bmod 10 = 3 \\ X_2 &= 326 \bmod 10 = 6 \\ X_3 &= 635 \bmod 10 = 5 \\ &\vdots \end{aligned}$$

Clearly the sequence produced by a congruential generator will eventually cycle and thus since there are at most m possible values, the maximum cycle length is m. Because computers use binary arithmetic, if we have $m = 2^k$ for some k, then taking $x \bmod m$ is very quick. An example of a good congruential generator is $m = 2^{32}$, A =1,664,525, and B = 1,013,904,223. An example of a bad congruential generator is RANDU, which was shipped with IBM computers in the 1970's. RANDU used $m = 2^{31}$, A =65,539, and $B = 0$.

18.1.2 *Seeding*

The number X_0 is called the seed. If you know the seed (as well as m, A, and B), then you can reproduce the whole sequence exactly. This is a very good idea from a scientific point of view; being able to repeat an experiment means that your results are verifiable.

To generate n pseudo-random numbers in R, use `runif(n)`. R does not use a congruential generator, but is still needs a seed to generate pseudo-random numbers. In R the command `set.seed(seed)` puts you at point `seed` (assumed integer) on the cycle of pseudo-random numbers. The current state of the random number generator is kept in the vector `.Random.seed`. You can save the value of `.Random.seed` and then use it to return to that point in the sequence of pseudo-random numbers. If the random number generator is not initialised before you start generating pseudo-random numbers, then R initialises it using a value taken from the system clock.

```
> set.seed(42)
> runif(2)

[1] 0.9148060 0.9370754

> RNG.state <- .Random.seed
> runif(2)

[1] 0.2861395 0.8304476

> set.seed(42)
> runif(4)

[1] 0.9148060 0.9370754 0.2861395 0.8304476

> .Random.seed <- RNG.state
> runif(2)

[1] 0.2861395 0.8304476
```

In order to be able to reproduce a sequence of pseudo-random numbers, you need to know the seed *and* the algorithm. To find out (and change) which algorithm R is using, use the function `RNGkind`. R allows you to use older versions of its pseudo-random number generator, so that simulation results obtained using older versions of R can still be verified.

18.2 Simulating discrete random variables

Let X be a discrete random variable taking values in the set $\{0, 1, \dots\}$ with cdf F and pmf p. The following snippet of code takes a uniform random variable U and returns a discrete random variable X with cdf F.

```
# given U ~ U(0,1)
X <- 0
while (F(X) < U) {
    X <- X + 1
}
```

When the algorithm terminates we have $F(X) \geq U$ and $F(X-1) < U$, that is $U \in (F(X-1), F(X)]$. Thus

$$\mathbb{P}(X = x) = \mathbb{P}(U \in (F(x-1), F(x)]) = F(x) - F(x-1) = p(x)$$

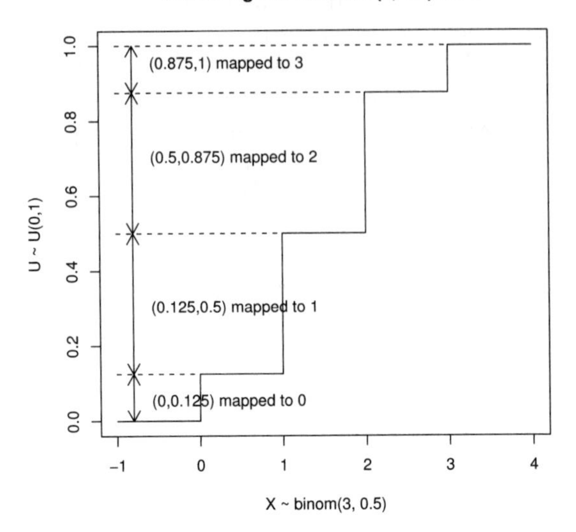

Figure 18.1 *Simulating a binom*$(3, 0.5)$ *rv by transforming a* $U(0,1)$ *rv.*

as required.

Figure 18.1 shows how this algorithm works, in the case $X \sim \text{binom}(3, 0.5)$. We see that F is used to map U to X. The algorithm can easily be generalised to any discrete distribution; see Exercise 8.

For simulating a finite rv R provides

```
sample(x, size, replace = FALSE, prob = NULL)
```

The inputs are

`x` A vector giving the possible values the rv can take;

`size` How many rv's to simulate;

`replace` Set this to `TRUE` to generate an iid sample, otherwise the rv's will be conditioned to be different from each other;

`prob` A vector giving the probabilities of the values in `x`. If omitted then the values in `x` are assumed to be equally likely.

18.2.1 Example: binomial

We present some code to simulate a binomial random variable as an example. We stress that R has superior binomial probability and simulation functions, compared with those that we present below. See `?rbinom` for more information.

If $X \sim \text{binom}(n,p)$ then it has pmf $p_X(x) = \binom{n}{x} p^x (1-p)^{n-x}$. The function `binom.cdf` below calculates the cdf F_X of X. The function `cdf.sim` takes as its first argument a function `F`, which is assumed to calculate the cdf of a non-negative integer valued random variable. `cdf.sim` also uses the argument `...` to pass parameters to the function `F`.

To simulate a single $\text{binom}(n,p)$ rv use `cdf.sim(binom.cdf, n, p)`.

```
binom.cdf <- function(x, n, p) {
  Fx <- 0
  for (i in 0:x) {
    Fx <- Fx + choose(n, i)*p^i*(1-p)^(n-i)
  }
  return(Fx)
}

cdf.sim <- function(F, ...) {
  X <- 0
  U <- runif(1)
  while (F(X, ...) < U) {
    X <- X + 1
  }
  return(X)
}
```

In the program above, suppose that U is close to 1. In this case we will need to calculate $F_X(x)$ for many values of x. But if we look at how `binom.cdf` is defined, each time we calculate $F_X(x)$ we recalculate $p_X(0), p_X(1), \ldots$, which is rather inefficient. We can avoid this by combining the loop in `cdf.sim`, which checks `F(X, ...) < U`, with the loop in `binom.cdf`, which calculates F_X. To improve the efficiency further we use a recursive formula to calculate $p_X(x)$, namely

$$p_X(x) = \frac{(n-x+1)p}{x(1-p)} p_X(x-1).$$

```
# program spuRs/resources/scripts/binom.sim.r

binom.sim <- function(n, p) {
  X <- 0
  px <- (1-p)^n
  Fx <- px
  U <- runif(1)
  while (Fx < U) {
    X <- X + 1
    px <- px*p/(1-p)*(n-X+1)/X
    Fx <- Fx + px
  }
  return(X)
```

```
}
```

To see that binom.sim works, observe that at the beginning of each cycle of the while loop we always have px equal to the pmf at X and Fx equal to the cdf at X. To verify numerically that binom.sim works, we generate a large sample using binom.sim, use it to estimate p_X, then compare the estimate with the known pmf. We use dbinom to calculate p_X (alternatively you could write your own function, as per Chapter 15, Exercise 12), and plot the output in Figure 18.2. The true values are indicated with a filled dot, and a plus sign is used for the estimates and their 95% confidence intervals.

```
# inputs
N <- 10000          # sample size
n <- 10             # rv parameters
p <- 0.7
set.seed(100)       # seed for RNG

# generate sample and estimate p
X <- rep(0, N)
for (i in 1:N) X[i] <- binom.sim(n, p)
phat <- rep(0, n+1)
for (i in 0:n) phat[i+1] <- sum(X == i)/N
phat.CI <- 1.96*sqrt(phat*(1-phat)/N)

# plot output
plot(0:n, dbinom(0:n, n, p), type="h", xlab="x", ylab="p(x)")
points(0:n, dbinom(0:n, n, p), pch=19)
points(0:n, phat, pch=3)
points(0:n, phat+phat.CI, pch=3)
points(0:n, phat-phat.CI, pch=3)
```

18.2.2 Sequences of independent trials

For random variables that are defined using a sequence of independent trials (the binomial, geometric, and negative binomial), we have alternative methods. Given a $U(0,1)$ rv U we can generate a Bernoulli rv B with parameter p using

```
# given U ~ U(0,1)
if (U < p) {B <- 1} else {B <- 0}
```

Thus, given n and p, to generate a binom(n, p) rv X we can use

```
X <- 0
for (i in 1:n) {
    U <- runif(1)
    if (U < p) X <- X + 1
}
```

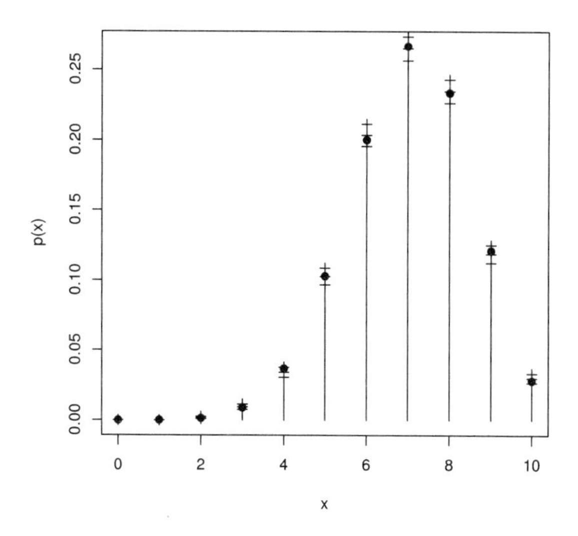

Figure 18.2 *Estimated and actual pmf for a binom*$(10, 0.7)$ *rv.*

Alternatively we can use that fact R coerces `TRUE` to 1 and `FALSE` to 0 to rewrite this in one line

```
X <- sum(runif(n) < p)
```

This is clearly much simpler than the approach of Example 18.2.1, and is faster. Its only disadvantage is that it uses more uniforms (n as opposed to 1), so that if you generate a lot of binomial random variables with this algorithm, then your 'random' numbers will start repeating themselves sooner. This is usually not a problem, however; the Mersenne-Twister (the random number generator used by R at the time of writing) has a cycle length of $2^{19937} - 1$, so even if we are using these a few hundred at a time, it will be a long while before our binomial random variables start cycling.

Given `p`, to generate a geom(p) rv `Y`, we can use

```
Y <- 0
success <- FALSE
while (!success) {
    U <- runif(1)
    if (U < p) {
        success <- TRUE
    } else {
        Y <- Y + 1
    }
}
```

The negative binomial distribution can be treated similarly: see Exercise 5.

18.3 Inversion method for continuous rv

In the following sections we study two general methods for simulating continuous distributions and also look at some techniques used to simulate the very important normal distribution.

Suppose that we are given $U \sim U(0,1)$ and want to simulate a continuous rv X with cdf F_X. Put $Y = F_X^{-1}(U)$ then we have

$$F_Y(y) = \mathbb{P}(Y \leq y) = \mathbb{P}(F_X^{-1}(U) \leq y) = \mathbb{P}(U \leq F_X(y)) = F_X(y).$$

That is, Y has the same distribution as X. Thus, if we can simulate a $U(0,1)$ rv, then we can simulate any continuous rv X for which we know F_X^{-1}. This is called the *inverse transformation method* or simply the *inversion method.* It is the continuous analogue of the method for simulating discrete random variables given in Section 18.2.

Another way of looking at this remarkable result is that, for any continuous rv X, $F_X(X) \sim U(0,1)$.

18.3.1 Example: uniform distribution

Consider $X \sim U(1,3)$. Verify that X has cdf $F_X(x) = 2(x-1)$ for $x \in (1,3)$ and thus that $F_X^{-1}(y) = 2y+1$ for $y \in (0,1)$. The inversion method therefore tells us to generate X using $2U+1$, where $U \sim U(0,1)$. Geometrically this result is clear: the factor of 2 stretches the $U(0,1)$ distribution from $(0,1)$ to $(0,2)$, and it is then translated to the right by 1. Figure 18.3 illustrates this transformation using a plot of F_X. Imagine a 'uniform rain' of observations on U falling on the interval $(0,1)$ on the vertical axis. The inverse cdf function converts this into a uniform rain on the interval $(1,3)$ on the horizontal axis, that is observations on $X \sim U(1,3)$.

18.3.2 Example: exponential distribution

If $X \sim \exp(\lambda)$ then the pdf is $f_X(x) = \lambda e^{-\lambda x}$, for $x > 0$, and by integrating we find

$$F_X(x) = \begin{cases} 0 & \text{for } x < 0; \\ 1 - e^{-\lambda x} & \text{for } x \geq 0. \end{cases}$$

Putting $y = F_X(x)$ we derive the inverse function as follows:

$$\begin{aligned} y &= 1 - e^{-\lambda x} \\ 1 - y &= e^{-\lambda x} \\ \log(1-y) &= -\lambda x \\ x &= -\frac{1}{\lambda}\log(1-y) \;=\; F_X^{-1}(y). \end{aligned}$$

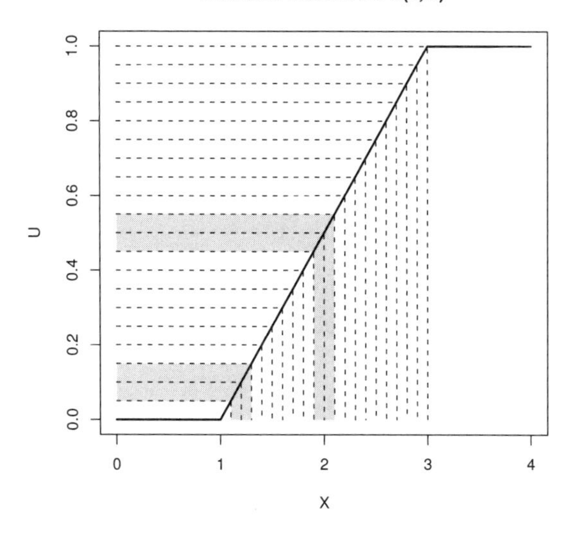

Figure 18.3 *Illustration of the inversion method. A 'uniform rain' of points on the vertical interval* $(0, 1)$ *becomes a uniform rain on the horizontal interval* $(1, 3)$.

So the inversion method generates $X \sim \exp(\lambda)$ by using $-\lambda^{-1} \log(1 - U)$ with $U \sim U(0, 1)$. It is easy to show that if $U \sim U(0, 1)$ then $1 - U \sim U(0, 1)$, so $-\lambda^{-1} \log(U) \sim \exp(\lambda)$.

Figure 18.4 illustrates the conversion of a uniform rain of points on the vertical interval $(0, 1)$ to an 'exponentially distributed rain' on the horizontal axis through this transformation. Notice that the values of U between 0.15 and 0.2 are transformed into a much smaller interval on the x-axis than those falling between 0.85 and 0.9 (the regions shaded in the figure).

18.4 Rejection method for continuous rv

The inversion method works well if we can find F^{-1} analytically. If not. we can use root-finding techniques to invert F numerically (see Exercise 16), but this can be time-consuming. An alternative method in this situation, which is often faster, is the rejection method.

To motivate the rejection method let us consider a simple example. Say we have a continuous random variable X with pdf f_X concentrated on the interval $(0, 4)$, as illustrated in Figure 18.5. We imagine 'sprinkling' points $P_1, P_2, \ldots$, uniformly at random under the density function. By sprinkling uniformly, we mean that a small target square under the pdf has the same chance of being hit wherever it is located. Our random points P_i are actually two-dimensional

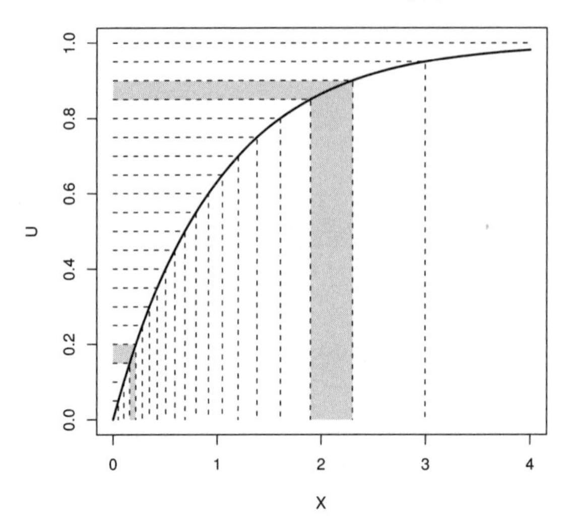

Figure 18.4 *Illustration of the inversion method. A 'uniform rain' of points on the vertical interval* $(0, 1)$ *becomes an 'exponentially distributed rain' on the horizontal axis.*

random variables (X_i, Y_i), where X_i and Y_i are the random coordinates of the i-th point.

Consider the distribution of X_1, the x coordinate of P_1. (Note that all X_i have the same distribution.) Let R be the shaded region under f_X between a and b, as shown in Figure 18.5, then

$$\begin{aligned}
\mathbb{P}(a < X_1 < b) &= \mathbb{P}(P_1 \text{ hits R}) \\
&= \frac{\text{Area of R}}{\text{Area under density}} \\
&= \frac{\int_a^b f_X(x)dx}{1} \\
&= \int_a^b f_X(x)dx.
\end{aligned}$$

Thus by the definition of the pdf, X_1 has the same distribution as X. So we can generate observations on X by taking the x coordinate of random points sprinkled under its pdf f_X. But how do we generate the points P_i uniformly under f_X? The answer is to generate points at random in the rectangle $[0, 4] \times [0, 0.5]$ (dotted in Figure 18.5), and then *reject* those that fall above the pdf, hence the name *rejection method.*

This method extends to any density with finite support that is bounded above. That is, $f_X(x) \leq k$ for all x and some constant k.

Rejection method (uniform envelope) Suppose that f_X is non-zero only on $[a, b]$, and $f_X \leq k$.

1. Generate $X \sim U(a, b)$ and $Y \sim U(0, k)$ independent of X (so $P = (X, Y)$ is uniformly distributed over the rectangle $[a, b] \times [0, k]$).
2. If $Y < f_X(X)$ then return X, otherwise go back to step 1.

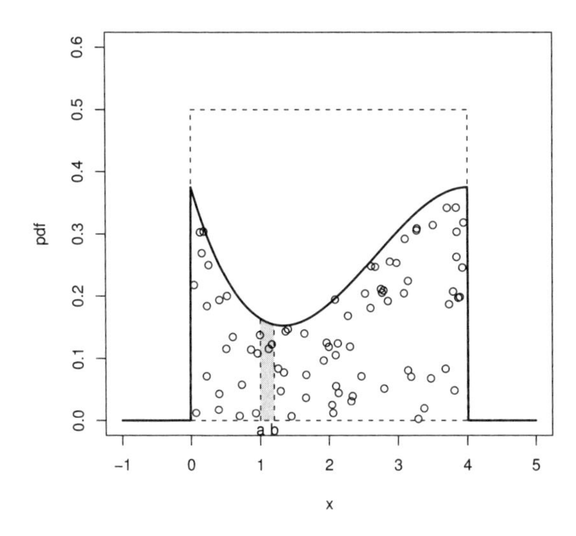

Figure 18.5 *Points uniformly distributed under a pdf.*

18.4.1 Example: triangular density

Consider the triangular pdf f_X defined as

$$f_X(x) = \begin{cases} x & \text{if } 0 < x < 1; \\ (2 - x) & \text{if } 1 \leq x < 2; \\ 0 & \text{otherwise.} \end{cases}$$

We apply the rejection method as follows:

```
# program spuRs/resources/scripts/rejecttriangle.r

rejectionK <- function(fx, a, b, K) {
  # simulates from the pdf fx using the rejection algorithm
  # assumes fx is 0 outside [a, b] and bounded by K
```

```
  # note that we exit the infinite loop using the return statement
  while (TRUE) {
    x <- runif(1, a, b)
    y <- runif(1, 0, K)
    if (y < fx(x)) return(x)
  }
}

fx<-function(x){
  # triangular density
  if ((0<x) && (x<1)) {
    return(x)
  } else if ((1<x) && (x<2)) {
    return(2-x)
  } else {
    return(0)
  }
}

# generate a sample
set.seed(21)
nreps <- 3000
Observations <- rep(0, nreps)
for(i in 1:nreps)   {
  Observations[i] <- rejectionK(fx, 0, 2, 1)
}

# plot a scaled histogram of the sample and the density on top
hist(Observations, breaks = seq(0, 2, by=0.1), freq = FALSE,
     ylim=c(0, 1.05), main="")
lines(c(0, 1, 2), c(0, 1, 0))
```

The output is given in Figure 18.6.

18.4.2 General rejection method

Our rejection method above uses a rectangular envelope to cover the target density f_X, then generates candidate points uniformly within the rectangle. However if the rectangle is infinite, then we cannot generate points uniformly within it, because it has infinite area. Instead we need a shape with finite area, within which we can simulate points uniformly.

Let X have pdf h and, given X, let $Y \sim U(0, kh(X))$ (so the range of Y depends on X), then (X, Y) is uniformly distributed over the region A defined by the curve kh above and 0 below. To see this we use conditional probability:

$$
\begin{aligned}
&\mathbb{P}((X,Y) \in (x, x+dx) \times (y, y+dy)) \\
&\quad = \mathbb{P}(Y \in (y, y+dy) \mid X \in (x, x+dx))\mathbb{P}(X \in (x, x+dx))
\end{aligned}
$$

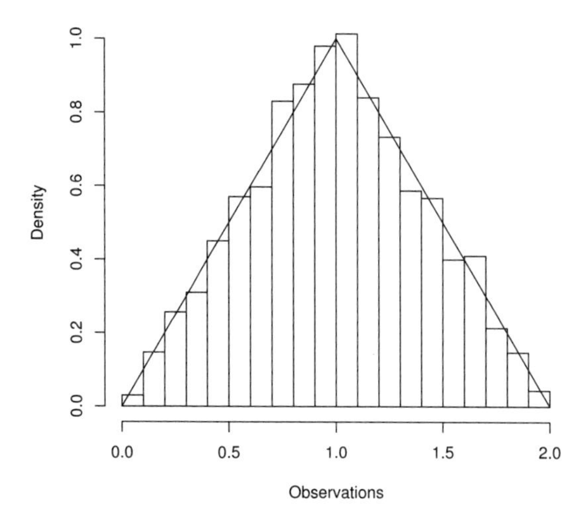

Figure 18.6 *Empirical pdf of the triangular distribution, simulated using the rejection method.*

$$
\begin{aligned}
&= \frac{dy}{kh(x)} h(x) dx \\
&= \frac{1}{k} dx dy.
\end{aligned}
$$

That is, the chance of being in a small rectangle of size $dx \times dy$ is the same anywhere in A. (We say that (X, Y) has a joint density, given by $\frac{1}{k} 1_{\{(x,y) \in A\}}$, where k is the area of A.)

Suppose we wish to simulate from the density f_X. Let h be a density we can simulate from, and choose k such that

$$k \geq k^* = \sup_x \frac{f_X(x)}{h(x)}.$$

Note that $k^* \geq 1$, with equality if and only if f_X and h are identical. Then kh forms an envelope for f_X, and we can generate points uniformly within this envelope. By accepting points below the curve f_X, we get the general rejection method:

General rejection method
To simulate from the density f_X, we assume that we have envelope density h from which you can simulate, and that we have some $k < \infty$ such that $\sup_x f_X(x)/h(x) \leq k$.

1. Simulate X from h.
2. Generate $Y \sim U(0, kh(X))$.
3. If $Y < f_X(X)$ then return X, otherwise go back to step 1.

18.4.3 Efficiency

The efficiency of the rejection method is measured by the expected number of times you have to generate a candidate point (X, Y). The area under the curve kh is k and the area under the curve f_X is 1, so the probability of accepting a candidate is $1/k$. Thus the number of times N we have to generate a candidate point has distribution $1+ \text{geom}(1/k)$, with mean $\mathbb{E}N = 1+(1-1/k)/(1/k) = k$. So, the closer h is to f_X, the smaller we can choose k, and the more efficient the algorithm is.

18.4.4 Example: gamma

For $m, \lambda > 0$ the $\Gamma(\lambda, m)$ density is $f(x) = \lambda^m x^{m-1} e^{-\lambda x}/\Gamma(m)$, for $x > 0$. There is no explicit formula for the cdf F or its inverse, so we will use the rejection method to simulate from f.

We will use an exponential envelope $h(x) = \mu e^{-\mu x}$, for $x > 0$. Using the inversion method we can easily simulate from h using $-\log(U)/\mu$, where $U \sim U(0, 1)$. To envelop f we need to find

$$k^* = \sup_{x>0} \frac{f(x)}{h(x)} = \sup_{x>0} \frac{\lambda^m x^{m-1} e^{(\mu-\lambda)x}}{\mu\Gamma(m)}.$$

Clearly k^* will be infinite if $m < 1$ or $\lambda \leq \mu$. For $m = 1$ the gamma is just an exponential. Thus we will assume $m > 1$ and choose $\mu < \lambda$. For $m \in (0, 1)$ the rejection method can still be used, but a different envelope is required.

To find k^* we take the derivative of the right-hand side above and set it to zero, to find the point where the maximum occurs. You can check that this is at the point $x = (m - 1)/(\lambda - \mu)$, which gives

$$k^* = \frac{\lambda^m (m-1)^{m-1} e^{-(m-1)}}{\mu(\lambda - \mu)^{m-1}\Gamma(m)}.$$

To improve efficiency we would like to choose our envelope to make k^* as small as possible. Looking at the formula for k^* this means choosing μ to make $\mu(\lambda - \mu)^{m-1}$ as large as possible. Setting the derivative with respect to

μ to zero, we see that the maximum occurs when $\mu = \lambda/m$. Plugging this back in we get $k^* = m^m e^{-(m-1)}/\Gamma(m)$.

We can now code up our rejection algorithm.

```
# program spuRs/resources/scripts/gamma.sim.r

gamma.sim <- function(lambda, m) {
  # sim a gamma(lambda, m) rv using rejection with an exp envelope
  # assumes m > 1 and lambda > 0
  f <- function(x) lambda^m*x^(m-1)*exp(-lambda*x)/gamma(m)
  h <- function(x) lambda/m*exp(-lambda/m*x)
  k <- m^m*exp(1-m)/gamma(m)
  while (TRUE) {
    X <- -log(runif(1))*m/lambda
    Y <- runif(1, 0, k*h(X))
    if (Y < f(X)) return(X)
  }
}

set.seed(1999)
n <- 10000
g <- rep(0, n)
for (i in 1:n) g[i] <- gamma.sim(1, 2)
hist(g, breaks=20, freq=F, xlab="x", ylab="pdf f(x)",
  main="theoretical and simulated gamma(1, 2) density")
x <- seq(0, max(g), .1)
lines(x, dgamma(x, 2, 1))
```

To check the function `gamma.sim` works we simulated a large sample, using the parameters $m = 2$ and $\lambda = 1$, and used them to estimate the density. The result, with the density plotted on top, is given in Figure 18.7.

18.5 Simulating normals

In this section we consider various ways to generate normal random variables. Historically the problem of simulating normal random variables has attracted a lot of attention because normal random variables are important, and because there is no one way of simulating them that is clearly best. To see what R uses, type `RNGkind()` then `?RNGkind`.

If $Z \sim N(0,1)$ then $\mu + \sigma Z \sim N(\mu, \sigma^2)$, so it is sufficient to be able to simulate standard $N(0,1)$ rv's.

18.5.1 Central Limit Theorem

The Central Limit Theorem (CLT) suggests an obvious approximate approach to simulating the normal, by averaging. Recall that for $U \sim U(0,1)$, $\mathbb{E}U = 1/2$

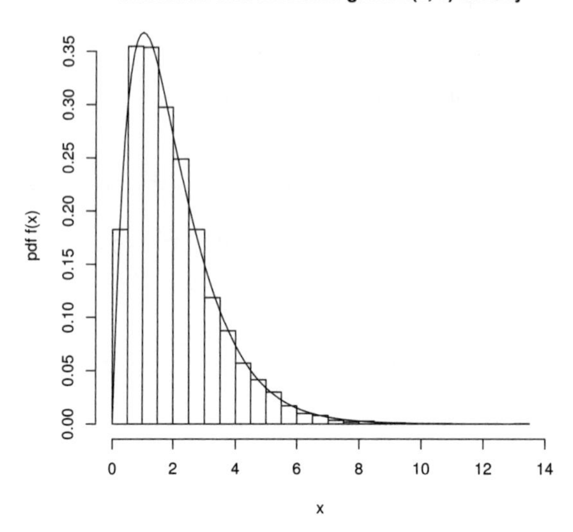

Figure 18.7 $\Gamma(1,2)$ *density estimated from a sample generated using the rejection method.*

and $\mathrm{Var}\, U = 1/12$, so if $U_1, \ldots, U_{12}$ are iid $U(0,1)$ then

$$Z = \left(\sum_{i=1}^{12} U_i\right) - 6$$

has mean 0 and variance 1, and thus by the CLT is (approximately) $N(0,1)$.

This generator works quite well in fact, but it is not difficult to do better and this approach is wasteful of uniforms.

18.5.2 Rejection with exponential envelope

If we chop a standard normal distribution in half and use only the positive side (scaled up by a factor of 2 to maintain a proper density), then we get the so-called 'half normal' density:

$$f_X(x) = \begin{cases} \sqrt{\frac{2}{\pi}} \exp\left(-\frac{1}{2}x^2\right) & \text{if } x > 0; \\ 0 & \text{otherwise.} \end{cases}$$

If $Z \sim N(0,1)$ then $|Z|$ has a half normal density. Conversely if X is half normal and $S = \pm 1$ with probability half each, independently of X, then

$$Z = SX \sim N(0,1).$$

We write $S \sim U\{-1,+1\}$ to indicate that the distribution of S is uniformly distributed over the finite set $\{-1,+1\}$.

We can use rejection to generate observations on the half normal X. Consider an exponential distribution with parameter $\lambda = 1$ as a possible envelope. That is, the envelope density is $h(x) = \exp(-x)$ for $x > 0$. One can easily check that

$$k^* = \sup_x \frac{f_X(x)}{h(x)} = \sup_x \sqrt{\frac{2}{\pi}} \exp(x - x^2/2) = \sqrt{\frac{2e}{\pi}}.$$

This gives us the following algorithm (here ϕ stands for the standard normal density):

Standard normal simulation using rejection

1. Generate $X \sim \exp(1)$ and $Y \sim U(0, \exp(-X)\sqrt{2e/\pi})$.
2. If $Y < \phi(X)$ then generate $S \sim U\{-1, +1\}$ and return $Z = SX$, otherwise go back to step 1.

It is possible to improve the efficiency of this algorithm a little. First, observe that to generate $X \sim \exp(1)$ we use $-\log(U)$ (using the inverse transform) where $U \sim U(0,1)$. Thus, $\exp(-X) = U$ and so $Y \sim U(0, U\sqrt{2e/\pi})$. Second, rather than generate $S \sim U\{-1, +1\}$ we note that if $Y < \phi(X)$, then $Y < \phi(X)/2$ with probability 1/2, independently of X. Incorporating these two refinements we get

Improved standard normal simulation using rejection.

1. Generate $U \sim U(0,1)$ and $Y \sim U(0, U\sqrt{2e/\pi})$
2. Put $X = -\log(U)$.
3. (a). If $Y < \phi(X)/2$ then return $Z = -X$,
   (b). Else if $\phi(X)/2 < Y < \phi(X)$ then return $Z = X$,
   (c). Else go back to step 1.

18.5.3 Box-Muller algorithm

Suppose $P = (X, Y)$ where X and Y are independent $N(0,1)$ rv's, then P is said to have a standard bivariate normal distribution. The Box-Muller algorithm works by simulating P in polar coordinates (R, Θ), then transforming these back to cartesian co-ordinates using $X = R\cos(\Theta)$ and $Y = R\sin(\Theta)$. Thus we generate two independent $N(0,1)$ rv's each time.

The derivation of the distribution of P in polar coordinates is not particularly straight-forward. It can be shown[1] that $R^2 \sim \exp(1/2)$ and $\Theta \sim U(0, 2\pi)$, independently of R. This gives us the following algorithm:

[1] The proof requires the transformation of a joint distribution function, using multivariate calculus.

Box-Muller simulation of standard normal

1. Generate $U_1, U_2 \sim U(0,1)$.
2. Set $\Theta = 2\pi U_1$ and $R = \sqrt{-2\log(U_2)}$.
3. Return $X = R\cos(\Theta)$ and $Y = R\sin(\Theta)$.

Calculating sines and cosines can be expensive (in terms of the time required), but there is a version of the Box-Muller algorithm that avoids this. Suppose that the point $Q = (A, B)$ is uniformly distributed over the unit circle. Let (S, Ψ) be the polar coordinates of Q, then one can show that $S^2 \sim U(0,1)$ and $\Psi \sim U(0, 2\pi)$, independently of S. Thus, $(\sqrt{-2\log(S^2)}, \Psi)$ has the same distribution as the polar coordinates of P, namely bivariate standard normal. The advantage of this representation is that it we can easily calculate X and Y from A and B. A little trigonometry gives us that

$$\begin{aligned} S^2 &= A^2 + B^2 \\ \cos(\Psi) &= A/S \\ \sin(\Psi) &= B/S \end{aligned}$$

so, for $R = \sqrt{-2\log(S^2)}$,

$$\begin{aligned} X &= R\cos(\Psi) = A\sqrt{\frac{-2\log(S^2)}{S^2}} \\ Y &= R\sin(\Psi) = B\sqrt{\frac{-2\log(S^2)}{S^2}}. \end{aligned}$$

We still need to generate Q, but this can be easily achieved using a rejection algorithm. Generate $U, V \sim U(-1,1)$ independently, then accept the point (U, V) if it is inside the unit circle, that is, if $U^2 + V^2 < 1$.

Putting these together we get the following:

Improved Box-Muller simulation of standard normal, with rejection step

1. Generate $U, V \sim U(-1,1)$.
2. Accept $S^2 = U^2 + V^2$ provided $S^2 < 1$ else return to step 1.
3. Set $W = \sqrt{-2\log(S^2)/S^2}$.
4. Return $X = UW$ and $Y = VW$.

18.6 Exercises

1. Express 45 in binary.

   Now express 45 mod 16 and 45 mod 17 in binary.

   What can you say about these three binary representations?

2. Find all of the cycles of the following congruential generators. For each cycle identify which seeds X_0 lead to that cycle.

   (a). $X_{n+1} = 9X_n + 3 \mod 11$.
   (b). $X_{n+1} = 8X_n + 3 \mod 11$.
   (c). $X_{n+1} = 8X_n + 2 \mod 12$.

3. Here is some pseudo-code of an algorithm for generating a sample $y_1, \ldots, y_k$ from the population $x_1, \ldots, x_n$ *without* replacement ($k \leq n$):

```
for (i in 1:k) {
  { Select j at random from 1:(n+1-i) }
  y[i] <- x[j]
  { Swap x[j] and x[n+1-i] }
}
```

   Implement this algorithm in R. (The built-in implementation is `sample`.)

4. Consider the discrete random variable with pmf given by:

$$\mathbb{P}(X = 1) = 0.1, \quad \mathbb{P}(X = 2) = 0.3, \quad \mathbb{P}(X = 5) = 0.6.$$

   Plot the cdf for this random variable.

   Write a program to simulate a random variable with this distribution, using the built-in function `runif(1)`.

5. How would you simulate a negative binomial random variable from a sequence of Bernoulli trials? Write a function to do this in R. (The built-in implementation is `rnbinom(n, size, prob)`.)

6. For $X \sim \text{Poisson}(\lambda)$ let $F(x) = \mathbb{P}(X \leq x)$ and $p(x) = \mathbb{P}(X = x)$. Show that the probability function satisfies

$$p(x+1) = \frac{\lambda}{x+1} p(x).$$

   Using this write a function to calculate $p(0), p(1), \ldots, p(x)$ and $F(x) = p(0) + p(1) + \cdots + p(x)$.

   If $X \in \mathbb{Z}_+$ is a random variable and `F(x)` is a function that returns the cdf F of X, then you can simulate X using the following program:

```
F.rand <- function () {
  u <- runif(1)
  x <- 0
  while (F(x) < u) {
    x <- x + 1
  }
  return(x)
}
```

   In the case of the Poisson distribution, this program can be made more efficient by calculating F just once, instead of recalculating it every time you call the function `F(x)`. By using two new variables, `p.x` and `F.x` for $p(x)$ and $F(x)$ respectively, modify this program so that instead of using the function F(x) it updates p.x and F.x within the while loop. Your program should have the form

```
F.rand <- function(lambda) {
  u <- runif(1)
  x <- 0
  p.x <- ?
  F.x <- ?
  while (F.x < u) {
    x <- x + 1
    p.x <- ?
    F.x <- ?
  }
  return(x)
}
```

You should ensure that at the start of the while loop you always have p.x equal to $p(x)$ and F.x equal to $F(x)$.

7. This exercise asks you to verify the function F.rand from Exercise 6. The idea is to use F.rand to estimate the Poisson probability mass function, and compare the estimates with known values. Let $X_1, \ldots, X_n$ be independent and identically distributed (iid) pois(λ) random variables, then we estimate $p_\lambda(x) = \mathbb{P}(X_1 = x)$ using

$$\hat{p}_\lambda(x) = \frac{|\{X_i = x\}|}{n}.$$

Write a program F.rand.test(n, lambda) that simulates n pois(λ) random variables and then calculates $\hat{p}_\lambda(x)$ for $x = 0, 1, \ldots, k$, for some chosen k. Have your program print a table giving $p_\lambda(x)$, $\hat{p}_\lambda(x)$ and a 95% confidence interval for $p_\lambda(x)$, for $x = 0, 1, \ldots, k$.

Finally, modify your program F.rand.test so that it also draws a graph of $\hat{p}$ and p, with confidence intervals, similar to Figure 18.2.

8. Suppose that X takes on values in the countable set $\{\ldots, a_{-2}, a_{-1}, a_0, a_1, a_2, \ldots\}$, with probabilities $\{\ldots, p_{-2}, p_{-1}, p_0, p_1, p_2, \ldots\}$. Suppose also that you are given that $\sum_{i=0}^{\infty} p_i = p$, then write an algorithm for simulating X.

   Hint: first decide whether or not $X \in \{a_0, a_1, \ldots\}$, which occurs with probability p.

9. Suppose that X and Y are independent rv's taking values in $\mathbb{Z}_+ = \{0, 1, 2, \ldots\}$ and let $Z = X + Y$.

   (a). Suppose that you are given functions X.sim() and Y.sim(), which simulate X and Y. Using these, write a function in R to estimate $\mathbb{P}(Z = z)$ for a given z.

   (b). Suppose that instead of X.sim() and Y.sim() you are given X.pmf(x) and Y.pmf(y), which calculate $\mathbb{P}(X = x)$ and $\mathbb{P}(Y = y)$ respectively.

Using these, write a function Z.pmf(z) to calculate $\mathbb{P}(Z = z)$ for a given z.

(c). Given Z.pmf(z) write a function in R to calculate $\mathbb{E}Z$.

Note that we may have $\mathbb{P}(Z = z) > 0$ for all $z \geq 0$. To approximate $\mu = \mathbb{E}Z$ numerically we use $\mu_n^{trunc} = \sum_{z=0}^{n-1} z\mathbb{P}(Z = z) + n\mathbb{P}(Z \geq n) = n - \sum_{z=0}^{n-1}(n - z)\mathbb{P}(Z = z)$. How can we decide how large n needs to be to get a good approximation?

Do you think this method of approximating $\mathbb{E}Z$ is better or worse than simulation?

10. Consider the following program, which performs a simulation experiment. The function X.sim() simulates some random variable X, and we wish to estimate $\mathbb{E}X$.

```
# set.seed(7)
# seed position 1
mu <- rep(0, 6)
for (i in 1:6) {
  # set.seed(7)
  # seed position 2
  X <- rep(0, 1000)
  for (j in 1:1000) {
    # set.seed(7)
    # seed position 3
    X[j] <- X.sim()
  }
  mu[i] <- mean(X)
}
spread <- max(mu) - min(mu)
mu.estimate <- mean(mu)
```

(a). What is the value of spread used for?

(b). If we uncomment the command set.seed(7) at seed position 3, then what is spread?

(c). If we uncomment the command set.seed(7) at seed position 2 (only), then what is spread?

(d). If we uncomment the command set.seed(7) at seed position 1 (only), then what is spread?

(e). At which position should we set the seed?

11. (a). Here is some code for simulating a discrete random variable Y. What is the probability mass function (pmf) of Y?

```
Y.sim <- function() {
  U <- runif(1)
  Y <- 1
  while (U > 1 - 1/(1+Y)) {
    Y <- Y + 1
```

```
  }
  return(Y)
}
```

Let N be the number of times you go around the while loop when `Y.sim()` is called. What is $\mathbb{E}N$ and thus what is the expected time taken for this function to run?

(b). Here is some code for simulating a discrete random variable Z. Show that Z has the same pmf as Y

```
Z.sim <- function() {
  Z <- ceiling(1/runif(1)) - 1
  return(Z)
}
```

Will this function be faster or slower that `Y.sim()`?

12. People arrive at a shoe store at random. Each person then looks at a random number of shoes before deciding which to buy.

(a). Let N be the number of people that arrive in an hour. Given that $\mathbb{E}N = 10$, what would be a good distribution for N?

(b). Customer i tries on X_i pairs of shoes they do not like before finding a pair they like and then purchase ($X_i \in \{0, 1, \ldots\}$). Suppose that the chance they like a given pair of shoes in 0.8, independently of the other shoes they have looked at. What is the distribution of X_i?

(c). Let Y be the total number of shoes that have been tried on, excluding those purchased. Supposing that each customer acts independently of other customers, give an expression for Y in terms of N and the X_i, then write functions for simulating N, X_i, and Y.

(d). What is $\mathbb{P}(Y = 0)$?

Use your simulation of Y to estimate $\mathbb{P}(Y = 0)$. If your confidence interval includes the true value, then you have some circumstantial evidence that your simulation is correct.

13. Consider the continuous random variable with pdf given by:

$$f(x) = \begin{cases} 2(x-1)^2 & \text{for } 1 < x \leq 2, \\ 0 & \text{otherwise.} \end{cases}$$

Plot the cdf for this random variable.

Show how to simulate a rv with this cdf using the inversion method.

14. Consider the continuous random variable X with pdf given by:

$$f_X(x) = \frac{\exp(-x)}{(1 + \exp(-x))^2} \qquad -\infty < x < \infty.$$

X is said to have a standard logistic distribution. Find the cdf for this random variable. Show how to simulate a rv with this cdf using the inversion method.

15. Let $U \sim U(0, 1)$ and let $Y = 1 - U$. Derive an expression for the cdf $F_Y(y)$ of Y in terms of the cdf of U and hence show that $Y \sim U(0, 1)$.

16. For a given u, adapt the bisection method from Chapter 10 to write a program to find the root of the function $\Phi(x) - u$ where $\Phi(x)$ is the cdf of the standard normal distribution. (You can evaluate Φ using numerical integration or by using the built-in R function.) Notice that the root satisfies $x = \Phi^{-1}(u)$.

    Using the inversion method, write a program to generate observations on a standard normal distribution. Compare the proportion of your observations that fall within the interval $(-1, 1)$ with the theoretical value of 68.3%.

17. The continuous random variable X has the following probability density function (pdf), for some positive constant c,

$$f(x) = \frac{3}{(1+x)^3} \text{ for } 0 \le x \le c.$$

    (a). Prove that $c = \sqrt{3} - 1$.
    (b). What is $\mathbb{E}X$? (Hint: $\mathbb{E}X = \mathbb{E}(X+1) - 1$.)
    (c). What is $\text{Var}\, X$? (Hint: start with $\mathbb{E}(X+1)^2$.)
    (d). Using the inversion method, write a function that simulates X.

18. The Cauchy distribution with parameter α has pdf

$$f_X(x) = \frac{\alpha}{\pi(\alpha^2 + x^2)} \qquad -\infty < x < \infty.$$

    Write a program to simulate from the Cauchy distribution using the inversion method.

    Now consider using a Cauchy envelope to generate a standard normal random variable using the rejection method. Find the values for α and the scaling constant k that minimise the probability of rejection. Write an R program to implement the algorithm.

CHAPTER 19

Monte-Carlo integration

The term Monte-Carlo is used to refer to techniques involving computer simulation, alluding to the games of chance played in the casinos of Monte Carlo. Monte-Carlo integration is numerical integration using simulation.

This chapter covers two simulation-based approaches to integration—the hit-and-miss method and the (improved) Monte-Carlo method—adding to the techniques that were introduced in Chapter 11. Again, our goal is to integrate a function for which the antiderivative is not known in closed form.

At the end of the chapter we give some comparative results on the errors of different numerical integration methods. We see that techniques like Simpson's rule work very well in one dimension, but are not efficient for calculating high-dimensional integrals $\int \cdots \int f(x_1, \ldots, x_d) dx_1 \cdots dx_d$. In contrast Monte-Carlo integration is not as good as Simpson's rule in one dimension but is relatively more efficient in higher dimensions.

19.1 Hit-and-miss method

We wish to calculate $I = \int_a^b f(x)dx$.

Let c and d be such that $f(x) \in [c, d]$ for all $x \in [a, b]$. Let A be the set bounded above by the curve and by the box $[a, b] \times [c, d]$, then $I = |A| + c(b - a)$. Thus if we can estimate $|A|$ then we can estimate I. A is illustrated as the shaded region in Figure 19.1.

To estimate $|A|$ imagine throwing darts at the box $[a, b] \times [c, d]$. On average the proportion that land under the curve will be given by the area of A over the area of the box, that is by $|A|/((b-a)(d-c))$, giving us a means of estimating $|A|$.

Take $X \sim U(a, b)$ and $Y \sim U(c, d)$, then (X, Y) is uniformly distributed over the box $[a, b] \times [c, d]$, and

$$\mathbb{P}((X,Y) \in A) = \mathbb{P}(Y \le f(X)) = \frac{|A|}{(b-a)(d-c)}.$$

Let $Z = 1_A(X, Y)$, that is $Z = 1$ if $Y \le f(X)$ and 0 otherwise, then $\mathbb{E}Z =$

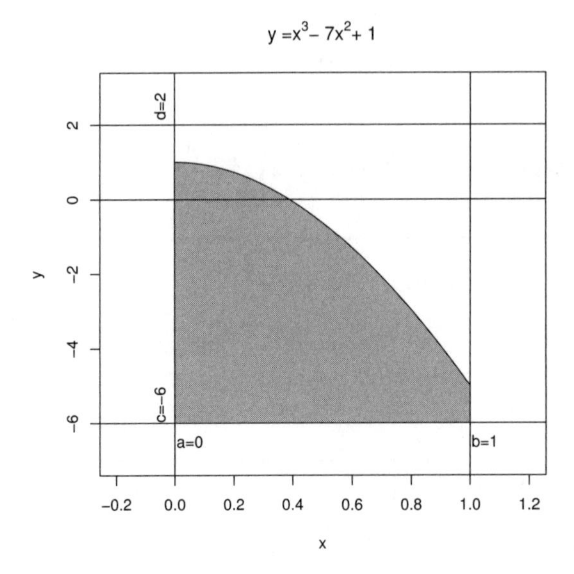

Figure 19.1 *The area of interest in the hit-and-miss method.*

$\mathbb{P}((X,Y) \in A)$ and we have

$$I = (\mathbb{E}Z)(b-a)(d-c) + c(b-a).$$

By simulating X and Y we can simulate Z and by repeatedly simulating Z, we can estimate $\mathbb{E}Z$ and thus I. Here is some code that implements the hit-and-miss method in R.

```
# program spuRs/resources/scripts/hit_miss.r

hit_miss <- function(ftn, a, b, f.min, f.max, n) {
  # Monte-Carlo integration using the hit and miss method
  # ftn is a function of one variable
  # [a, b] is the range of integration
  # f.min and f.max are bounds on ftn over the range [a, b]
  # that is f.min <= ftn(x) <= f.max for all x in [a, b]
  # n is the number of samples used in the estimation
  # that is the number of calls made to the function ftn
  Z.sum <- 0
  for (i in 1:n) {
    X <- runif(1, a, b)
    Y <- runif(1, f.min, f.max)
    Z <- (ftn(X) >= Y)
    Z.sum <- Z.sum + Z
    # cat("X =", X, "Y =", Y, "Z =", Z, "Z.sum =", Z.sum, "\n")
  }
  I <- (b - a)*f.min + (Z.sum/n)*(b - a)*(f.max - f.min)
```

```
  return(I)
}
```

We apply the method to estimate

$$\begin{aligned}\int_0^1 (x^3 - 7x^2 + 1)dx &= (x^4/4 - 7x^3/3 + x)|_0^1 \\ &= -13/12 = -1.0833 \text{ (to 4 decimal places).}\end{aligned}$$

Taking the min and max of each term we see that on $[0, 1]$ the function is bounded below by $c = 0 - 7 + 1 = -6$ and above by $d = 1 + 0 + 1 = 2$.

```
> source('../scripts/hit_miss.r')
> f <- function(x) x^3 - 7*x^2 + 1
> hit_miss(f, 0, 1, -6, 2, 10)

[1] -1.2

> hit_miss(f, 0, 1, -6, 2, 100)

[1] -0.88

> hit_miss(f, 0, 1, -6, 2, 1000)

[1] -0.784

> hit_miss(f, 0, 1, -6, 2, 10000)

[1] -1.0912

> hit_miss(f, 0, 1, -6, 2, 100000)

[1] -1.08928

> hit_miss(f, 0, 1, -6, 2, 1000000)

[1] -1.084752
```

We see that the number of repetitions n needs to be very large just to get just two decimal places accuracy.

Here is a vectorised version of the previous program. We have added a line to plot the successive approximations to the integral. The output is given in Figure 19.2.

```
hit_miss2 <- function(ftn, a, b, c, d, n) {
  # Monte-Carlo integration using the hit & miss method
  # vectorised version
  X <- runif(n, a, b)
  Y <- runif(n, c, d)
  Z <- (Y <= sapply(X, ftn))
  I <- (b - a)*c + (cumsum(Z)/(1:n))*(b - a)*(d - c)
  plot(1:n, I, type = "l")
  return(I[n])
}
```

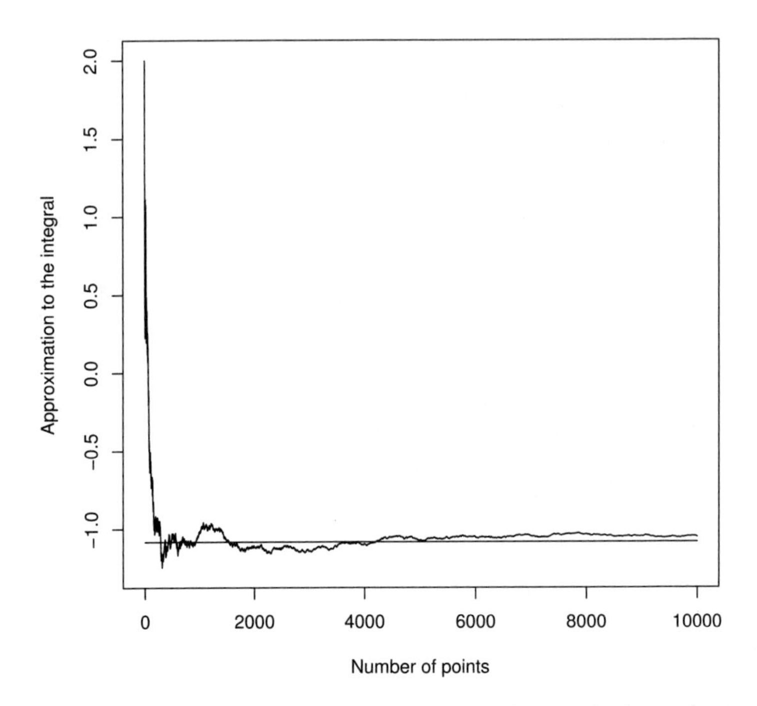

Figure 19.2 *Successive approximations to the integral using the hit-and-miss method.*

```
> source('hit_miss2.r')
> hit_miss2(f, 0, 1, -6, 2, 10000)

[1] -1.052

> lines(c(1, 10000), c(-13/12, -13/12))
```

19.2 (Improved) Monte-Carlo integration

Hit-and-miss Monte-Carlo converges very slowly. In this section we give a better Monte-Carlo integration technique, which is the technique people usually refer to as 'Monte-Carlo integration'. When we say one Monte-Carlo technique is better than another we mean that, using the same number of function calls, it has smaller variance. Bear in mind that because our estimates are based on random samples, they are themselves random variables.

Again we consider the integral $I = \int_a^b f(x)dx$. From Riemann's definition of the integral we have

$$I = \lim_{n\to\infty} \sum_{i=0}^{n-1} f(a+i(b-a)/n)(b-a)/n.$$

Here the term $f(a+i(b-a)/n)(b-a)/n$ approximates the integral from $a+i(b-a)/n$ to $a+(i+1)(b-a)/n$ by the area of a rectangle of width $(b-a)/n$ and height $f(a+i(b-a)/n)$.

Now consider the random variable X_n, which takes values in the set $\{a, a+(b-a)/n, a+2(b-a)/n, \ldots, a+(n-1)(b-a)/n\}$ with equal probability $1/n$, then

$$\begin{aligned} \mathbb{E}f(X_n) &= \sum_{i=0}^{n-1} f(a+i(b-a)/n)\mathbb{P}(X_n = a+i(b-a)/n) \\ &= \sum_{i=0}^{n-1} f(a+i(b-a)/n)/n. \end{aligned}$$

Thus, $I = \lim_{n\to\infty} \mathbb{E}f(X_n)(b-a)$.

19.2.1 Lemma

$X_n \xrightarrow{\text{d}} U(a,b)$ as $n \to \infty$. That is, for any $x \in [a,b]$,

$$\mathbb{P}(X_n \le x) \to \mathbb{P}(U \le x) = \frac{x-a}{b-a} \text{ as } n \to \infty,$$

where $U \sim U(a,b)$.

Proof. Computing the left-hand side above we get

$$\begin{aligned} \mathbb{P}(X_n \le x) &= \frac{|\{i : a+i(b-a)/n \le x\}|}{n} \\ &= \frac{|\{i \le (x-a)n/(b-a)\}|}{n} \\ &= \left(1 + \left\lfloor \frac{x-a}{b-a} n \right\rfloor\right) \frac{1}{n} \\ &\to \frac{x-a}{b-a} \text{ as } n \to \infty \end{aligned}$$

as required. (Note that we start counting at 0.)

Using the lemma we get

$$\begin{aligned} I &= \lim_{n\to\infty} \mathbb{E}f(X_n)(b-a) \\ &= \mathbb{E}f(\lim_{n\to\infty} X_n)(b-a) \\ &= \mathbb{E}f(U)(b-a) \text{ where } U \sim U(a,b). \end{aligned}$$

(Strictly speaking we need to justify exchanging the limit and expectation. Although we won't do this here, it can be done provided f is bounded and continuous.) Thus if $U_1, \ldots, U_n$ are an iid sample of $U(a,b)$ random variables, then our estimate of I is

$$\hat{I} = \frac{1}{n} \sum_{i=1}^{n} f(U_i)(b-a).$$

The following function performs Monte-Carlo integration of the function `ftn` over the interval $[a, b]$.

```
mc.integral <- function(ftn, a, b, n) {
  # Monte-Carlo integral of ftn over [a, b] using a sample of size n
  u <- runif(n, a, b)
  x <- sapply(u, ftn)
  return(mean(x)*(b-a))
}
```

19.2.2 Accuracy in higher dimensions

The big-O notation is used to describe how fast a function grows. We say $f(x)$ is $O(x^{-\alpha})$ if $\limsup_{x\to\infty} f(x)/x^{-\alpha} = \limsup_{x\to\infty} f(x)x^{\alpha} < \infty$.

Let d be the dimension of our integral and n the number of function calls used, then the accuracy of the different numerical integration techniques we have seen is as follows:

Method	Error
Trapezoid	$O(n^{-2/d})$
Simpson's rule	$O(n^{-4/d})$
Hit-and-miss Monte-Carlo	$O(n^{-1/2})$
Improved Monte-Carlo	$O(n^{-1/2})$

We see that the size of the error for the Monte-Carlo methods does not depend on d and that, asymptotically, they are preferable when $d > 8$.

19.3 Exercises

1. Suppose that X and Y are iid $U(0,1)$ random variables.

   (a). What is $\mathbb{P}((X,Y) \in [a,b] \times [c,d])$ for $0 \le a \le b \le 1$ and $0 \le c \le d \le 1$?

   Based on your previous answer, what do you think you should get for $\mathbb{P}((X,Y) \in A)$, where A is an arbitrary subset of $[0,1] \times [0,1]$?

   (b). Let $A = \{(x,y) \in [0,1] \times [0,1] : x^2 + y^2 \le 1\}$. What is the area of A?

(c). Define the rv Z by

$$Z = \begin{cases} 1 & \text{if } X^2 + Y^2 \leq 1, \\ 0 & \text{otherwise.} \end{cases}$$

What is $\mathbb{E}Z$?

(d). By simulating Z, write a program to estimate π.

2. Which is more accurate, the hit-and-miss method or the improved Monte-Carlo method? Suppose that $f : [0,1] \to [0,1]$ and we wish to estimate $I = \int_0^1 f(x)\,dx$.

   Using the hit-and-miss method, we obtain the estimate

   $$\hat{I}_{HM} = \frac{1}{n}\sum_{i=1}^{n} X_i,$$

   where $X_1, \ldots, X_n$ are an iid sample and $X_i \sim \text{binom}(1, I)$ (make sure you understand why this is the case).

   Using the improved Monte-Carlo method, we obtain the estimate

   $$\hat{I}_{MC} = \frac{1}{n}\sum_{i=1}^{n} f(U_i),$$

   where $U_1, \ldots, U_n$ are an iid sample of $U(0,1)$ random variables.

   The accuracy of the hit-and-miss method can be measured by the standard deviation of $\hat{I}_{HM}$, which is just $1/\sqrt{n}$ times the standard deviation of X_1. Similarly the accuracy of the basic Monte-Carlo method can be measured by the standard deviation of $\hat{I}_{MC}$, which is just $1/\sqrt{n}$ times the standard deviation of $f(U_1)$.

   Show that

   $$\text{Var}\, X_1 = \int_0^1 f(x)\,dx - \left(\int_0^1 f(x)\,dx\right)^2,$$

   and that

   $$\text{Var}\, f(U_1) = \int_0^1 f^2(x)\,dx - \left(\int_0^1 f(x)\,dx\right)^2.$$

   Explain why (in this case at least) the improved Monte-Carlo method is more accurate than the hit-and-miss method.

3. The previous exercise gave a theoretical comparison of the hit-and-miss and improved Monte-Carlo method. Can you verify this experimentally?

   Repeat the example of Section 19.1 using the improved Monte-Carlo method. How many function calls are required to get 2 decimal places accuracy?

4. The trapezoidal rule for approximating the integral $I = \int_0^1 f(x)\,dx$ can be broken into two steps

   Step 1: $I = \sum_{i=0}^{n-1}$ (Area under the curve from i/n to $(i+1)/n$);

Step 2: Area under the curve from i/n to $(i+1)/n \approx \frac{1}{2}(f(i/n) + f((i+1)/n)) \times \frac{1}{n}$.

In two dimensions the integral $I = \int_0^1 \int_0^1 f(x,y)\,dx\,dy$ can be broken down as

$$\sum_{i=0}^{n-1}\sum_{j=0}^{n-1} (\text{Volume under the surface above the square } [i/n,(i+1)/n] \times [j/n,(j+1)/n]).$$

(a). By analogy with the trapezoidal method, suggest a method for approximating the volume under the surface above the square $[i/n,(i+1)/n] \times [j/n,(j+1)/n]$, and thus a method for approximating the two-dimensional integral.

(b). Can you suggest a two-dimensional analogue for the improved Monte-Carlo algorithm?

CHAPTER 20

Variance reduction

Previously we have used $\overline{X} = \sum_{i=1}^{n} X_i/n$ to estimate μ, where $X_1, \ldots, X_n$ are an iid sample with mean μ. This chapter introduces several innovations to the way we sample $X_1, \ldots, X_n$ that can dramatically increase the accuracy of our estimate when judiciously applied. We cover antithetic sampling, importance sampling, and correction using control variates.

Consider using simulation to estimate the parameters of a distribution. Due to the inherent randomness, our estimates will vary from one simulation run to the other. Naturally it is desirable to reduce this variability as much as possible, thus improving the reliability of the resulting estimates. We can always reduce overall variability by increasing the number of simulation trials, but this can take too long. The variance reduction techniques we introduce sometimes offer substantial efficiency gains by working smarter not harder, using an understanding of the structure of the simulation task at hand.

Antithetic sampling reframes our estimate as a sum of negatively correlated random variables, using the fact that negative correlation reduces the variance of a sum.

Importance sampling involves placing samples where they will be most beneficial, that is, where the underlying variability is high.

Control variates uses a variable Y with known parameters to control another variable X with unknown parameters. Instead of working with what might be a highly variable X directly, we work with the residual variability of $X - Y$.

20.1 Antithetic sampling

Suppose we are interested in estimating a parameter θ and we have two unbiased estimators X and Y, with finite variances $\sigma_X^2 = \sigma_Y^2 = \sigma^2$. Clearly $Z = (X+Y)/2$ is also unbiased, and is also a candidate estimator for θ, but is it any better than X or Y?

Given X, Y, and Z are all unbiased, we can compare them using their variances (the smaller the better). We have

$$\operatorname{Var} Z = \tfrac{1}{4}\operatorname{Var} X + \tfrac{1}{4}\operatorname{Var} Y + \tfrac{1}{2}\operatorname{Cov}(X,Y) = \tfrac{1}{2}(\sigma^2 + \operatorname{Cov}(X,Y)).$$

If X and Y are independent then, compared to X or Y on their own, the variance of Z decreases by a factor of two, corresponding to the fact that we are using twice as many sample points. However if X and Y are negatively correlated, or *antithetic*, then $\mathrm{Cov}(X, Y)$ will be negative and $\mathrm{Var}\, Z$ even smaller. That is, there is more to be gained from averaging negatively correlated estimates than from averaging independent estimates (and averaging positively correlated averages is relatively worse).

We motivate the idea of antithetic variates by considering an interesting modification to a famous problem in geometric probability.

20.1.1 Example: Buffon's needle and cross

In 1733 the Compte de Buffon calculated the probability q that a needle of length l, thrown at random onto a table ruled with parallel lines of distance $d \geq l$ apart, would not intersect a line. If we let $p = 1 - q$ then Buffon showed that $p = 2l/\pi d$, thus simulation of this experiment offers a way to estimate p and hence π.

From now we assume $l = d$, so $p = 2/\pi$. Let N be the total number of intersections in n independent throws of the needle, then $N \sim \mathrm{binom}(n, p)$. Let $\hat{p} = N/n$ be our estimator for p, then $\mathrm{Var}\, \hat{p} = p(1-p)/n$.

Suppose now that instead of throwing a single needle, you throw a pair of needles, fixed at right angles at their centres, to form a cross. For convenience we will imagine that the cross is composed of a red and a black needle. Let N_R and N_B be the total number of intersections in n throws of the cross, for the red and black needle, respectively. Clearly N_R and N_B each have the same distribution as N, namely $\mathrm{binom}(n, p)$. It should also be clear that N_R and N_B are negatively correlated: if the red needle lies roughly parallel to the lines on the table then it is unlikely to intersect any, but the black needle will be nearly perpendicular to the lines on the table and thus very likely to intersect one.

If we put $\hat{p}_R = N_R/n$ and $\hat{p}_B = N_B/n$, then we use the average $\hat{p}_C = (N_R + N_B)/2n$ to estimate p. Note that $\hat{p}_C$ uses n tosses of the cross, but, because N_R and N_B are antithetic, it will have smaller variance than $\hat{p}$ based on $2n$ tosses of a needle.

20.1.2 General antithetic variate technique

Say we wish to estimate $\theta = \mathbb{E}(Z)$, with $\mathrm{Var}\,(Z) = \sigma^2$.

We can estimate θ using the average of $2n$ independent observations on Z. That is, using $\hat{\theta}_1 = \sum_{i=1}^{2n} Z_i/2n$. We have $\mathrm{Var}\,(\hat{\theta}_1) = \sigma^2/2n$.

Alternatively, suppose we can generate n independent pairs of observations

(X_i, Y_i), where X_i and Y_i have the same distribution as Z, but are negatively correlated. We estimate θ using the unbiased estimate $\hat{\theta}_a = (\overline{X} + \overline{Y})/2$, which has variance

$$\begin{aligned} \text{Var}\,(\hat{\theta}_a) &= \frac{1}{4}(\text{Var}\,(\overline{X}) + \text{Var}\,(\overline{Y}) + 2\text{Cov}(\overline{X}, \overline{Y})) \\ &= \frac{\sigma^2}{2n} + \frac{1}{2n}\text{Cov}(X_i, Y_i) \\ &= \frac{\sigma^2}{2n}\left(1 + \rho(X_i, Y_i)\right), \end{aligned}$$

where $\rho(X_i, Y_i)$ is the correlation of X_i and Y_i. Thus we have a $-100\rho\%$ variance reduction compared to $\hat{\theta}_1$, attributable to the negative covariance.

20.1.3 *Example: improved Monte-Carlo integration*

Consider $\theta = \int_0^1 g(u)du$ where g is an increasing function on $[0, 1]$. Our usual improved Monte-Carlo estimate, based on $2n$ observations, is

$$\hat{\theta}_1 = \sum_{i=1}^{2n} g(U_i)$$

where $U_1, \ldots, U_{2n}$ are iid $U(0, 1)$.

Put $(X_i, Y_i) = (g(U_i), g(1 - U_i))$ then, using the fact that $1 - U_i \sim U(0, 1)$, X_i and Y_i have the same distribution. Thus $\mathbb{E}(X_i) = \mathbb{E}(Y_i) = \theta$ and we can form the antithetic estimate:

$$\hat{\theta}_a = \frac{1}{2n}\left(\sum_{i=1}^{n} g(U_i) + \sum_{i=1}^{n} g(1 - U_i)\right).$$

For this to be useful we need that $\text{Cov}(X_i, Y_i)$ is negative. Note that $\hat{\theta}_a$ uses only n uniforms, $U_1, \ldots, U_n$, but requires $2n$ calls to the function g. That is, $\hat{\theta}_1$ and $\hat{\theta}_a$ make the same number of function calls and thus should take approximately the same amount of time to run.

Since g is increasing we can find u^* in $[0, 1]$ so that $g(1 - u) > \theta$ if $u < u^*$ and $g(1 - u) < \theta$ if $u > u^*$. Thus

$$\begin{aligned} \text{Cov}(X_i, Y_i) &= \text{Cov}(g(U_i), g(1 - U_i)) \\ &= \mathbb{E}(g(U_i) - \theta)(g(1 - U_i) - \theta) \\ &= \mathbb{E}g(U_i)(g(1 - U_i) - \theta) \\ &< g(u^*)\int_0^{u^*}(g(1 - u) - \theta)du + g(u^*)\int_{u^*}^{1}(g(1 - u) - \theta)du \\ &= 0. \end{aligned}$$

That is, X_i and Y_i are antithetic, as required. By symmetry the same result also holds for g decreasing.

The following R program enables you to compare the variances of $\hat{\theta}_1$ and $\hat{\theta}_a$ empirically, for a function of your choice (we initially consider $g(x) = 1 - x^2$). We put $n = 50$, then calculate each estimator $N = 5000$ times and form the sample variance.

Note that we have written vectorised code, which eschews explicit loops for computational efficiency. The `colMeans` function computes the means of each column of a matrix.

```
> g <- function(x) 1 - x^2
> N <- 5000
> n <- 50
> u_1 <- matrix(runif(2 * n * N), ncol = N)
> theta_1 <- colMeans(g(u_1))
> u_a <- matrix(runif(n * N), ncol = N)
> theta_a <- 0.5 * (colMeans(g(u_a)) + colMeans(g(1 - u_a)))
> var1 <- var(theta_1)
> vara <- var(theta_a)
> reduction <- 100 * (var1 - vara)/var1
> cat("Variance theta_1 is", var1, "\n")

Variance theta_1 is 0.0009080722

> cat("Variance theta_a is", vara, "\n")

Variance theta_a is 0.0001112241

> cat("Variance reduction is", reduction, "percent \n")

Variance reduction is 87.75163 percent
```

We can show theoretically that $\mathrm{Var}(\hat{\theta}_1) = 2/45n$ and $\mathrm{Var}(\hat{\theta}_a) = 1/180n$, corresponding to an 87.5% variance reduction.

There is an interesting geometric interpretation of the antithetic estimator in this case. We note that $\int_0^1 g(x)dx = \int_0^1 g(1-x)dx$, so we can rewrite θ as

$$\theta = \int_0^1 \frac{1}{2}\left(g(x) + g(1-x)\right) dx.$$

Write the improved Monte-Carlo estimator for θ in this form and you will see it is identical to $\hat{\theta}_a$. So the antithetic approach in this case is equivalent to replacing g by h where $h(x) = (g(x) + g(1-x))/2$. The function h averages $g(x)$ with its mirror image around $x = 1/2$. If h is less variable than g, this results in a variance reduction. If g is already symmetric around $1/2$, there is no reduction. If h is constant then the estimator is constant with 100% variance reduction. (Try $g(x) = \sin(\pi x)$ or $g(x) = 1 - x$ in the program.)

20.1.4 Antithetic pairs through inversion

In general, to generate antithetic pairs (X_i, Y_i), $i = 1, \ldots, n$, where X_i and Y_i have the same cdf F but are negatively correlated, we can use the inverse transformation method. For $U_i \sim U(0,1)$ we set $X_i = F^{-1}(U_i)$ and $Y_i = F^{-1}(1 - U_i)$, so both have cdf F, and $\mathrm{Cov}(X_i, Y_i) < 0$ follows immediately from the the previous section on putting $g = F^{-1}$.

20.2 Importance sampling

We will motivate importance sampling by introducing a generalisation of the improved Monte-Carlo integration method considered in Section 19.2.

Previously we observed that if $\theta = \int_a^b \phi(x)\frac{1}{b-a}dx$, then $\theta = \mathbb{E}\phi(U)$, where $U \sim U(a,b)$. We generalise this by permitting a distribution other than the uniform. That is, suppose that we wish to estimate $\theta = \int_a^b \phi(x)f(x)dx$, where f is a pdf with support $[a, b]$ (possibly infinite), then we have

$$\theta = \int_a^b \phi(x)f(x)dx = \mathbb{E}\phi(X) \text{ where } X \text{ has pdf } f.$$

Thus if $X_1, \ldots, X_n$ is a random sample from f, then an unbiased estimator of θ is

$$\hat{\theta} = \frac{1}{n}\sum_{i=1}^{n} \phi(X_i).$$

The variance of $\hat{\theta}$ is

$$\mathrm{Var}\,\hat{\theta} = \frac{1}{n}\left(\int_a^b \phi(x)^2 f(x)dx - \theta^2\right),$$

which cannot be calculated exactly unless we already know θ.

This more general formulation of Monte-Carlo integration allows us to evaluate the integral of a function h by thinking of it as the product ϕf, for a suitably chosen pdf f. The choice of f can make a big difference to $\mathrm{Var}\,\hat{\theta}$, as the next example shows.

20.2.1 Example: evaluating a simple integral three different ways

Assume we wish to evaluate the integral

$$\theta = \int_0^1 (1 - x^2)dx = \frac{2}{3}$$

(whose value we luckily already know) using a Monte-Carlo method.

Method 1: Our improved Monte-Carlo method uses uniformly distributed points over the interval $[0, 1]$. We obtain the estimator

$$\hat{\theta}_1 = \frac{1}{n}\sum_{i=1}^{n}\phi(X_i) = \frac{1}{n}\sum_{i=1}^{n}(1 - X_i^2)$$

where the X_i are $U(0, 1)$. A simple calculation shows $\text{Var}\,(\hat{\theta}_1) = 4/45n$.

To apply the generalised Monte-Carlo method we ask: 'Are all points in $[0, 1]$ equally important in evaluating the integral?' The function is larger near zero so this region makes a proportionately larger contribution to the area that we want to estimate. Hence we speculate whether we can reduce the variability of our Monte-Carlo estimator by choosing the distribution of our random evaluation points to somehow match the shape of the integrand.

Method 2: Our first generalised Monte-Carlo estimator reformulates the integral as

$$\theta = \int_0^1 \frac{2}{3}\frac{3}{2}(1 - x^2)dx$$

which corresponds to setting $\phi(x) = \frac{2}{3}$ and $f(x) = \frac{3}{2}(1 - x^2)$, which is a pdf over $[0, 1]$. We then have

$$\hat{\theta}_2 = \frac{1}{n}\sum_{i=1}^{n}\phi(X_i) = \frac{2}{3},$$

where $X_1, \ldots, X_n$ is an iid sample from f.

In this case we have made the distribution of our random points exactly match the shape of the function being integrated. And our estimator is equal to θ exactly with no variability at all! ($\text{Var}\,\hat{\theta}_2 = 0$.) Of course there is no real gain here, as we needed to know θ to verify that our choice of f was a proper density, but this illustrates our objective. Since in real cases of interest θ is unknown, we simply try as best we can to match the function with a known pdf. In this example let us consider a simple triangular distribution for f, to give some emphasis to the higher values near zero.

Method 3: Our second general Monte-Carlo estimator reformulates the integral as

$$\theta = \int_0^1 (1 - x^2)dx = \int_0^1 \frac{1}{2}(1 + x)2(1 - x)dx$$

which corresponds to setting $\phi(x) = \frac{1}{2}(1 + x)$ and $f(x) = 2(1 - x)$, which is a pdf over $[0, 1]$. We then have

$$\hat{\theta}_3 = \frac{1}{n}\sum_{i=1}^{n}\frac{1}{2}(1 + X_i)$$

where the X_i, $i = 1, \ldots, n$, are an iid sample with the new f distribution. A simple calculation yields $\text{Var}\,(\hat{\theta}_3) = 1/72n$. Hence the variability of the original improved Monte-Carlo estimator has been reduced by a factor of 6.4.

To put this example into the general framework of *importance sampling*, consider again the expectation

$$\theta = \int \phi(x) f(x) dx = \mathbb{E}\phi(X),$$

where X has pdf f. Now imagine choosing a density g that shadows ϕf as closely as possible. If Y has pdf g then we have

$$\begin{aligned} \mathbb{E}\phi(X) &= \int \phi(x) f(x) dx \\ &= \int \frac{\phi(x) f(x)}{g(x)} g(x) dx \\ &= \int \psi(x) g(x) dx \\ &= \mathbb{E}\psi(Y), \end{aligned}$$

where $\psi(x) = \phi(x) f(x)/g(x) = w(x)\phi(x)$.

Let $Y_1, \ldots, Y_n$ be an iid sample with pdf g, then our importance sampling estimator for $\mathbb{E}\psi(Y)$ is

$$\hat{\theta}_g = \frac{1}{n} \sum_{i=1}^{n} \psi(Y_i) = \frac{1}{n} \sum_{i=1}^{n} w(Y_i)\phi(Y_i),$$

where $w(x) = f(x)/g(x)$. This estimator can be thought of as a weighted version of the original improved Monte-Carlo estimator, where the weights compensate for the fact that we are sampling from g rather than f. We have that $\hat{\theta}_g$ is an unbiased estimator with

$$\text{Var}\,(\hat{\theta}_g) = \frac{1}{n} \text{Var}\,\psi(Y_1).$$

Clearly the better g shadows ϕf then the closer ψ is to a constant, and the greater the variance reduction achieved.

20.2.2 Example: standard normal tail probability

Suppose we want to estimate the probability

$$\theta = \mathbb{P}(Z > 2) = \int_2^{\infty} f(x) dx = 0.02275$$

where Z has a standard normal distribution with pdf $f(x) = e^{-x^2/2}/\sqrt{2\pi}$.

Method 1: We think of the integral as

$$\theta = \int_2^{\infty} f(x) dx = \int_{-\infty}^{\infty} \phi(x) f(x) dx$$

where $\phi(x)$ equals 1 if $x > 2$ and 0 otherwise. Then, if $X_1, \ldots, X_n$ are an iid sample from f, we use

$$\hat{\theta}_1 = \frac{1}{n}\sum_{i=1}^{n}\phi(X_i) = \frac{N}{n}$$

where $N \sim \text{binom}(n, \theta)$. So we are simply estimating the probability by generating normal observations and counting the proportion greater than 2. $\text{Var}(\hat{\theta}_1) = \theta(1-\theta)/n = 0.0223/n$.

Method 2: To benefit from importance sampling we need to choose a density that is similar to the tail of the standard normal above two and zero below two. One standard method is to use a shifted version of the original density, relocated towards the important values. In this case we choose a half-normal density shifted to start at two, that is

$$g(x) = \begin{cases} 0 & \text{if } x < 2; \\ \sqrt{\frac{2}{\pi}}e^{-(x-2)^2/2} & \text{if } x > 2. \end{cases}$$

If X has pdf g, we obtain

$$\begin{aligned} \theta &= \mathbb{E}\left(\frac{f(X)}{g(X)}\right) = \frac{e^2}{2}\mathbb{E}(e^{-2X}), \text{ and} \\ \hat{\theta}_g &= \frac{1}{n}\sum_{i=1}^{n}\frac{e^2}{2}e^{-2X_i}, \end{aligned}$$

where the X_i are an iid sample from g. By construction $\hat{\theta}_g$ is unbiased and

$$\text{Var}(\hat{\theta}_g) = \frac{1}{n}\text{Var}\left(\frac{1}{2}e^2 e^{-2X_1}\right) = \frac{1}{n}\left(\frac{e^4}{4}\mathbb{E}(e^{-4X_1}) - \theta^2\right).$$

A simple integration shows $\mathbb{E}(e^{-4X_1}) = 2\mathbb{P}(Z > 4)$ so $\text{Var}(\hat{\theta}_g) = 0.000347/n$. In this case the variance has been reduced by a factor of around 64. One intuitive way of understanding this variance reduction is that $\hat{\theta}_g$ uses more information from each X_i than $\hat{\theta}_1$, which only notes whether the value is greater or smaller than 2.

20.2.3 Example: standard normal central probability

Let $Z \sim N(0, 1)$ and consider

$$\theta = \int_0^1 e^{-x^2/2}dx = \sqrt{2\pi}\mathbb{P}(0 < Z < 1).$$

A simple estimator for θ is $\hat{\theta}_1 = \sqrt{2\pi}\hat{p}$, where $\hat{p}$ is the proportion of n iid standard normals in $(0, 1)$. We have $\hat{p} \sim \text{binom}(n, \theta/\sqrt{2\pi})/n$, so $\hat{\theta}_1$ has variance $\theta(\sqrt{2\pi} - \theta)/n = 1.413/n$.

To apply importance sampling, note that by using a second-order Taylor expansion of $e^{x^2/2}$ about 0, we have

$$e^{-x^2/2} = \frac{1}{e^{x^2/2}} \approx \frac{1}{1+x^2/2} = h(x).$$

Thus we choose our importance sampling density $g \propto h(x)$. This is a truncated form of a non-standard Cauchy density. Since

$$\int_0^x \frac{1}{1+u^2/2} du = \sqrt{2} \arctan\left(\frac{x}{2}\right)$$

our density g has cdf

$$G(x) = \frac{\arctan(x/\sqrt{2})}{\arctan(1/\sqrt{2})} \text{ for } x \in (0,1).$$

We can generate observations on g using the inverse transformation method with $G^{-1}(u) = \sqrt{2}\tan(u \arctan(1/\sqrt{2}))$.

Our importance sampling estimator is

$$\hat{\theta}_g = \frac{1}{n} \sum_{i=1}^{n} e^{-X_i^2/2} \sqrt{2} \arctan\left(\frac{1}{\sqrt{2}}\right) \left(1 + \frac{X_i^2}{2}\right)$$

where $X_i = G^{-1}(U_i)$ and $U_1, \ldots, U_n$ are iid $U(0,1)$.

The following (vectorised) R code estimates the variance reduction using importance sampling with density g. We calculate $\hat{\theta}_1$ and $\hat{\theta}_g$ N times, each time with a sample size of n.

```
> Ginv <- function(u) {
+     sqrt(2) * tan(u * atan(1/sqrt(2)))
+ }
> Psi <- function(x) {
+     exp(-(x^2)/2) * sqrt(2) * atan(1/sqrt(2)) * (1 + (x^2)/2)
+ }
> N <- 10000
> n <- 50
> u_a <- matrix(runif(n * N), ncol = N)
> theta_a <- colMeans(Psi(Ginv(u_a)))
> var1 <- 1.413/n
> vara <- var(theta_a)
> reduction <- 100 * (var1 - vara)/var1
> cat("Variance theta_1 is", var1, "\n")

Variance theta_1 is 0.02826

> cat("Variance theta_a is", vara, "\n")

Variance theta_a is 8.433276e-06

> cat("Variance reduction is", reduction, "% \n")
```

```
Variance reduction is 99.97016 %
```

In this example importance sampling reduces the variance by a factor of approximately 3400.

20.3 Control variates

Like antithetic variates, control variates take advantage of covariance. The difference is that control variates obtain a variance reduction from positive covariance rather than negative covariance. The basic idea is to use one variable Y with known mean μ to control another variable X with unknown mean θ. Suppose that $\text{Cov}(X, Y) > 0$, then we define the 'controlled' version of X to be

$$X^* = X - \alpha(Y - \mu)$$

where $\alpha > 0$ is some constant. Clearly $\mathbb{E}(X^*) = \theta$ so X^* is an unbiased estimator of θ, and

$$\begin{aligned} \text{Var}\,(X^*) &= \text{Var}\,(X) + \alpha^2\text{Var}\,(Y) - 2\alpha\text{Cov}(X, Y - \mu) \\ &= \text{Var}\,(X) - \alpha(2\text{Cov}(X, Y) - \alpha\text{Var}\,(Y)). \end{aligned}$$

We have $\text{Var}\,X^* < \text{Var}\,X$ if and only if $2\text{Cov}(X, Y) - \alpha\text{Var}\,(Y) > 0$, or equivalently $0 < \alpha < 2\text{Cov}(X, Y)/\text{Var}\,(Y)$.

As $f(\alpha) = \text{Var}\,(X^*)$ is a parabola, it is minimised by choosing α such that $f'(\alpha) = 0$. That is

$$\alpha = \alpha^* = \frac{\text{Cov}(X, Y)}{\text{Var}\,(Y)}.$$

Hence the minimum value of $\text{Var}\,(X^*)$ is

$$\begin{aligned} f(\alpha^*) &= \text{Var}\,(X) - \frac{\text{Cov}(X, Y)^2}{\text{Var}\,(X)\text{Var}\,(Y)}\text{Var}\,(X) \\ &= \text{Var}\,(X)(1 - \rho^2) \end{aligned}$$

where $\rho = \frac{\text{Cov}(X,Y)}{\sqrt{\text{Var}\,(X)\text{Var}\,(Y)}}$ is the correlation coefficient (which may be familiar to the reader as the residual variability in a linear regression of X on Y). The resulting variance reduction is $100\rho^2\%$.

20.3.1 Example: standard normal central probability

We revisit Example 20.2.3. We are estimating

$$\theta = \int_0^1 e^{-x^2/2} dx$$

Let $X = \hat{\theta}$ be the previously derived importance sampling estimator. That is

$$X = \hat{\theta} = \frac{1}{n}\sum_{i=1}^{n} e^{-T_i^2/2}\sqrt{2}\arctan\left(\frac{1}{\sqrt{2}}\right)\left(1+\frac{T_i^2}{2}\right)$$
$$= \frac{1}{n}\sum_{i=1}^{n}\psi_1(T_i), \text{ say,}$$

where the T_i, $i = 1, \ldots, n$, are an iid sample with density g as before. A second-order Taylor expansion of $e^{-x^2/2}$ near 0 gives $e^{-x^2/2} \approx 1 - x^2/2$. Accordingly we define

$$\mu = \int_0^1 \left(1 - \frac{x^2}{2}\right) dx = \frac{5}{6},$$

and choose as our control variate Y the estimator $\hat{\mu}$ for μ based on the same importance sampling distribution g as used to estimate θ. That is

$$Y = \hat{\mu} = \frac{1}{n}\sum_{i=1}^{n}\left(1-\frac{T_i^2}{2}\right)\sqrt{2}\arctan\left(\frac{1}{\sqrt{2}}\right)\left(1+\frac{T_i^2}{2}\right)$$
$$= \frac{1}{n}\sum_{i=1}^{n}\psi_2(T_i), \text{ say.}$$

By construction $\mathbb{E}Y = \mu$, and we can ensure a positive correlation between X and Y by using the same T_i to generate them. For $\alpha > 0$ we form X^*, the controlled version of X, as

$$X^* = \hat{\theta}_c = X - \alpha\left(Y - \frac{5}{6}\right)$$
$$= \hat{\theta} - \alpha\left(\hat{\mu} - \frac{5}{6}\right).$$

Note this equation makes it clear that the control variate estimator $\hat{\theta}_c$ is the original estimator $\hat{\theta}$ plus a correction term. For example if the control variable $\hat{\mu}$ exceeds its known mean, the positive correlation would suggest that $\hat{\theta}$ might also be high, so the estimate is corrected down.

Note also that the optimal choice of α, namely $\mathrm{Cov}(X, Y)/\mathrm{Var}\, Y = \mathrm{Cov}(\psi_1(T_1), \psi_2(T_1))/\mathrm{Var}\,\psi_2(T_1)$, is not known, but we can estimate it using a sample covariance and variance based on the simulation, as is done in the following vectorised R code. Note the use of the `colSums` function, analogous to the `colMeans` function.

```
> Ginv <- function(u){
+   sqrt(2)*tan(u*atan(1/sqrt(2)))
+ }
> psi1 <- function(x){
+   exp(-(x^2)/2)*sqrt(2)*atan(1/sqrt(2))*(1+(x^2)/2)
+ }
> psi2 <- function(x){
```

```
+    (1-(x^2)/2)*sqrt(2)*atan(1/sqrt(2))*(1+(x^2)/2)
+ }
> N <- 10000  # Number of estimates of each type
> n <- 50     # Sample size
> commonG <- matrix(Ginv(runif(n*N)), ncol=N)
> p1g <- psi1(commonG)
> p2g <- psi2(commonG)
> theta_hat <- colMeans(p1g)
> mu_hat <- colMeans(p2g)
> samplecov <- colSums((p1g - theta_hat) * (p2g - 5/6))/n
> samplevar <- colSums((p2g - 5/6)^2)/n
> alphastar <- samplecov/samplevar
> theta_hat_c <- theta_hat - alphastar*(mu_hat - 5/6)
> var1 <- var(theta_hat)
> varc <- var(theta_hat_c)
> reduction<-100*(var1-varc)/var1
> cat("Variance theta_hat is", var1, "\n")

Variance theta_hat is 8.22984e-06

> cat("Variance theta_hat_c is", varc, "\n")

Variance theta_hat_c is 3.069343e-08

> cat("Variance reduction is", reduction, "percent \n")

Variance reduction is 99.62705 percent
```

We find the controlled version is significantly less variable. Relative to the naive approach of looking at the proportion of standard normals in $(0, 1)$, the combination of control variates and importance sampling has reduced the variance by a factor of order 10^6.

20.4 Exercises

1. Write a program to calculate the Monte-Carlo integral of a function `ftn(x)`, using antithetic sampling, then use it to estimate

$$B(z, w) = \int_0^1 x^{z-1}(1-x)^{w-1}\,dx, \text{ for } z = 0.5, w = 2.$$

   $B(z, w)$ is called the beta function, and is finite for all $z, w > 0$.

2. Suppose that X has a continuous cdf F, and that F^{-1} is known. Let $U_1, \ldots, U_n$ be iid $U(0, 1)$ rv's and put $X_i = F^{-1}(U_i)$, then we can estimate $\mu = \mathbb{E}X$ and $\sigma^2 = \operatorname{Var} X$ using $\overline{X} = n^{-1}\sum_i X_i$ and $S^2 = (n-1)^{-1}\sum_i (X_i - \overline{X})^2$, respectively.

   (a). Show that if $U \sim U(0, 1)$, then

$$\operatorname{Cov}(F^{-1}(U), F^{-1}(1-U)) \leq 0.$$

(b). Show how to use antithetic sampling to improve our estimate of μ.

(c). Suppose that the distribution of X is symmetric about μ. Show that antithetic sampling will *not* improve our estimate of σ^2.

3. Consider the integral

$$I = \int_0^1 \sqrt{1-x^2}\,dx.$$

(a). Estimate I using Monte-Carlo integration.

(b). Estimate I using antithetic sampling, and compute an estimate of the percentage variance reduction achieved by using the antithetic approach.

(c). Approximate the integrand by a straight line and use a control variate approach to estimate the value of the integral. Estimate the resulting variance reduction achieved.

(d). Use importance sampling to estimate the integral I. Try using three different importance sampling densities, and compare their effectiveness.

4. Suppose that the rv X has mean μ and can be simulated. Further, suppose that f is a non-linear function, and that we wish to estimate $a = \mathbb{E}f(X)$ using simulation.

Using $g(x) = f(\mu) + (x-\mu)f'(\mu)$ and tuning parameter $\alpha = 1$, estimate a using control variates. That is, if $X_1, \ldots, X_n$ are an iid sample distributed as X, show that for $\alpha = 1$, the controlled estimate of a is

$$\frac{1}{n}\sum_{i=1}^{n} f(X_i) - (\overline{X} - \mu)f'(\mu).$$

Furthermore, using the fact that for x close to μ, $g(x) \approx f(x)$, show that the controlled estimate can be written approximately as

$$\frac{1}{n}\sum_{i=1}^{n} f(X_i) - f(\overline{X}) + f(\mu).$$

Finally, derive the optimal (theoretical) value of α.

5. Daily demand for a newspaper is approximately gamma distributed, with mean 10,000 and variance 1,000,000. The newspaper prints and distributes 11,000 copies each day. The profit on each newspaper sold is \$1, and the loss on each unsold newspaper is \$0.25.

(a). Express the expected daily profit as a finite integral, then estimate it using both Simpson's method and Monte-Carlo integration.

(b). Improve your Monte-Carlo estimate using importance sampling and/or a control variate.

(c). For m integer valued, a $\Gamma(\lambda, m)$ rv can be written as the sum of m iid $\exp(\lambda)$ rv's. Using this approach to simulate gamma rv's, estimate the expected daily profit using antithetic sampling.

6. Consider estimating $I = \int_0^1 g(x)dx$ by improved Monte-Carlo integration. We showed in Section 20.1.3, that using antithetic variates is equivalent to replacing g by $h(x) = (g(x) + g(1-x))/2$, which averages g with its mirror image around $x = 1/2$. Further variance reduction may be possible by iterating this process on subintervals, as illustrated below.

(a). Let $g(x) = x^4$. Write an R program to calculate the improved Monte-Carlo estimator $\hat{I}$ of I, and to estimate its variance.

(b). Repeat (a) using antithetic variates, and compute the variance reduction achieved.

(c). Using the fact that $h(x) = h(1-x)$, verify that

$$I = \int_0^{1/2} (g(x) + g(1-x))dx.$$

Then verify that over this subinterval you can again replace the integrand by a function which averages its value with the value of its mirror image around $x = 1/4$. Hence verify that

$$I = \int_0^{1/4} (g(x) + g(1-x) + g((1/2) - x) + g((1/2) + x))dx$$

Use this to estimate I and calculate the resulting variance reduction.

CHAPTER 21

Case studies

21.1 Introduction

In this chapter we present three case studies: extended examples intended to demonstrate some of our simulation techniques. Simulation is ubiquitous in science, so trying to list all the areas where it appears would be an endless task. To give you a taste, here are some (but not all) of the areas where simulation is being employed in the authors' home department.

- Spin systems: big lattices of interacting molecules.
- Granular materials: how do grains of dirt move about when you put a weight (like a building) on them?
- Molecular geometry: the shape of complex molecules has an important effect on how they act.
- Stock markets: how should we value financial instruments such as bonds, options, etc?
- Health care: modelling and then optimising patient care.
- Telecommunications: optimal design of communication networks.
- Carbon modelling: where is all the carbon, and how will it affect global warming?
- Forest management: where, when, and what should I plant?

The development of new simulation techniques is a scientific field in its own right. As computer power increases, numerical simulation and optimisation techniques become more sophisticated and more widely applicable. Here is a list (not exhaustive) of some of the simulation topics we have not been able to cover in this book. The interested reader is encouraged to explore!

- Stochastic processes: simulating and analysing systems that evolve over time. That is, instead of having independent samples, our random variables are *dependent*. Discrete event simulation is one of the most important methodologies in this area.
- Markov Chain Monte-Carlo: the simulation technique that underpins modern Bayesian statistics.

- Stochastic optimisation: using a stochastic (random) process to optimise a function. Techniques include simulated annealing, genetic algorithms, cross-entropy, ant-heaps, and many more.
- Bootstrapping: a very clever statistical technique for extracting information from a sample by resampling.
- Meta-modelling: using a simpler but faster simulation to approximate a complex but slow simulation.
- Perfect simulation: how to reach an asymptotic limit in a finite amount of time.

21.2 Epidemics

The science of epidemiology, the study of the spread of disease, includes mathematical/statistical models of how disease spreads. In this section we look at some of these models and investigate their behaviour using simulation.

21.2.1 SIR model

SIR stands for Susceptible, Infected, and Removed. In this model we suppose that individuals can be one of three types: susceptible if they have not yet caught the disease, infected if they currently have the disease, and removed if they have had the disease and have since recovered (and are now immune) or died. In our following descriptions, we will use the type labels—susceptible, infected, and removed—as shorthand to describe individuals of that type. We measure time in discrete steps. At each time step, each infected can infect susceptibles or can recover/die, at which point the infected is removed.

Let $S(t)$, $I(t)$ and $R(t)$ be the number of susceptible, infected and removed individuals at time t. At each time step each infected has probability α of infecting each susceptible. (This assumes that each infected has equal contact with all susceptibles. This is called a *mixing* assumption.) At the end of each time step, after having had a chance to infect people, each infected has probability β of being removed.

We take initial conditions

$$\begin{aligned} S(0) &= N; \\ I(0) &= 1; \\ R(0) &= 0. \end{aligned}$$

Note that the total population is $N+1$ and this remains fixed. That is $S(t) + I(t) + R(t) = N + 1$ for all t.

Each time step t the chance that a susceptible remains uninfected is $(1-\alpha)^{I(t)}$.

That is, each infected must fail to pass on the infection to the susceptible. Thus,

$$S(t+1) \sim \text{binom}(S(t), (1-\alpha)^{I(t)}).$$

As each infected has a chance $\boldsymbol{\beta}$ of being removed, we have

$$R(t+1) \sim R(t) + \text{binom}(I(t), \boldsymbol{\beta}).$$

Given $S(t+1)$ and $R(t+1)$ we get $I(t+1)$ from the total population

$$I(t+1) = N + 1 - R(t+1) - S(t+1).$$

These rules are enough to write a simulation of an SIR process.

```
# program spuRs/resources/scripts/SIRsim.r

SIRsim <- function(a, b, N, T) {
  # Simulate an SIR epidemic
  # a is infection rate, b is removal rate
  # N initial susceptibles, 1 initial infected, simulation length T
  # returns a matrix size (T+1)*3 with columns S, I, R respectively
  S <- rep(0, T+1)
  I <- rep(0, T+1)
  R <- rep(0, T+1)
  S[1] <- N
  I[1] <- 1
  R[1] <- 0
  for (i in 1:T) {
    S[i+1] <- rbinom(1, S[i], (1 - a)^I[i])
    R[i+1] <- R[i] + rbinom(1, I[i], b)
    I[i+1] <- N + 1 - R[i+1] - S[i+1]
  }
  return(matrix(c(S, I, R), ncol = 3))
}
```

In Figure 21.1 we plot $S(t)$, $I(t)$, and $R(t)$ for four separate simulations, with $\alpha = 0.0005$ and $\boldsymbol{\beta} = 0.1, 0.2, 0.3$, and 0.4. We see that as $\boldsymbol{\beta}$ increases, the size of the epidemic decreases.

To see the range of behaviour possible for a single choice of α and $\boldsymbol{\beta}$, we plot several realisations of the simulation on the same graph: see Figure 21.2. We see that epidemics either die out quickly or else grow to be quite large.

It would be nice to know exactly how α and $\boldsymbol{\beta}$ affect the size of an epidemic. Using simulation we can estimate $\mathbb{E}S(T)$ for different values of α and $\boldsymbol{\beta}$ and see how it varies. The following program does this for $\alpha \in [0.0001, 0.001]$ and $\beta \in [0.1, 0.5]$ and plots the results on a 3D-graph. (See Section 7.7 for guidance on 3D-plotting.) The output is given in Figure 21.3.

```
# program spuRs/resources/scripts/SIR_grid.r
# discrete SIR epidemic model
#
```

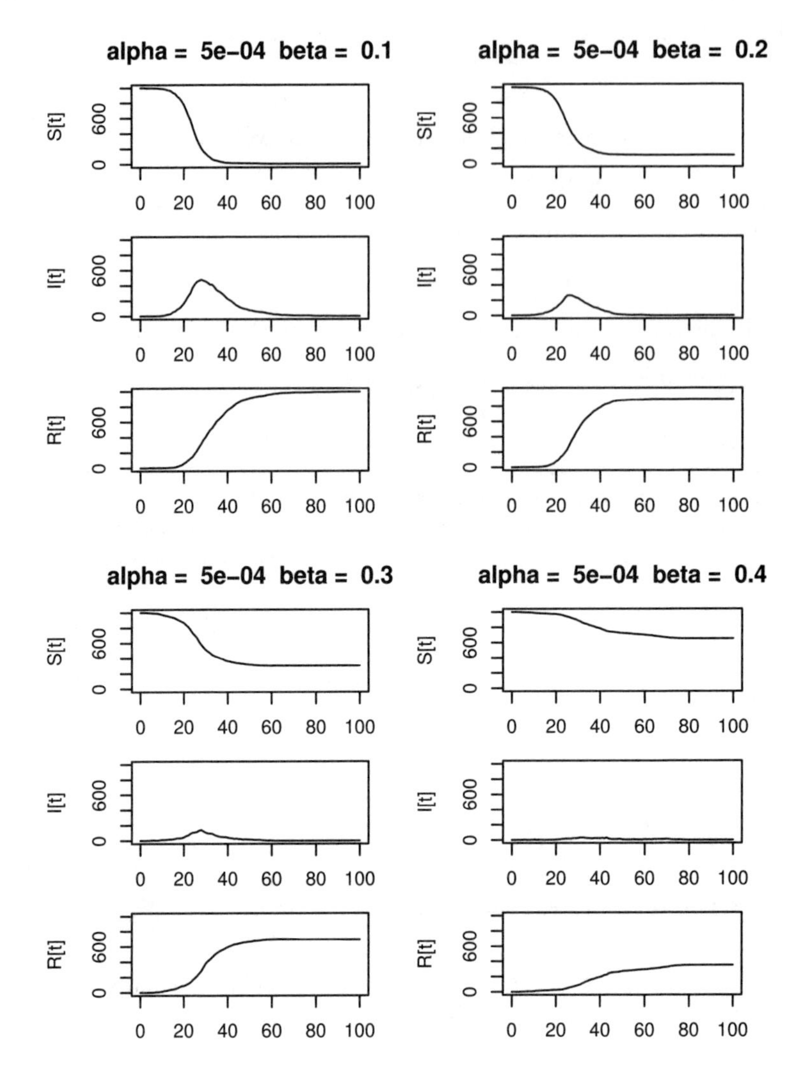

Figure 21.1 *Simulations of an SIR epidemic with* $\alpha = 0.0005$ *and* $\beta = 0.1, 0.2, 0.3,$ *and* 0.4.

```
# initial susceptible population N
# initial infected population 1
# infection probability a
# removal probability b
#
# estimates expected final population size for different values of
# the infection probability a and removal probability b
# we observe a change in behaviour about the line Na = b
```

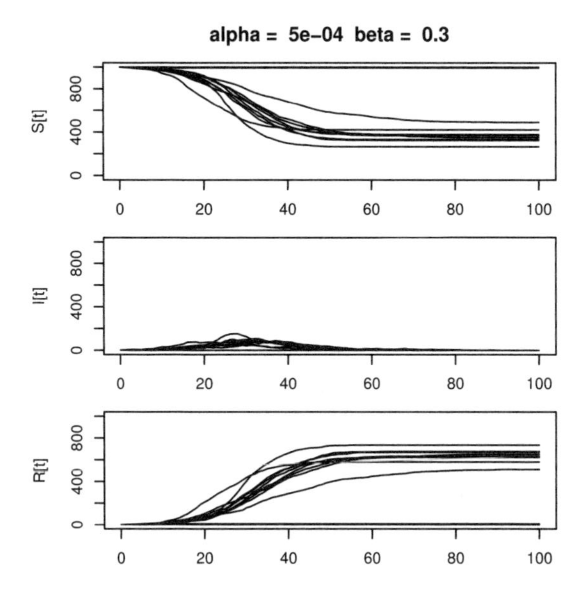

Figure 21.2 *Twenty realisations of an SIR epidemic with* $\alpha = 0.0005$ *and* $\boldsymbol{\beta} = 0.3$.

```
# (Na is the expected number of new infected at time 1 and
# b is the expected number of infected who are removed at time 1)

SIR <- function(a, b, N, T) {
  # simulates SIR epidemic model from time 0 to T
  # returns number of susceptibles, infected and removed at time T
  S <- N
  I <- 1
  R <- 0
  for (i in 1:T) {
    S <- rbinom(1, S, (1 - a)^I)
    R <- R + rbinom(1, I, b)
    I <- N + 1 - S - R
  }
  return(c(S, I, R))
}

# set parameter values
N <- 1000
T <- 100
a <- seq(0.0001, 0.001, by = 0.0001)
b <- seq(0.1, 0.5, by = 0.05)

n.reps <- 400 # sample size for estimating E S[T]
f.name <- "SIR_grid.dat" # file to save simulation results
```

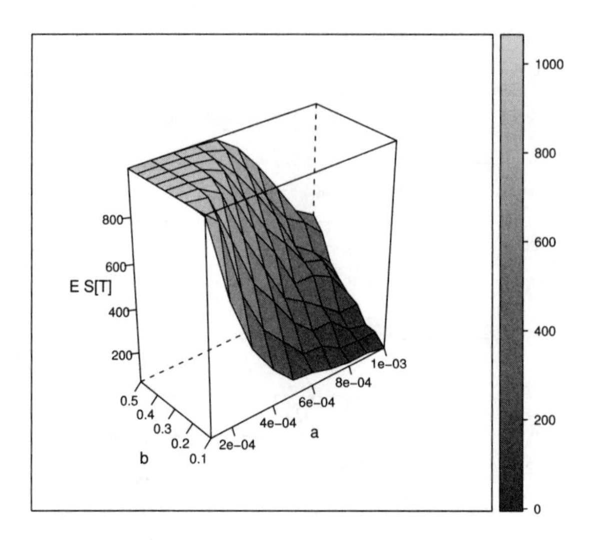

Figure 21.3 *Average epidemic size for various infection rates* α *and removal rates* $\boldsymbol{\beta}$.

```
# estimate E S[T] for each combination of a and b
write(c("a", "b", "S_T"), file = f.name, ncolumns = 3)
for (i in 1:length(a)) {
  for (j in 1:length(b)) {
    S.sum <- 0
    for (k in 1:n.reps) {
      S.sum <- S.sum + SIR(a[i], b[j], N, T)[1]
    }
    write(c(a[i], b[j], S.sum/n.reps), file = f.name,
      ncolumns = 3, append = TRUE)
  }
}

# plot estimates in 3D
g <- read.table(f.name, header = TRUE)
library(lattice)
print(wireframe(S_T ~ a*b, data = g, scales = list(arrows = FALSE),
                aspect = c(.5, 1), drape = TRUE,
                xlab = "a", ylab = "b", zlab = "E S[T]"))
```

We observe a change in behaviour about the line $N\alpha = \boldsymbol{\beta}$. $N\alpha$ is the expected number of new infected at time 1 and $\boldsymbol{\beta}$ is the expected number of infected who are removed at time 1. When $N\alpha > \beta$ then we nearly always get a big epidemic, but when $N\alpha \leq \beta$ the size of the epidemic drops away sharply.

For more insight into why this threshold occurs, we look at a class of models called *branching processes.*

21.2.2 Branching processes

An epidemic has the potential to be large if, in its early stages, $\mathbb{E}$(new infected) $>$ $\mathbb{E}$(infected removed). For a general epidemic, calculating $\mathbb{E}$(new infected) is difficult because individuals interact:

- Finite population size means infected individuals are 'competing' for individuals to infect;
- Spatial restrictions restrict contact between infected and susceptible.

The SIR model ignores spatial interactions but does model the finite population. Branching processes ignore the finite population restriction as well. This results in a simpler but hopefully still useful model. The branching process can be viewed as a model for the early stages of an epidemic.

Branching processes are typically described in terms of births and population growth rather than infection. Let Z_n be the size of the population at generation/time n. At each time step every individual independently gives birth to a random number of offspring, with distribution X, then dies. (You can include the case where the individual does not die by adding 1 to X.) Put $Z_0 = 1$ then we have

$$Z_{n+1} = X_{n,1} + \cdots + X_{n,Z_n},$$

where $X_{n,i}$ is the i-th family size in generation n. The $X_{n,i}$ are iid with the same distribution as X.

If you just look at the infected, then the first step of an SIR epidemic is the same as the first step of a branching process, with $X_{0,1} = A + B$ where $A \sim \text{binom}(N, \alpha)$ are the new infected and $B \sim \text{binom}(1, 1 - \beta)$ is 1 if the initial infected is not removed and 0 otherwise. Note that $\mathbb{E}X = N\alpha + 1 - \beta$ so the condition for an epidemic to grow, $N\alpha > \beta$, is equivalent to $\mathbb{E}X > 1$.

Here is some code for simulating and plotting a branching process. It makes use of the construct ... for passing a variable number of inputs to a function.

```
# Program spuRs/resources/scripts/bp.r
# branching process simulation

bp <- function(gen, rv.sim, ...) {
  # population of a branching process from generation 0 to gen
  # rv.sim(n, ...) simulates n rv's from the offspring distribution
  # Z[i] is population at generation i-1; Z[1] = 1
```

```
  Z <- rep(0, gen+1)
  Z[1] <- 1
  for (i in 1:gen) {
    if (Z[i] > 0) {
      Z[i+1] <- sum(rv.sim(Z[i], ...))
    }
  }
  return(Z)
}

bp.plot <- function(gen, rv.sim, ..., reps = 1, logplot = TRUE) {
  # simulates and plots the population of a branching process
  # from generation 0 to gen; rv.sim(n, ...) simulates n rv's
  # from the offspring distribution
  # the plot is repeated reps times
  # if logplot = TRUE then the population is plotted on a log scale
  # Z[i,j] is population at generation j-1 in the i-th repeat
  Z <- matrix(0, nrow = reps, ncol = gen+1)
  for (i in 1:reps) {
    Z[i,] <- bp(gen, rv.sim, ...)
  }
  if (logplot) {
    Z <- log(Z)
  }
  plot(c(0, gen), c(0, max(Z)), type = "n", xlab = "generation",
    ylab = if (logplot) "log population" else "population")
  for (i in 1:reps) {
    lines(0:gen, Z[i,])
  }
  return(invisible(Z))
}
```

Figure 21.4 gives some sample output where we took $X \sim \text{binom}(2, 0.6)$. There are 20 simulations over 20 generations. Note that in half the simulations the population has died out, in the other half it appears to be growing exponentially. The command used was `bp.plot(20, rbinom, 2, .6, 20, logplot = F)`.

What is the relationship between the offspring distribution X and the growth of the process? To investigate this question we fixed T then used simulation to estimate $\log \mathbb{E}Z_T$ for a number of different X and then plotted this against $\mu = \mathbb{E}X$. We put $T = 50$ and $X \sim \text{binom}(2, p)$ for $p \in [.3, .6]$. Here is the code we used; the output is given in Figure 21.5. Note that values of $\log(0)$ $(= -\infty)$ are not plotted.

```
# program spuRs/resources/scripts/bp_grid.r

bp.sim <- function(gen, rv.sim, ...) {
  # population of a branching process at generation gen
```

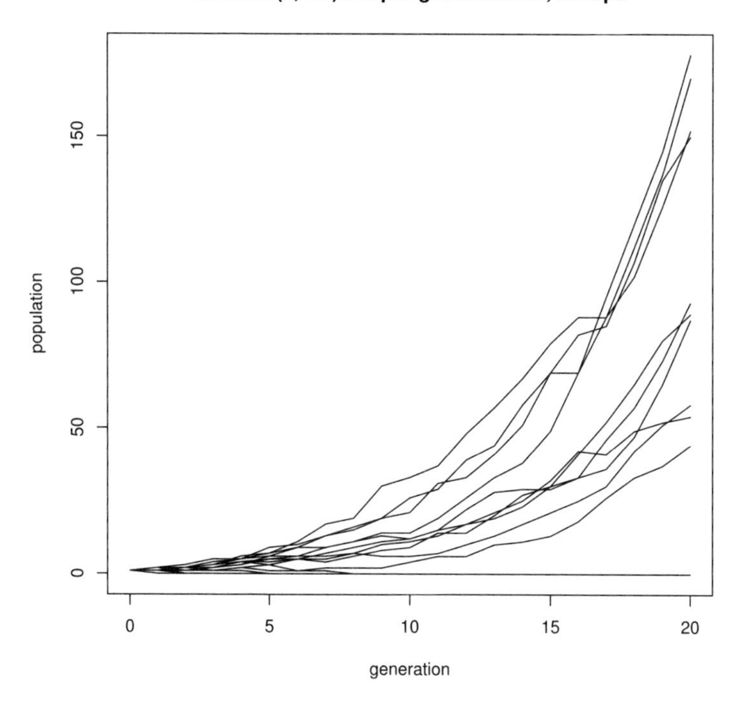

Figure 21.4 *Several realisations of a branching process.*

```
  # rv.sim(n, ...) simulates n rv's from the offspring distribution
  Z <- 1
  for (i in 1:gen) {
    if (Z > 0) {
      Z <- sum(rv.sim(Z, ...))
    }
  }
  return(Z)
}


# set parameter values
gen <- 50
size <- 2
prob <- seq(0.3, 0.6, by = 0.01)
n.reps <- 100 # sample size for estimating E Z

# estimate E Z for each value of prob
mu <- rep(0, length(prob))
Z.mean <- rep(0, length(prob))
```

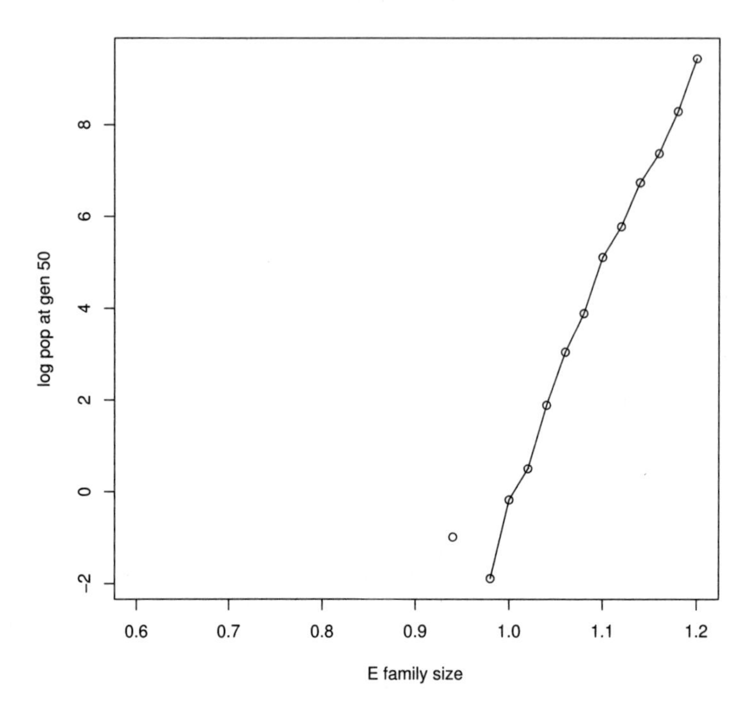

Figure 21.5 *Expected population at time T agianst the expected family size.*

```
for (i in 1:length(prob)) {
  Z.sum <- 0
  for (k in 1:n.reps) {
    Z.sum <- Z.sum + bp.sim(gen, rbinom, size, prob[i])
  }
  mu[i] <- size*prob[i]
  Z.mean[i] <- Z.sum/n.reps
}

# plot estimates
# note that values of log(0) (= -infinity) are not plotted
plot(mu, log(Z.mean), type = "o",
     xlab = "E family size", ylab = paste("log pop at gen", gen))
```

There is a quite convincing linear relationship between $\mathbb{E}X$ and $\log \mathbb{E}Z_T$, with an x-intercept at 1. That is, for some constant $c = c(T)$, we have

$$\begin{aligned} \log \mathbb{E}Z_T &\approx c(\mathbb{E}X - 1) \\ \mathbb{E}Z_T &\approx e^{c(\mathbb{E}X-1)}. \end{aligned}$$

Thus if $\mathbb{E}X > 1$ then $\mathbb{E}Z_T > 1$ but if $\mathbb{E}X < 1$ then $\mathbb{E}Z_T < 1$.

Because the branching process is a relatively simple model, we can prove some exact results for it. In particular it is possible to show that

$$\mathbb{E}Z_n = (\mathbb{E}X)^n. \tag{21.1}$$

So if $\mathbb{E}X > 1$ the process grows exponentially (on average), while if $\mathbb{E}X < 1$ then it dies out exponentially fast (on average). This agrees with our previous observations of the SIR process.

A useful exercise is to verify the relationship (21.1) using simulation.

21.2.3 Forest fire model

The forest fire model incorporates spatial interactions. Like the SIR model we suppose that we have a population made up of susceptible (unburnt), infected (on fire), and removed (burnt out) individuals. The difference is that the individuals are placed on a grid and an infected individual can only infect a susceptible individual if they are neighbours. We define the neighbours of a point (x, y) to be the eight points $(x-1, y-1)$, $(x-1, y)$, $(x-1, y+1)$, $(x, y-1)$, $(x, y+1)$, $(x+1, y-1)$, $(x+1, y)$, and $(x+1, y+1)$ (smaller or larger neighbourhoods can also be considered).

We take time in discrete steps. At each step an infected individual has a probability α of infecting each of its susceptible neighbours. Thus for a susceptible individual, the probability of remaining uninfected is $(1-\alpha)^x$ where x is the number of infected neighbours. After having had a chance to infect its neighbours, an individual is removed with probability β.

We restrict our forest fire to a grid of size $N \times N$. Let X_t be a matrix of size $N \times N$ representing the population at time t. We put

$$X_t(i,j) = \begin{cases} 2 & \text{if the individual at } (i,j) \text{ is susceptible;} \\ 1 & \text{if the individual at } (i,j) \text{ is infected;} \\ 0 & \text{if the individual at } (i,j) \text{ is removed.} \end{cases}$$

Here is some code for simulating the forest fire model and printing the results. An example of the output is provided in Figure 21.6. If you play around with this for a while you will see that we still see a threshold below which the fire rarely gets going but above which there is a chance that it can grow quite large. Again there is a balance between how fast new infections appear and how fast infected individuals are removed.

```
# program: spuRs/resources/scripts/forest_fire.r
# forest fire simulation
rm(list = ls())

neighbours <- function(A, i, j) {
  # calculate number of neighbours of A[i,j] that are infected
```

```
  # we have to check for the edge of the grid
  nbrs <- 0
  # sum across row i - 1
  if (i > 1) {
    if (j > 1) nbrs <- nbrs + (A[i-1, j-1] == 1)
    nbrs <- nbrs + (A[i-1, j] == 1)
    if (j < ncol(A)) nbrs <- nbrs + (A[i-1, j+1] == 1)
  }
  # sum across row i
  if (j > 1) nbrs <- nbrs + (A[i, j-1] == 1)
  nbrs <- nbrs + (A[i, j] == 1)
  if (j < ncol(A)) nbrs <- nbrs + (A[i, j+1] == 1)
  # sum across row i + 1
  if (i < nrow(A)) {
    if (j > 1) nbrs <- nbrs + (A[i+1, j-1] == 1)
    nbrs <- nbrs + (A[i+1, j] == 1)
    if (j < ncol(A)) nbrs <- nbrs + (A[i+1, j+1] == 1)
  }
  return(nbrs)
}

forest.fire.plot <- function(X) {
  # plot infected and removed individuals
  for (i in 1:nrow(X)) {
    for (j in 1:ncol(X)) {
      if (X[i,j] == 1) points(i, j, col = "red", pch = 19)
      else if (X[i,j] == 0) points(i, j, col = "grey", pch = 19)
    }
  }
}

forest.fire <- function(X, a, b, pausing = FALSE) {
  # simulate forest fire epidemic model
  # X[i, j] = 2 for susceptible; 1 for infected; 0 for removed

  # set up plot
  plot(c(1,nrow(X)), c(1,ncol(X)), type = "n", xlab = "", ylab = "")
  forest.fire.plot(X)

  # main loop
  burning <- TRUE
  while (burning) {
    burning <- FALSE
    # check if pausing between updates
    if (pausing) {
      input <- readline("hit any key to continue")
    }

    # update
```

```
    B <- X
    for (i in 1:nrow(X)) {
      for (j in 1:ncol(X)) {
        if (X[i, j] == 2) {
          if (runif(1) > (1 - a)^neighbours(X, i, j)) {
            B[i, j] <- 1
          }
        } else if (X[i, j] == 1) {
          burning <- TRUE
          if (runif(1) < b) {
            B[i, j] <- 0
          }
        }
      }
    }
    X <- B

    # plot
    forest.fire.plot(X)
  }

  return(X)
}

# spark
set.seed(3)
X <- matrix(2, 21, 21)
X[11, 11] <- 1
# big fires
#X <- forest.fire(X, .1, .2, TRUE)
X <- forest.fire(X, .2, .4, TRUE)
# medium fires
#X <- forest.fire(X, .07, .2, TRUE)
#X <- forest.fire(X, .1, .4, TRUE)
# small fires
#X <- forest.fire(X, .05, .2, TRUE)
#X <- forest.fire(X, .07, .4, TRUE)
```

Clearly as α increases and/or $\boldsymbol{\beta}$ decreases, the chance of a large fire will increase. Like the SIR and branching process models, we imagine that there will be a threshold above which large fires become much more likely. For example, suppose that the fire is burning along a straight front. Along the front each susceptible tree is adjacent to three burning trees, so the probability of catching on fire is $1-(1-\alpha)^3$. Thus, given burning trees are removed with probability $\boldsymbol{\beta}$, we might conjecture that the fire will grow if $1-(1-\alpha)^3 > \boldsymbol{\beta}$.

As it turns out, this conjecture understates the chance of a large fire. The reason is that fire fronts are not straight, and an irregular front will move faster than a straight front. Even a front that starts straight will quickly

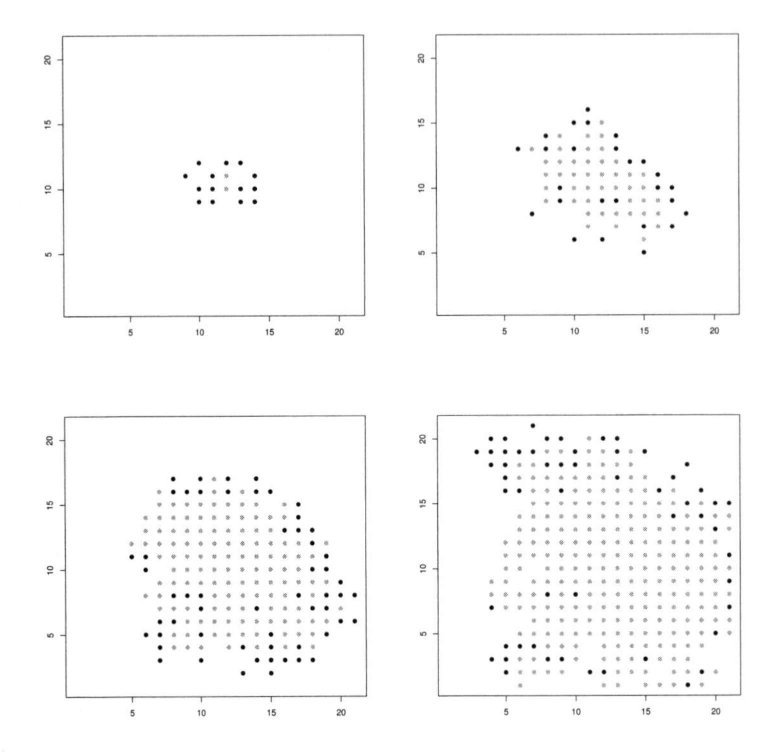

Figure 21.6 *Simulation of a forest fire epidemic at times 5, 10, 15, and 20. Infected individuals are dark grey and removed individuals light grey. Here* $\alpha = 0.2$ *and* $\beta = 0.4$*; we started with a single infected individual in the centre of the grid.*

contort, which one can easily see in the simulation, using the following initial condition.

```
X <- matrix(2, 21, 21)
X[21,] <- 1
```

21.3 Inventory

To meet demand in time and compete in the market, a company needs to keep stock in hand. The purpose of inventory theory is to determine rules or policies that minimise the cost of running an inventory system, while meeting customer demand. The following are possible costs associated with an inventory system

1. *Ordering and setup cost:* This includes the cost of paperwork and billing associated with an order, and may include overheads on the cost of delivery. If the product is produced internally, this cost may also include the cost of setting up and shutting down a machine in a production system.

The *Lead time* is the length of time between when an order is placed and when the order arrives.

2. *Purchasing cost:* For outsourced products this will include per-item transportation costs as well as the cost of the product. For goods produced internally, this includes the cost of raw materials and labour.
3. *Holding cost:* This is the cost of holding one unit of inventory for one period of time. If the period is one year then it is the annual holding cost. This cost can include insurance costs, the cost of renting space, security costs, loss due to spoilage, and the effects of inflation.
4. *Shortage cost:* When a demand cannot be met in time, a shortage is said to have occurred. There are two possible cases:

   (a) The customers accept delivery on a later date, which is called a *backlogged demand*;

   (b) The customers refuse to have the delivery on a later date, which is called a *lost sale.*

   In the second case the shortage cost is primarily the lost revenue. In the first case the shortage cost includes penalties paid for late delivery. In both cases the shortage cost can also include a component that represents lost future sales due to the lack of service shortage represents.

21.3.1 Continuous Review Inventory Model

The continuous review inventory model makes the following assumptions.

1. The inventory system is under *continuous review*, which means that sales are recorded when they occur so that the level of inventory in the system $I(t)$ is known at all times t.
2. The demand is a Poisson process with a rate of D items per year.
3. The lead time L is a known constant.
4. There is an ordering cost of K and a price per unit of p.
5. The unit holding cost is h per year.
6. Shortage results in lost sales, with a shortage cost of s per item.

We suppose that the inventory policy (or the ordering policy) is a so-called (q, r) policy. That is, when the inventory level is r (reorder point), an order of size q is placed, which will arrive after a lead time of L. Our objective is to choose q and r to minimise cost.

The expected demand over a lead time period is LD. Hence, if we reorder when $I(t) = r$ the expected minimum inventory level will be $m = r - LD$. The quantity m is called the *safety stock*. We will assume that $m \geq 0$, that is $r \geq LD$, and that $q \geq r$.

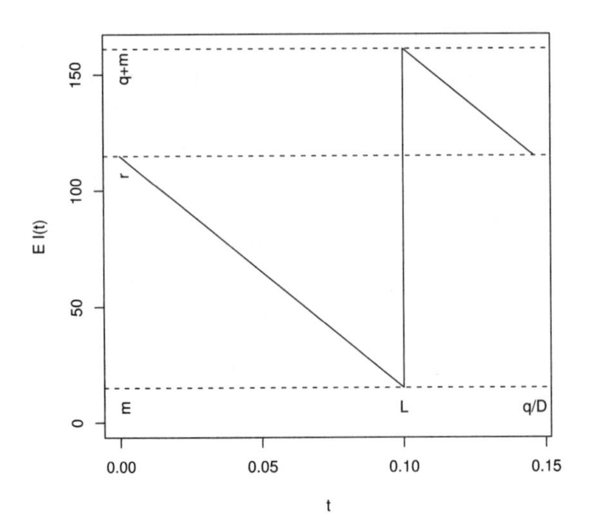

Figure 21.7 *Expected inventory level under a (q, r) policy, over a single cycle.*

We would like to estimate $c(q, r)$, which is the expected cost per unit time of running the system (that is, the annual cost), and then choose q and r to minimise it. However we have to be careful what we mean by 'cost per unit time', because the costs change as the level of inventory changes. The way around this problem is to consider *cycles*. Define a cycle to be the time from one reorder point to the next, when the stock is at level r. With a little thought you should see that these cycle lengths are *independent*.[1] Let C be the running costs and T the length of a single cycle, then we define

$$c(q, r) = \mathbb{E}\left(\frac{C}{T}\right) \approx \frac{\mathbb{E}C}{\mathbb{E}T}.$$

In Figure 21.7 we plot the expected inventory level over a single cycle. The expected demand is D per year. Thus the graph of $\mathbb{E}I(t)$ will decrease from r to m with a constant slope of $-D$, jump to $q + m$, then decrease to r with slope $-D$. We see immediately that $\mathbb{E}T = q/D$.

To estimate $\mathbb{E}C$ we split the cost into four parts—holding cost, ordering cost, purchasing cost, and shortage cost—and consider each in turn.

1. Put $I(0) = r$. Noting that $m = r - LD$, the expected holding cost over a

[1] In fact, our inventory system is an example of a *renewal process*, and the reorder points are known as *renewal times*.

single cycle is

$$\begin{aligned}\mathbb{E}\int_0^T hI(t)\,dt &\approx h\int_0^{q/D} \mathbb{E}I(t)\,dt \\ &= h\int_0^{q/D} (m+Dt)\,dt \\ &= h\left(\frac{q^2}{2D}+\frac{(r-LD)q}{D}\right).\end{aligned}$$

Here we have approximated T by $\mathbb{E}T$.

2. The ordering cost per cycle is exactly K.
3. The purchasing cost per cycle is exactly pq.
4. To calculate the shortage cost we note that demand during the lead time has a Poisson(DL) distribution. To simplify things we will approximate the demand during the lead time by a continuous distribution with probability density function $f(x)$. Given this the expected shortage during the lead time will be

$$n(r) = \int_r^\infty (x-r)f(x)\,dx$$

and so the expected shortage cost will be $sn(r)$.

Putting these together, the expected cost per unit time under a (q, r) policy is (approximately)

$$c(q,r) = h\left(r - LD + \frac{q}{2}\right) + \frac{KD}{q} + pD + \frac{sDn(r)}{q}.$$

Theorem A necessary condition for $c(q, r)$ to be minimised is that q and r satisfy the equations

$$q = \sqrt{\frac{2D(K+sn(r))}{h}} \quad \text{and} \quad 1 - F(r) = \frac{qh}{sD}, \tag{21.2}$$

where $F(r) = \int_0^r f(x)dx$.

Proof. We note that

$$\frac{\partial c(q,r)}{\partial q} = \frac{h}{2} - \frac{KD}{q^2} - \frac{sDn(r)}{q^2}$$

and

$$\frac{\partial c(q,r)}{\partial r} = h + \frac{sDn'(r)}{q}$$

where

$$\begin{aligned}n'(r) &= \frac{d}{dr}\left(\int_r^\infty xf(x)dx - r\int_r^\infty f(x)dx\right) \\ &= -rf(r) - \int_r^\infty f(x)dx + rf(r)\end{aligned}$$

$$= \quad F(r) - 1.$$

Setting $\partial c(q,r)/\partial q = \partial c(q,r)/\partial r = 0$ gives the result.

The *service level* α is the probability of not running out of stock in any given cycle, namely $F(r)$. In practice, rather than solve the above equations for q and r, what practitioners often do is specify the required service level beforehand, based on perceived customer requirements. Typically we take $\alpha = 0.95$ or 0.99. Having specified α and thus r, the expected cost is now a function of q alone, which we minimise in the usual way. We have

$$c(q) = h\left(r - LD + \frac{q}{2}\right) + \frac{KD}{q} + pD + \frac{sDn(r)}{q}.$$

From this the optimal value of q is

$$q^* = \sqrt{\frac{2D(K + sn(r))}{h}} \approx \sqrt{\frac{2KD}{h}},$$

noting that for α close to 1, $n(r)$ will be small. This last value is known as the *Economic Order Quantity* (EOQ) in the inventory literature.

For example, suppose that $D = 1000$ per year, $L = 0.1$ years, $K = 1000$, $p = 100$, $h = 100$ per year, and $s = 200$. Let X be the demand during the lead time, then $X \sim \text{pois}(100) \approx N(100, 100)$. Using the normal approximation we have $f(x) = \frac{1}{\sqrt{200\pi}} \exp(-(x - 100)^2/200)$.

If we specify a service level of $\alpha = 0.95$, then r satisfies $\mathbb{P}(X \leq r) = 0.95$. To calculate the left-hand side we can use Simpson's rule for numerical integration. To solve the equation we can use the Newton-Raphson algorithm. Let $F(x) = \int_{-\infty}^{x} f(u)du = 0.5 + \int_{100}^{x} f(u)du$ (the second form avoids having an infinite domain to integrate over).

```
> rm(list = ls())
> source("../scripts/simpson.r")
> f <- function(x) exp(-(x - 100)^2/200)/sqrt(200 * pi)
> F <- function(x) {
+     if (x > 100)
+         return(0.5 + simpson(f, 100, x))
+     else if (x < 100)
+         return(0.5 - simpson(f, x, 100))
+     else return(0.5)
+ }
> source("../scripts/newtonraphson.r")
> g <- function(r) c(F(r) - 0.95, f(r))
> r <- newtonraphson(g, 100)

At iteration 1 value of x is: 111.2798
At iteration 2 value of x is: 115.0524
At iteration 3 value of x is: 116.3077
```

```
At iteration 4 value of x is: 116.4469
At iteration 5 value of x is: 116.4485
At iteration 6 value of x is: 116.4485
Algorithm converged
```

Rounding to the nearest integer we get $r = 116$. Using the EOQ to approximate q we have $q = \sqrt{2KD/h} = 141$ (rounding to the nearest integer).

How good are these values of q and r? To find out, we solve the Equations (21.2) and compare the answers.

Let A be any 2×2 non-singular matrix, then the optimal $(q, r)^T$ is a fixed point of the equation

$$G\begin{pmatrix} q \\ r \end{pmatrix} = A\begin{pmatrix} q^2h - 2D(K + s\,n(r)) \\ (1 - F(r))sD - qh \end{pmatrix} + \begin{pmatrix} q \\ r \end{pmatrix}.$$

If we can choose A so that G is a *contraction*, then we can obtain the fixed point by iterating G. We say G is a contraction if there exists $\delta \in (0, 1)$ such that, for any vectors $\mathbf{x}$ and $\mathbf{y}$, $\|G(\mathbf{x}) - G(\mathbf{y})\| \leq \delta\|\mathbf{x} - \mathbf{y}\|$. In this case, putting $\mathbf{x}_n = G(\mathbf{x}_{n-1})$ we have

$$\begin{aligned} \|\mathbf{x}_{n+1} - \mathbf{x}_n\| &= \|G(\mathbf{x}_n) - G(\mathbf{x}_{n-1})\| \\ &\leq \delta\|\mathbf{x}_n - \mathbf{x}_{n-1}\| \\ &\leq \delta^n\|\mathbf{x}_1 - \mathbf{x}_0\| \;\rightarrow\; 0 \text{ as } n \rightarrow \infty. \end{aligned}$$

It follows that for any k, $\|\mathbf{x}_{n+k} - \mathbf{x}_n\| \leq \delta^n\|\mathbf{x}_1 - \mathbf{x}_0\|/(1 - \delta)$, and thus that $\mathbf{x}_n$ converges, to $\mathbf{x}_*$ say (this is Cauchy's convergence criterion). Since G is continuous,

$$\mathbf{x}_* = \lim_{n\to\infty} \mathbf{x}_{n+1} = \lim_{n\to\infty} G(\mathbf{x}_n) = G(\lim_{n\to\infty} \mathbf{x}_n) = G(\mathbf{x}_*).$$

That is, $\mathbf{x}_*$ is a fixed point of G.

To calculate G we need $n(r) = \int_r^\infty (x - r)f(x)dx$. Using the change of variables $y = (x - 100)^2/2$, we can rewrite n as

$$n(r) = \sqrt{50/\pi}\exp(-(r - 100)^2/200) - (r - 100)(1 - F(r)).$$

This reformulation has the advantage that it does not require an integral over an infinite domain.

After some trial and error, it turns out that a suitable A is

$$\begin{pmatrix} -1/50,000 & 0 \\ 0 & 1/50,000 \end{pmatrix}.$$

Using $(141, 116)^T$ as the starting point for our iteration, G does indeed converge to a fixed point:

```
> n <- function(r) {
+   return(sqrt(50/pi)*exp(-(r - 100)^2/200) - (r - 100)*(1 - F(r)))
+ }
```

```
> G <- function(x) {
+   q <- x[1]
+   r <- x[2]
+   A <- matrix(c(-1, 0, 0, 1), 2, 2)/50000
+   return( A %*% c(100*q^2 - 2000*(1000 + 200*n(r)),
+                   (1 - F(r))*200000 - 100*q) + c(q, r) )
+ }
> tol <- 1e-3
> x <- c(141, 116)
> x.diff <- 1
> while (x.diff > tol) {
+   x.old <- x
+   x <- G(x)
+   x.diff <- sum(abs(x - x.old))
+ }
> x

         [,1]
[1,] 145.9390
[2,] 114.5481
```

Rounding to the nearest integer we get $q = 146$ and $r = 115$. Comparing our two solutions, first note that the service level corresponding to $r = 115$ is $F(r) = 0.933$ (to 3 significant figures), a little lower than the level of 0.95 we initially assumed. Calculating the annual cost we have

$$c(141, 116) = 116,071.9, \quad c(146, 115) = 116,050.8.$$

So in this case using the EOQ gave a reasonable approximation to the optimal value of q.

21.3.2 Simulated inventory level

The cost per unit time $c(q, r)$, derived in the previous section, incorporated some simplifying assumptions. In particular we assumed

$$\mathbb{E}\left(\frac{C}{T}\right) \approx \frac{\mathbb{E}C}{\mathbb{E}T}$$

and

$$\mathbb{E}\int_0^T hI(t)\,dt \approx h\int_0^{q/D} \mathbb{E}I(t)\,dt.$$

We also assumed that the demand during the lead time could be approximated by a continuous distribution.

To judge how much of an effect these simplifying assumptions have on $c(q, r)$, we use simulation to provide an independent estimate. We will use a technique called *discrete event simulation*.

Let $I(t)$ be the level of stock (that is, inventory) at time t, and $c(t)$ the accumulated costs at time t. The triple $(t, I(t), c(t))$ describes the *state* of our system. Discrete event simulation updates the state only when certain events occur. In our case the relevant events are *purchases* and the arrival of *new stock*.

Suppose that at the previous event the state was $(u, I(u), c(u))$ and that the next event after time u happens at time v.

If the new event is a purchase then $I(v) = \max\{I(u) - 1, 0\}$. Updating the costs is more complex:

- Over the time interval $(u, v]$ the holding costs have increased by $I(u)(v-u)$;
- If $I(u) = 0$ then there will be a shortage cost of s;
- If $I(v) = r$ then there will be a reordering cost of $K + qp$.

If the new event is the arrival of new stock, then $I(v) = I(u) + q$ and $c(v) = c(u) + I(u)(v - u)$.

We maintain a list of events and when they will occur. We update this list every time an event occurs, by removing the event that has just occurred, and adding any new events we now know about. In our case, if the event at time v is a purchase, then we generate a new purchase event at time $v + A$, where $A \sim \exp(D)$. That is, A is the time between arrivals in a Poisson process of rate D. Moreover, if the stock level drops to r then we generate a new stock arrival event at time $v + L$. The arrival of new stock does not trigger any new events. (By assumption $q > r$, so we know that it is never necessary to order new stock immediately.)

Once we have defined rules for updating the state for each type of event, and for generating new events, the simulation has the following simple form (pseudo-code):

```
initialise state and event list
while (stopping condition not met) {
  get next event
  if event type = a
    update state and event list
  else if event type = b
    update state and event list
  else ...
}
```

The event list We will implement the event list as a `list` in R. Each element will itself be a list, with two named elements: `type` and `time`. We will assume that the elements of the event list are ordered according to their `time` components.

Given this structure, to get the next event we just need the first element of the event list:

```
current.event <- event.list[[1]]
event.list <- event.list[-1]
```

Inserting a new event into the event list requires more work, as we need to preserve the ordering. Here is a function to do this for us:

```
add_event <- function(event.list, new.event) {
  # add new.event to event.list
  N <- length(event.list)
  if (N == 0) return(list(new.event))
  # find position n of new.event
  n <- 1
  while ((n <= N) && (new.event$time > event.list[[n]]$time)) {
    n <- n + 1
  }
  # add new.event to event.list
  if (n == 1) {
    event.list <- c(list(new.event), event.list)
  } else if (n == N + 1) {
    event.list <- c(event.list, list(new.event))
  } else {
    event.list <- c(event.list[1:(n-1)], list(new.event), event.list[n:N])
  }
  return(event.list)
}
```

In our case the event list will only ever contain the next purchase event and sometimes also the next stock arrival event.

Here is our program for simulating an inventory system. To simulate a single cycle we put $I(0) = r$ and $c(0) = 0$ and then run the simulation until the next time $I(t) = r$.

```
# program: spuRs/resources/scripts/inventory_sim.r

rm(list=ls())
set.seed(1939)
source("../scripts/add_event.r")

# inputs
# system parameters
D <- 1000
L <- 0.1
K <- 1000
p <- 100
h <- 100
s <- 200
# control parameters
q <- 146
r <- 115
```

```
# initialise system and event list
n <- 0  # number of events so far
t <- 0  # time
stock <- r
costs <- 0
event.list <- list(list(type = "purchase", time = rexp(1, rate = D)))
event.list <- add_event(event.list, list(type = "new stock", time = L))
# initialise stopping condition
time.to.stop <- FALSE
# simulation
while (!time.to.stop) {
  # get next event
  current.event <- event.list[[1]]
  event.list <- event.list[-1]
  n <- n + 1
  # update state and event list according to type of current event
  if (current.event$type == "purchase") {
    # update system state
    t[n+1] <- current.event$time
    if (stock[n] > 0) {  # reduce inventory, update holding costs
      costs[n+1] <- costs[n] + h*stock[n]*(t[n+1] - t[n])
      stock[n+1] <- stock[n] - 1
    } else {             # lost sale
      costs[n+1] <- costs[n] + s
      stock[n+1] <- stock[n]
    }
    # generate next purchase
    new.event <- list(type = "purchase", time = t[n+1] + rexp(1, rate = D))
    event.list <- add_event(event.list, new.event)
    # check for end of cycle
    if (stock[n+1] == r) {
      # order more stock
      new.event <- list(type = "new stock", time = t[n+1] + L)
      event.list <- add_event(event.list, new.event)
      costs[n+1] <- costs[n+1] + K + q*p
    }
  } else if (current.event$type == "new stock") {
    # update system state
    t[n+1] <- current.event$time
    costs[n+1] <- costs[n] + h*stock[n]*(t[n+1] - t[n])
    stock[n+1] <- stock[n] + q
  }
  # check stopping condition
  if (stock[n+1] == r) time.to.stop <- TRUE
}
```

It is worthwhile to plot the stock (inventory) level over a single cycle, and compare with the expected stock level, which was the basis of the analysis used in Section 21.3.1.

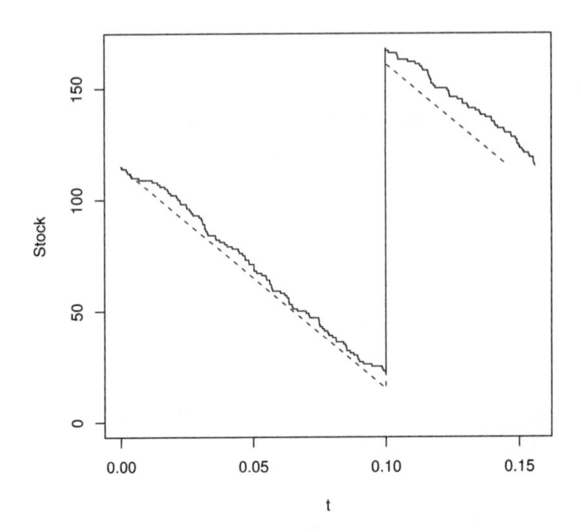

Figure 21.8 *Simulated and expected stock (inventory) level for a continuous review inventory model.*

```
plot(t, stock, type = "s", ylim=c(0, max(stock)))
lines(c(0, L, L, q/D), c(r, r-L*D, q+r-L*D, r), lty=2, col="red")
```

The output is given in Figure 21.8. We see that qualitatively the simulated stock level looks a lot like the expected stock level.

To estimate $c(q, r)$ we need to run the simulation for several cycles. The program `inventory_sim.r` incrementally updates the state vectors `t`, `stock` and `costs` at each event. This gives us a complete record of the process, but is too slow for simulating more than a few cycles. Thus to estimate $c(q, r)$ we rewrite the program so that it only keeps the current state, not the whole history. We also need to change the stopping condition, so that we stop after a fixed number of cycles. Finally, for each cycle we need to record the observed value of C/T. The rewritten program can be found as `inventory2_sim.r` in the `resources/scripts` directory within the `spuRs` archive.

Simulating 1000 cycles we obtained $\hat{c}(q, r) = 116,338.3$ with a 95% CI of (115,826.9, 116,849.8). Our approximation from Section 21.3.1 was 116,050.8, which sits comfortably in the confidence interval.

21.3.3 A two-stage inventory system

Our approximation of $c(q, r)$ for a continuous review inventory system worked quite well. Unfortunately such an analysis becomes much harder for more complex systems, and we have to rely more on simulation.

Consider an inventory system with a retail store and a depot. The store sells items one at a time, keeps a small amount of inventory on site, and frequently orders replacement stock from the depot. The depot supplies batches of stock to the store, keeps a large amount of inventory, and infrequently orders large quantities of replacement stock. Delivery from the depot to the store should be quite quick, but the lead time for deliveries to the depot could be quite large. Such systems are used when storage at the store is expensive, but storage at the depot is cheap.

In practice a depot will often serve several stores, however we will restrict ourselves to a single store.

The parameters needed to describe this two-stage system are

- D demand (at store);
- L_1, L_2 lead time for store and depot;
- K_1, K_2 ordering/delivery cost for store and depot;
- p per item cost (depot only);
- h_1, h_2 holding cost per item per unit time at store and depot;
- s shortage cost (at store);
- q_1, q_2 order quantities for store and depot;
- r_1, r_2 reorder point for store and depot.

Using discrete event simulation, we describe the state of the system using the variables

- Time t;
- Inventory at the store I_1;
- Inventory at the depot I_2;
- Cumulative cost c;

and we have the following events

- Purchase at the store;
- Stock arrives at the store;
- Stock arrives at the depot.

There is a complication to the two-stage system that does not appear in the simple continuous review model. It is possible that when the store orders stock from the depot, the depot is empty. We cannot treat this as a lost sale, rather the order has to be backlogged, then filled when the depot gets new stock. A convenient way to deal with backlogged orders is to create a new event

- Backlogged order

When the depot fails to fill an order we just create a backlogged order at some predetermined time b in the future. That is, we wait time b then try again.

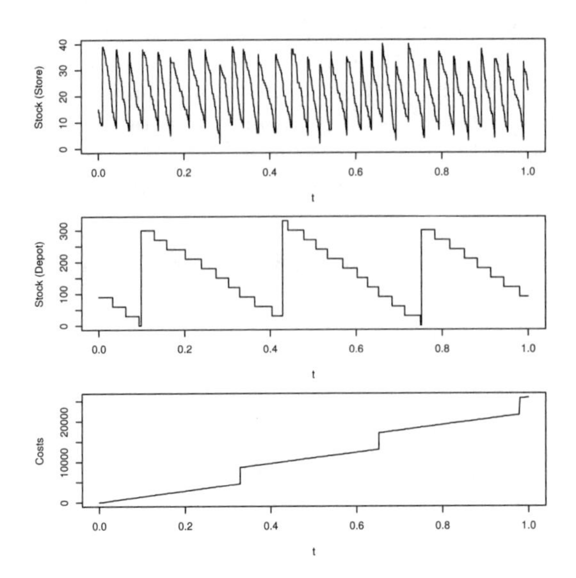

Figure 21.9 *A simulation of the two-stage inventory system, showing the level of stock (inventory) at the store and depot, and cumulative costs.*

Pseudo-code Rather than give a full implementation of the two-stage inventory system here, we will map out a suitable structure using pseudo-code. The interested reader can find a working version as `inventory_2stage_sim.r` in the `spuRs` archive, and some sample output is given in Figure 21.9, using a plausible set of parameter values.

The two-stage inventory system is still a renewal process, though the cycles are now more complex. Observe that when the inventory level at the depot reaches the reorder point r_2, it must be that the inventory level at the store has just reached r_1, because the only time we take stock from the depot is when it is ordered by the store. At this point we also know all there is to know about coming events: there will be a stock arrival at the store after a lead time of L_1; a stock arrival at the depot after a lead time of L_2; the time to the next purchase event is exponentially distributed, with rate D; and there will be no pending backlogged order, because we know the depot has just filled an order. Thus the point where the inventory levels I_1 and I_2 hit r_1 and r_2 is a renewal point, and marks the start/finish of independent cycles.

For this example, even though we have a renewal structure, instead of running the simulation for a given number of cycles, we have chosen to run it for a fixed length of time, T say. The basic structure of our program is as follows:

```
# initialise state variables
t <- 0
```

```
I1 <- r1
I2 <- r2
c <- 0
# initialise event list
create empty event list
add stock_arrival_at_store event at time L1
add stock_arrival_at_depot event at time L2
add purchase event at time X ~ exp(D)
# run the simulation
while (t < T) {
  t.old <- t
  get next event from event list
  if (next event is a purchase) {
    # update state and event list for a purchase
    ...
  } else if (next event is a stock_arrival_at_store) {
    # update state and event list for a stock_arrival_at_store
    ...
  } else if (next event is a stock_arrival_at_depot) {
    # update state and event list for a stock_arrival_at_depot
    ...
  } else { # next event is a backlogged_order
    # update state and event list for a backlogged_order
    ...
  }
}
```

With each event we need to adjust the time and add accumulated holding costs to c, other changes to the state variables and event list depend on the event in question. We consider the purchase event first:

```
# update state and event list for a purchase
# update time
t <- new event time
# update holding costs
c <- c + h1*I1*(t - t.old) + h2*I2*(t - t.old)
# update stock level
if (I1 > 0) {
  I1 <- I1 - 1
} else {
  # incur shortfall cost
  c <- c + s
}
# check store reorder level
if (I1 == r1) {
  # order from depot
  ...
}
# schedule next purchase
add purchase event at time t + X where X ~ exp(D)
```

The process of making an order from the depot requires some thought, as it will affect the level of stock at the depot, which means we also need to check the depot reorder point. Moreover, if the depot does not have enough stock to fill the order, then we have to generate a backlogged order. We will assume that the level of stock at the depot is always a multiple of q_1, which means that r_2 must also be a multiple of q_1. The advantage of this assumption is that we know to reorder only when $I_2 = r_2$, rather than when $I_2 \leq r_2$. If we reorder whenever $I_2 \leq r_2$ we can make several orders while we are waiting for the first one to arrive. (A more general way of dealing with this issue is to include in the state description a logical variable that indicates whether or not the store is waiting for an order to arrive.)

```
# order from depot
if (I2 >= q1) {
  # depot can fill order
  I2 <- I2 - q1
  c <- c + K1
  add stock_arrival_at_store event at time t + L1
  # check depot reorder level
  if (I2 == r2) {
    # order from supplier
    c <- c + K2 + q2*p
    add stock_arrival_at_depot event at time t + L2
  }
} else {
  # depot cannot fill order
  add backlogged_order event at time t + d
}
```

A backlogged order event involves updating the state, then attempting an order from the depot, as above.

```
# update state and event list for a backlogged_order
# update time
t <- new event time
# update holding costs
c <- c + h1*I1*(t - t.old) + h2*I2*(t - t.old)
# order from depot
...
```

The stock arrival events are both straightforward.

```
# update state and event list for a stock_arrival_at_store
# update time
t <- new event time
# update holding costs
c <- c + h1*I1*(t - t.old) + h2*I2*(t - t.old)
# update stock
I1 <- I1 + q1
```

```
# update state and event list for a stock_arrival_at_depot
# update time
t <- new event time
# update holding costs
c <- c + h1*I1*(t - t.old) + h2*I2*(t - t.old)
# update stock
I2 <- I2 + q2
```

Putting all these bits together we get our complete program. The process of breaking down a problem into smaller manageable tasks is sometimes called top-down programming or top-down refinement, and is an important technique for dealing with large problems. In this case we have used what is called an event based viewpoint to structure the problem, but there are other possibilities, such as the *process based* viewpoint, the *activity based* viewpoint, or the *three-phase* approach. For further reading on the topic of discrete event simulation, have a look at the book 'Computer Simulation in Management Science', by Mike Pidd, or 'Simulation Modelling and Analysis', by Law and Kelton.

21.4 Seed dispersal

Plant ecologists who perform research in plant propagation are often interested in how far plant seeds disperse from a parent plant. Information about dispersal enables ecologists to make predictions about the ability of an invasive species to colonise a new area, for example.

One of the first questions we can ask is, 'what is the mean displacement of a seed from the parent plant?' In order to frame this question in the context of a model, we can ask, 'what is the distribution of (R, Θ), the polar coordinates of the displacement from the parent plant of a randomly chosen seed?' To collect suitable data to answer these questions, the ecologists install seedtraps in lines that extend out from the parent plant (see Figure 21.10). These lines are called *transects*. After a specified amount of time (for example, a single flowering season), the seeds in each trap are counted; these seed counts at given distances form the experimental data that we have to work with.

We will assume that the seed rain is radially symmetric around the plant, although this is usually untrue. This assumption is called *isotropy*. An immediate consequence of the assumption is that $\Theta \sim U(0, 2\pi)$, independently of R. Moreover, the dispersal of seeds along each transect will be identically distributed, so it is sufficient for us to restrict our attention to a single transect.

Let T be the distance from the parent plant of a seed chosen at random *from the transect*. Importantly, T has a different distribution to R, which is what we really want to know. The reason is that the seeds in the closer traps are over weighted relative to the seeds in the remote traps, because their traps subtend a greater angle than do the remote traps. That is, the near traps sample a

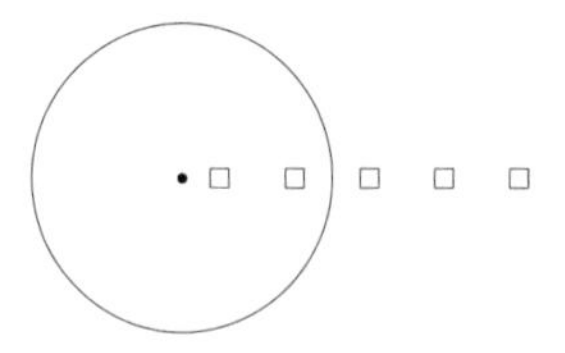

Figure 21.10 *Transect of seedtraps from plant; squares represent seed traps, the circle represents the median of the overall seed shadow, the black dot is the focal parent plant.*

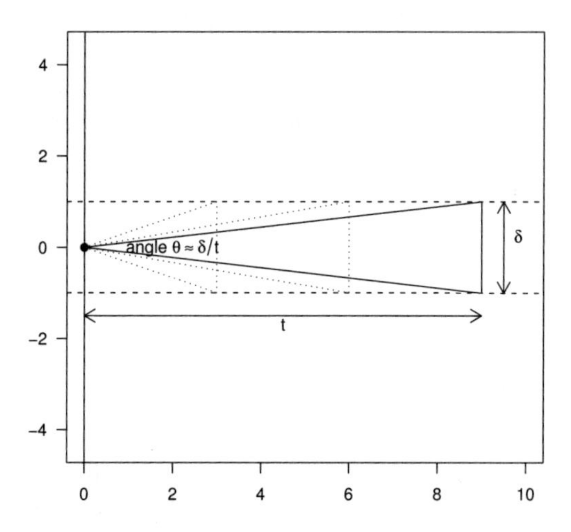

Figure 21.11 *Relating distance along the transect t to the radial distance r.*

larger slice of the circular seed rain than do the remote traps. The situation is illustrated in Figure 21.11. Suppose the traps have width δ (assumed small), then seeds that fall at distance t on the transect will have polar coordinates (r, α), where $r = t$ and $-\theta/2 < \alpha < \theta/2$, for θ such that

$$t \sin \theta = \delta.$$

If δ/t is small then so is θ, in which case $\sin \theta \approx \theta$ and we get

$$-\frac{\delta}{2t} < \alpha < \frac{\delta}{2t}.$$

That is, the overcounting of seeds at distance t along the transect is inversely proportional to t.

A further problem is that we do not actually observe T. Suppose that trap i covers area $[x_i - \epsilon/2, x_i + \epsilon/2] \times [-\delta/2, \delta/2]$, for $j = 1, \ldots, k$. Using the trap centres as our displacements, we observe a discretised version of T, call it T^*, where

$$\mathbb{P}(T^* = x_i) = \frac{\mathbb{P}(x_i - \epsilon/2 < T < x_i + \epsilon/2)}{\sum_{j=1}^{k} \mathbb{P}(x_j - \epsilon/2 < T < x_j + \epsilon/2)}.$$

In practice, if the traps are regularly spaced and reasonably close together (relative to the range of observations), we will just treat our observations of T^* as if they are observations of T. That is, we will ignore this problem.

Let $t_1, \ldots, t_n$ be our sample from T. A probability density function can be used to represent the relative number of seeds that are located along the transect, as a function of distance from the parent plant. We will refer to this as the *transect* pdf. For the moment, we shall assume that the transect pdf follows the exponential function; that is, for $0 \leq t < \infty$ and $\tau > 0$,

$$f_T(t) = \tau e^{-t\tau},$$

where τ is the rate parameter. The expected mean and variance of T in terms of the parameters of the model are $1/\tau$ and $1/\tau^2$ respectively, and we can estimate τ using $\hat{\tau} = 1/\bar{t}$, where $\bar{t}$ is the mean distance from the seeds to the plant.

Fitting the transect pdf is straightforward, but how does this give us the density of R, which we call the *radial pdf*? Let f_R be the pdf of R, then from Figure 21.11 we see that

$$\begin{aligned} f_T(t)\,dt &= \mathbb{P}(t < T < t + dt) \\ &\approx \mathbb{P}\left(t < R < t + dt \text{ and } -\frac{\delta}{2t} < \Theta < \frac{\delta}{2t}\right) \\ &= f_R(t)\,dt\,\frac{\delta}{2t} \quad \text{as } R \text{ and } \Theta \text{ are independent.} \end{aligned}$$

That is

$$f_R(r) \propto r f_T(r). \tag{21.3}$$

The approximation step above comes from putting $\sin\Theta \approx \Theta$. The approximation becomes exact in the limit as $\delta \to 0$.

Thus, in the case $T \sim \exp(\tau)$ we have $f_R(r) = kre^{-r\tau}$, for $0 \leq r < \infty$ and $\tau > 0$, where k is some normalising constant, chosen so that the density function integrates to 1. Using integration by parts it is easy to check that $k = \tau^2$, so

$$f_R(r) = r\tau^2 e^{-r\tau}.$$

This is the gamma distribution, with shape of 2 and rate of τ. Thus, the mean and variance of the distance that a seed travels, determined radially, are $\mathbb{E}R = 2/\tau$ and $\operatorname{Var} R = 2/\tau^2$, respectively. If we mistakenly use the exponential distribution instead of the gamma distribution, then our model for the seeds will place them too close to the plant and insufficiently variable.

In short, if we measure a transect of seedtraps and fit an exponential distribution to the numbers using $\hat{\tau} = 1/\bar{t}$, then to model the seed rain in two dimensions, we use a $\Gamma(\tau, 2)$ distribution for the radial distance and an independent $U(0, 2\pi)$ distribution for the angle.

Note, by integrating both sides of Equation 21.3, we can deduce that in general

$$f_R(r) = \frac{r f_T(r)}{\mathbb{E}(T)}. \tag{21.4}$$

That is, $\mathbb{E}(T)$ is the appropriate rescaling factor for the length weighted radial distribution.

21.4.1 Simulating the radial distance R

Consider now the problem of simulating the process by which a plant species colonises a new area. If we model this at the level of individual plants, then we need to be able to simulate where the seeds of each plant land. That is, we need to be able to simulate (R, Θ). Of course, we also need to know how many seeds a plant produces and when, how long the plant lives, and the chance that a seed will successfully germinate, which will depend on where it lands, but these are questions for another time.

As we have seen, if we know the transect density f_T then we can obtain the radial density f_R using Equation 21.4. However the exact functional form of the transect density may not be known. What we would like is a general technique which, assuming we can simulate T, allows us to simulate R. For example, if we wanted to make no assumptions at all about the distribution of T, we could simulate T directly from the observations $t_1, \ldots, t_n$. That is, put $\mathbb{P}(T = t_i) = 1/n$ for $i = 1, \ldots, n$. A more sophisticated approach would be to use a non-parametric estimate of f_T, but this is beyond the scope of this book.

We now demonstrate that it is possible to simulate f_R using f_T and rejection sampling. That is, we can simulate f_R without knowing its closed-form expression, just so long as we know f_T. Suppose that the range of T is bounded by a. That is $0 \leq T \leq a$. Take $U \sim U(0, a)$ independently of T then define

$$S = T \,|\, T > U.$$

That is, for $r \in [0, a]$,

$$\mathbb{P}(S \leq r) = \mathbb{P}(T \leq r \,|\, T > U).$$

To calculate the right-hand side probability we need the following version of the Law of Total Probability, which we give here without proof. For any random variables X and Y, with Y continuous, and any set $A \subset \mathbb{R}^2$, we have

$$\mathbb{P}((X, Y) \in A) = \int_y \mathbb{P}((X, y) \in A \,|\, Y = y) f_Y(y) dy.$$

In our case, noting that T and U are independent, we have

$$\begin{aligned}
\mathbb{P}(T \le r \,|\, T > U) &= \frac{\mathbb{P}(U < T \le r)}{\mathbb{P}(T > U)} \\
&= \frac{\int_0^a \mathbb{P}(U < t \le r) f_T(t)dt}{\int_0^a \mathbb{P}(t > U) f_T(t)dt} \\
&= \frac{\int_0^r (t/a) f_T(t)dt}{\int_0^a (t/a) f_T(t)dt} \\
&= \frac{\int_0^r f_R(t)dt}{\int_0^a f_R(t)dt} \\
&= F_R(t) \;=\; \mathbb{P}(R \le t).
\end{aligned}$$

That is, $S = T \,|\, T > U$ has the same distribution as R, the radial displacement. Moreover, if we can simulate T then we can simulate S easily using a rejection algorithm. Suppose that `T.sim()` simulates T, then to simulate S (or equivalently R), we can use the function below. The argument a gives an upper bound on the range of T.

```
R.sim <- function(a) {
  while (TRUE) {
    U <- runif(1, 0, a)
    T <- T.sim()
    if (T > U) return(T)
  }
}
```

Recall from Equation 21.3 that compared to R, displacements as measured by T are over represented by a factor proportional to the inverse distance from the origin. Our rejection algorithm *thins out* our observations of T, by a factor inversely proportional to the distance from the origin, negating the over representation caused by measuring along a transect.

Our definition of S required the range of T to be bounded. In practice, provided we are prepared to live with some occasional errors, if T has an unbounded range then we just take a large enough that $\mathbb{P}(T > a) \le \epsilon$, for some small ϵ. For example, if $T \sim \exp(1/2)$ then $\mathbb{P}(T > a) = \exp(-a/2)$, so for $\epsilon = 0.0001$ we get $a \ge -2\log\epsilon = 18.42$ (to 2 decimal places). What happens is that simulated values of T greater than a are always accepted, rather than being thinned.

To test that the distributions of S and R really are the same, we consider a case where we know what the distribution of R is, and compare that with an empirical estimate of the density of S, calculated from a simulated sample. We take the case $T \sim \exp(1/2)$ and $R \sim \Gamma(1/2, 2)$ and put $a = 20$. The output of our simulation experiment is given in Figure 21.12. In the left panel, we show the transect pdf (solid line) and the analytically computed radial pdf (dotted line). In the right panel, we include the analytical radial pdf again, and add an empirical estimate of the density of S. The two are extremely close.

To simulate S we use a vectorised version of `R.sim`. This has the advantage of speeding up the simulation, but the disadvantage that we don't know exactly how many observations of S we are going to get. In this case, provided we get enough observations to estimate the density of S, this is not a problem.

```
# program spuRs/resources/scripts/seed-test.r

# set up two plots side-by-side
par(las=1, mfrow=c(1,2), mar=c(4,5,0,2))

# graph f_R and f_T on the LHS plot
curve(dgamma(x, shape=2, rate=1/2), from=0, to=20,
      ylim=c(0, dexp(0, rate=1/2)), lty=2,
      xlab="r", ylab=expression(paste(f[T](r), " and ", f[R](r))))
curve(dexp(x, rate=1/2), add=TRUE)
abline(h=0, col="grey")

# generate T, U, and S samples for case 1
T <- rexp(1000000, rate=1/2)
U <- runif(1000000, min=0, max=20)
S <- T[T > U]

# graph estimate of f_S and f_R on the RHS plot
hist(S, breaks=seq(0, max(S)+0.5, 0.5), freq=FALSE,
     xlim=c(0,20), ylim=c(0, dexp(0, rate=1/2)),
     main="", xlab="r",
     ylab=expression(paste(f[R](r), " and ", hat(f)[S](r))),
     col="lightgrey", border="darkgrey")
curve(dgamma(x, shape=2, rate=1/2), add=TRUE)
```

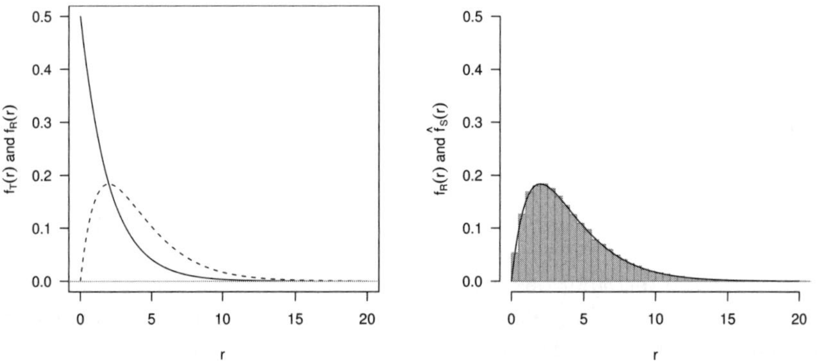

Figure 21.12 *Acceptance/rejection sampling for seed shadows using an exponential transect pdf.*

To conclude this example, we consider a couple of other cases where the radial pdf of f_R can be obtained analytically and can also be easily simulated, thereby providing a further test of our rejection algorithm.

In the first case we suppose that the transect pdf f_T is a lognormal density with parameters μ and σ^2. That is,

$$f_T(x \mid \mu, \sigma^2) = \frac{1}{\sqrt{2\pi}x\sigma} e^{-(\log x - \mu)^2/(2\sigma^2)}.$$

The radial pdf is thus

$$f_R(r) \propto e^{-(\log r - \mu)^2/(2\sigma^2)}.$$

This suggests deriving the distribution of $\log R$ (by applying the transformation theory from Section 14.5.2), which gives $\log R \sim N(\mu + \sigma^2, \sigma^2)$. Hence, R is still lognormal but with parameters $\mu + \sigma^2$ and σ^2.

In the second case we use a Weibull distribution, with parameters a and b, which has density

$$f_T(x \mid a, b) = \frac{a}{b}\left(\frac{x}{b}\right)^{a-1} \exp\left(-\left(\frac{x}{b}\right)^a\right).$$

Thus, for a Weibull transect with parameters $a = 2$ and $b = 2$ say, we have the radial pdf

$$f_R(r) \propto x^2 e^{-x^2/4},$$

which we recognise as a chi distribution with three degrees of freedom, scaled by a factor of $\sqrt{2}$. We write $R \sim \sqrt{2}\chi_3$. A χ random variable with k degrees of freedom is defined as the square root of a χ^2_k random variable.

To estimate f_R we use code very similar to the code that created Figure 21.12. The only substantial changes are in simulating T, the distribution of the transect pdf (Figure 21.13).

```
> # set up two plots side-by-side
> par(las=1, mfrow=c(1,2), mar=c(4,5,3,2))
> # Construct a graphic for the Lognormal transect pdf
> T <- rlnorm(1000000, meanlog = 0.5, sdlog = 0.55)
> U <- runif(1000000, min=0, max=20)
> S <- T[T > U]
> hist(S, breaks=seq(0, max(S)+0.5, 0.125), freq=FALSE,
+      xlim=c(0,7),  ylim=c(0, dexp(0, rate=1/2)),
+      main="Lognormal", xlab="r",
+      ylab=expression(paste(f[R](r), " and ", hat(f)[S](r))),
+      col="lightgrey", border="darkgrey")
> curve(dlnorm(x, meanlog = 0.5, sdlog = 0.55), add=TRUE, lty=2)
> curve(dlnorm(x, meanlog = 0.8025, sdlog = 0.55), add=TRUE)
> # Construct a graphic for the Weibull transect pdf
> T <- rweibull(1000000, shape=2, scale=2)
> U <- runif(1000000, min=0, max=20)
```

```
> S <- T[T > U]
> hist(S, breaks=seq(0, max(S)+0.5, 0.125), freq=FALSE,
+      xlim=c(0,7),  ylim=c(0, dexp(0, rate=1/2)),
+      main="Weibull", xlab="r",
+      ylab=expression(paste(f[R](r), " and ", hat(f)[S](r))),
+      col="lightgrey", border="darkgrey")
> curve(dweibull(x, shape=2, scale=2), add=TRUE, lty=2)
> curve((1/(2*sqrt(pi)))*x^2*exp(-(x^2)/4),add=TRUE)
```

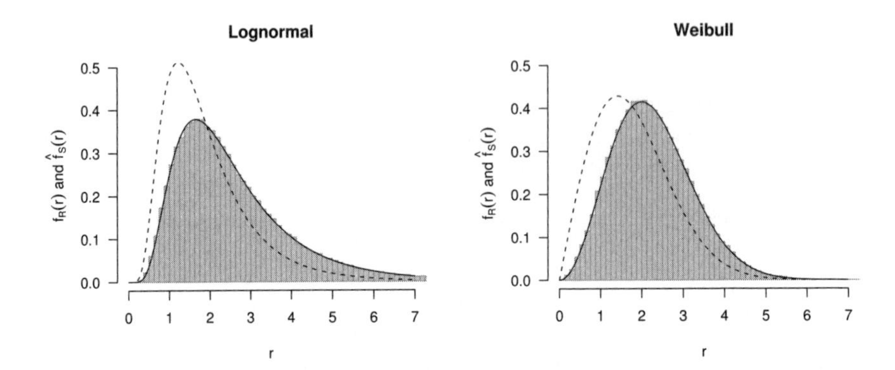

Figure 21.13 *Acceptance/rejection sampling for seed shadows using lognormal and Weibull transect pdfs. The dotted lines represent the transect pdfs, the shaded histograms represent the radial pdfs from simulation, and the solid lines are the exact radial pdfs we derived.*

21.4.2 Object-oriented programming implementation

In Section 8.4 (on object-oriented programming, or OOP), we developed a `trapTransect` class. In this section we will construct a `transectHolder` class that contains one or more `trapTransect` objects, and methods to fit a nominated pdf to the seed distances along the transect, and to simulate random seed locations from the fitted transect pdf using rejection sampling as outlined above.

Here is the S3 `trapTransect` constructor function from Section 8.4, and two methods, `print` and `mean`. Recall that the seed data are stored as seed counts at given distances.

```
> trapTransect <- function(distances, seed.counts, trap.area = 0.0001) {
+   if (length(distances) != length(seed.counts))
+     stop("Lengths of distances and counts differ.")
+   if (length(trap.area) != 1) stop("Ambiguous trap area.")
+   trapTransect <- list(distances = distances,
+                        seed.counts = seed.counts,
```

```
+                        trap.area = trap.area)
+   class(trapTransect) <- "trapTransect"
+   return(trapTransect)
+ }
> print.trapTransect <- function(x, ...) {
+   str(x)
+ }
> mean.trapTransect <- function(x, ...) {
+   return(weighted.mean(x$distances, w=x$seed.counts))
+ }
```

We imagine a situation in which we have a large number of plants, around each of which a transect of seedtraps has been installed. Critically, the number of seedtraps may vary by plant. For example, the seedtrap count might change by plant height. In order to store the observations for later analysis, we need a storage device that does not require that everything be the same length, for which a `list` is ideal. That is, a `transectHolder` should contain a list of `trapTransect` objects (and possibly some other bits and pieces).

We wish to have a function `fitDistances`, that takes a `transectHolder` object and fits a nominated pdf to the observed (transect) seed distances. That is, we assume that the transect distribution is the same for each plant, and fit a single pdf to all of our observations. There are various ways we could do this; our approach is to fit the nominated pdf to each transect, then summarise across the transects by taking the mean of the computed parameters. This strategy is not optimal, but is not unreasonable.

We could hide the `fitDistances` function inside the `transectHolder` constructor function, which we present below, but it might be useful to try different models, so we prefer the `fitDistances` function to be readily available to the user. The work of actually fitting a pdf is performed using maximum likelihood by the convenient `fitdistr` function from the `MASS` package: we just need to get our data into the correct format. To this end we define `getDistances`, which takes a `trapTransect` object and returns a vector of seed distances.

```
> fitDistances <- function(x, family=NULL) {
+   # x$transects is a list of trapTransect objects
+   # family is a string giving the name of a pdf
+   require(MASS)  # we need this package for the fitdistr() function
+   getDistances <- function(y) {
+     rep(y$distances, y$seed.counts)
+   }
+   getEstimates <- function(distance) {
+     fitdistr(distance, family)$estimate
+   }
+   distances <- lapply(x$transects, getDistances)
+   parameter.list <- lapply(distances, getEstimates)
+   parameters <- colMeans(do.call(rbind, parameter.list))
```

```
+   return(parameters)
+ }
```

Observe the function `do.call`, which accepts as arguments a function name and a list of arguments for the function, and calls the function using the list of arguments.

As before, for the sake of brevity, we omit useful checks for correct object class, and we omit informative behaviour in the case of an empty transect. Note that even though we are operating upon arbitrary numbers of objects, we never need to invoke a loop. Instead we vectorise all the operations using the graceful `lapply` function.

We now give a constructor for our `transectHolder` class. Rather than just create a list of `trapTransect` objects, we will add infrastructure in order to simplify its use for our purposes. The extra infrastructure that we will add is the automatic fitting of a nominated pdf to the transects, and the ability to simulate from the fitted model.

```
> transectHolder <- function(..., family="exponential") {
+   transectHolder <- list()
+   transectHolder$transects <- list(...)
+   distname <- tolower(family)
+   transectHolder$family <- family
+   transectHolder$parameters <-  fitDistances(transectHolder, distname)
+   transectHolder$rng <- switch(distname,
+                              "beta" = "rbeta",
+                              "chi-squared" = "rchisq",
+                              "exponential" = "rexp",
+                              "f" = "rf",
+                              "gamma" = "rgamma",
+                              "log-normal" = "rlnorm",
+                              "lognormal" = "rlnorm",
+                              "negative binomial" = "rnbinom",
+                              "poisson" = "rpois",
+                              "weibull" = "rweibull",
+                              NULL)
+   if (is.null(transectHolder$rng))
+     stop("Unsupported distribution")
+   class(transectHolder) <- "transectHolder"
+   return(transectHolder)
+ }
```

This simple constructor again omits checks for suitable arguments, successful fitting of the probability density function, and so on. The list of `trapTransect` objects is stored as `transectHolder$transects`. `fitDistances` is used to estimate the parameters for the nominated pdf; the pdf family and fitted parameters are stored as `transectHolder$family` and `transectHolder$parameters`. `transectHolder$rng` stores the name of the function that will be used to simulate from the fitted transect distribution. When complete, the object is then assigned the `transectHolder` class.

A function to print the object might look like this.

```
> print.transectHolder <- function(x, ...){
+   print(paste("This object of class transectHolder contains ",
+               length(x$transects), " transects.", sep=""))
+   str(x)
+ }
```

We can construct a function to simulate n random seed locations, given a `transectHolder` object, using the generic function `simulate`.

```
> methods(simulate)

[1] simulate.lm*

   Non-visible functions are asterisked

> simulate

function (object, nsim = 1, seed = NULL, ...)
UseMethod("simulate")
<environment: namespace:stats>
```

We need to write a version that will be specific to our class. We make sure that we match the argument names of the generic function.

```
> simulate.transectHolder <- function(object, nsim=1, seed=NULL, ...) {
+   if (!is.null(seed)) set.seed(seed)
+   distances <- c()
+   while(length(distances) < nsim) {
+     unfiltered <- do.call(object$rng,
+                           as.list(c(10*nsim, object$parameters)))
+     filter <- runif(10*nsim, 0, max(unfiltered))
+     distances <- c(distances, unfiltered[unfiltered > filter])
+     }
+   distances <- distances[1:nsim]
+   angles <- runif(nsim, 0, 2*pi)
+   return(data.frame(distances = distances,
+                     angles = angles,
+                     x = cos(angles) * distances,
+                     y = sin(angles) * distances))
+ }
```

Notice that, using `do.call`, we directly invoke the random-number generator and pass to it the estimated parameters, without knowing what distribution it is, nor how many parameters it requires.

We now demonstrate the construction of a `transectHolder` object, using data that mimics the structure of a field experiment, and simulate five random seedlings using the distribution fitted to the trap data.

```
> transect.1 <- trapTransect(distances = 1:4,
+                            seed.counts = c(4, 3, 2, 0))
> transect.2 <- trapTransect(distances = 1:3,
+                            seed.counts = c(3, 2, 1))
> transect.3 <- trapTransect(distances=(1:5)/2,
+                            seed.counts = c(3, 4, 2, 3, 1))
> allTraps <- transectHolder(transect.1, transect.2, transect.3,
+                              family="Weibull")
> allTraps
```

```
[1] "This object of class transectHolder contains 3 transects."
List of 4
 $ transects :List of 3
  ..$ :List of 3
  .. ..$ distances  : int [1:4] 1 2 3 4
  .. ..$ seed.counts: num [1:4] 4 3 2 0
  .. ..$ trap.area  : num 1e-04
  .. ..- attr(*, "class")= chr "trapTransect"
  ..$ :List of 3
  .. ..$ distances  : int [1:3] 1 2 3
  .. ..$ seed.counts: num [1:3] 3 2 1
  .. ..$ trap.area  : num 1e-04
  .. ..- attr(*, "class")= chr "trapTransect"
  ..$ :List of 3
  .. ..$ distances  : num [1:5] 0.5 1 1.5 2 2.5
  .. ..$ seed.counts: num [1:5] 3 4 2 3 1
  .. ..$ trap.area  : num 1e-04
  .. ..- attr(*, "class")= chr "trapTransect"
 $ family    : chr "Weibull"
 $ parameters: Named num [1:2] 2.37 1.8
  ..- attr(*, "names")= chr [1:2] "shape" "scale"
 $ rng       : chr "rweibull"
 - attr(*, "class")= chr "transectHolder"
```

```
> simulate(allTraps, 5, seed = 123)
```

```
  distances   angles          x          y
1 1.9707469 3.769842 -1.5944481 -1.1582653
2 0.7456877 2.091192 -0.3707734  0.6469754
3 2.8898740 3.070046 -2.8824807  0.2065838
4 1.4863482 5.997136  1.4259521 -0.4193945
5 1.6207433 3.034165 -1.6114000  0.1737775
```

This brief demonstration concludes the first phase of development of our class.

We are now able to test our earlier conjecture (that the displacement pdf f_R and the transect pdf f_T are related by $f_R(r) \propto r f_T(r)$), for a wider range of transect distributions, using simulation. We can proceed as follows.

1. Choose one of the available transect distributions, and simulate a two-dimensional seed shadow using the acceptance sampling algorithm.
2. Using only the random points located within a fixed-width transect, compare the quantiles with the original transect distribution.

In short, we should be able to recover our original transect distribution by the correct simulation followed by sampling along a transect.

We simulate from the Weibull distribution, with arbitrary but known shape and scale parameters, and discretise the random numbers to mimic the process of sampling for seedtraps. The pdf of the Weibull density that we used is as follows:

$$f(x) = \frac{a}{b}\left(\frac{x}{b}\right)^{a-1} \exp\left(-\left(\frac{x}{b}\right)^{a}\right)$$

where the shape parameter is a and scale parameter is b.

```
> simulated.seed.points <- table(round(rweibull(1000, shape = 2,
+     scale = 5)))[-1]
```

We drop the first measure to remove zeros from the observations. We use these simulated seedtrap points to construct a transect, and store that in a `transectHolder`, in the process fitting the Weibull pdf to the seedtrap data.

```
> simulated.transect <-
+   trapTransect(distances = as.numeric(names(simulated.seed.points)),
+                seed.counts = simulated.seed.points)
> simulated.holder <- transectHolder(simulated.transect, family="Weibull")
```

Finally we simulate a new site, using the fitted model, and select only those points that are in our new transect, which for convenience's sake we will not discretise. We arbitrarily set our transect as being $x > 0$ and $-0.5 < y < 0.5$. If our conjecture is correct then the distribution of these points should be close to the original fitted density, which is Weibull.

```
> good.site <- simulate(simulated.holder, 100000)
> good.points <- good.site$x[abs(good.site$y) < 0.5 & good.site$x > 0]
```

We compare the distributions using the following code, with output in Figure 21.14. Rather than try to visually compare simulated and theoretical distributions, here we provide a scatterplot of the simulated and theoretical quantiles, called a quantile-quantile plot. If the simulated distribution matches the theoretical distribution well, then the simulated quantiles should line up well with the theoretical quantiles. The comparison seems favourable; the simulated distribution along the transect matches the theoretical distribution quite well.

```
> par(las = 1)
> quantiles <- (1:99)/100
> plot(quantile(good.points, probs = quantiles), do.call(qweibull,
+     c(list(quantiles), unlist(simulated.holder$parameters))),
+     xlab = "Simulated Quantiles", ylab = "Theoretical Quantiles")
> abline(0, 1, col = "darkgrey")
```

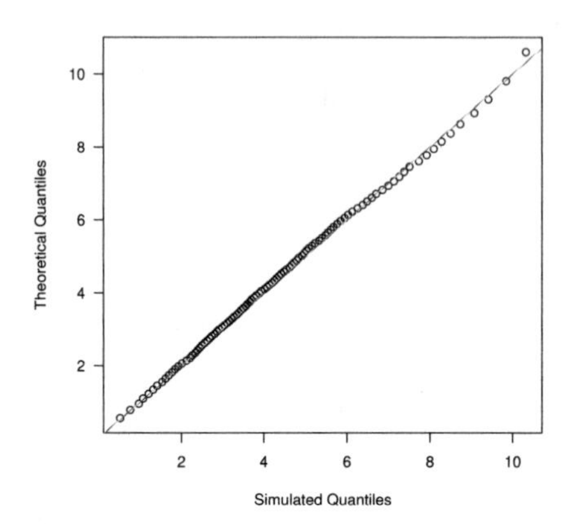

Figure 21.14 *Quantile-quantile plot of observed and theoretical seedling distance distributions.*

To conclude our development, we will add functions to compute the mean and standard deviation of the transect distance, using all the transects contained in the `transectHolder` object. We will assume that each of the plants should have equal weight. Because each object contained within `transectHolder$transects` is a `trapTransect`, we can reuse the `mean.trapTransect` function that we have already written.

```
> mean.transectHolder <- function(x) {
+     mean(sapply(x$transects, mean))
+ }
```

Note that the call to `mean` within the `sapply` function will automatically deploy `mean.trapTransect` if the objects to which the function is being applied are of class `trapTransect`. If at any time we need to use a different function for the mean of `trapTransect` objects, all we have to do is rewrite `mean.trapTransect`.

`sd` is *not* a generic function, so although we can create `sd.transectHolder`, it will not be automatically used in place of `sd` if the latter is called. An explicit call to the function `sd.transectHolder` is needed, as below.

```
> var.trapTransect <- function(x) {
```

```
+     return(var(rep(x$distances, x$seed.counts)))
+ }
> sd.transectHolder <- function(x) {
+     sqrt(mean(sapply(x$transects, var.trapTransect)))
+ }
```

We can now invoke these functions to find the mean of the means of the transect seed distances, and the quadratic mean of the standard deviations of the transect seed distances.

```
> mean(allTraps)

[1] 1.584046

> sd.transectHolder(allTraps)

[1] 0.7746426
```

CHAPTER 22

Student projects

This chapter presents a suite of problems that can be tackled by students. They are less involved than the case studies that were detailed in the preceding chapter, but more substantial than the exercises that we have included in each chapter.

22.1 The level of a dam

In this assignment we will model the changing level (or height) of water in a dam (Figure 22.1). The minimum level is 0 and the maximum is $h_{\max}$. The level increases when rain falls in the catchment area and decreases as a result of evaporation and use. We will ignore any loss due to leaks or seepage.

22.1.1 Height and volume

Volume Let $A(h)$ be the cross-sectional area of the dam at height h.

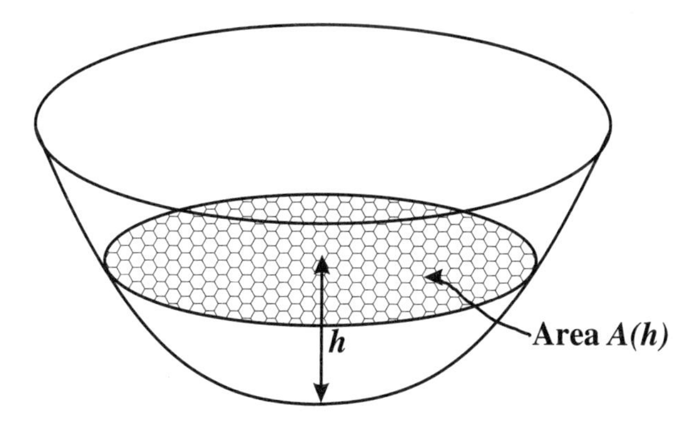

Figure 22.1 *An idealized dam.*

The volume of water contained by the dam when it is filled to level h is

$$V(h) = \int_0^h A(u)\, du.$$

Write a function `volume(h, hmax, ftn)` that returns $V(h)$ for $h \in [0, h_{\max}]$, where `hmax` is $h_{\max}$ and `ftn` is a function of a single variable which is assumed to return $A(h)$. For $h < 0$ your function should return 0, and for $h > h_{\max}$ it should return $V_{\max} = V(h_{\max})$.

Use at least 100 subdivisions when calculating the integral numerically.

Height If the current level of the dam is h and the volume of the dam changes by an amount v, then the level of the dam becomes $u = H(h, v)$ where u satisfies

$$V(u) = V(h) + v.$$

Note that if the right-hand side of this equation is $> V_{\max}$ or < 0, then this equation has no solution. In this case we take $u = h_{\max}$ or $u = 0$, respectively.

Using a root-finding algorithm, write a function `height(h, hmax, v, ftn)` that returns $H(h, v)$, where `hmax` is $h_{\max}$ and `ftn` is a function of a single variable that is assumed to return $A(h)$.

Use a tolerance of `1e-6` in your root-finding algorithm.

Test case Suppose that the dam is bowl-shaped with profile given by the equation $y = \pi x^2$. That is, the dam has the shape obtained by rotating the curve $y = \pi x^2$ about the y-axis (Figure 22.2).

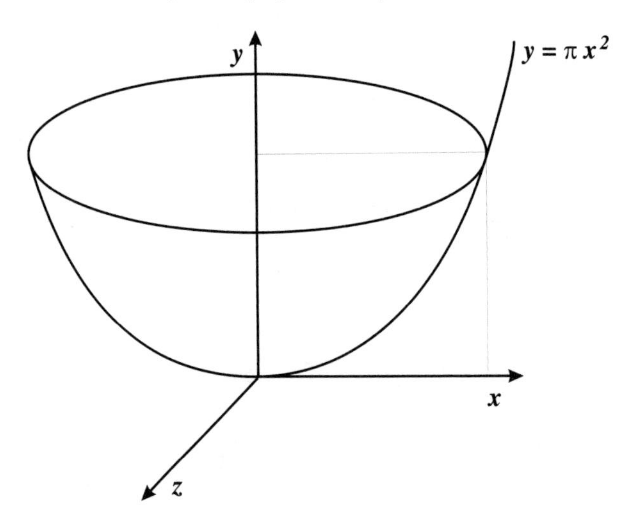

Figure 22.2 *A schematic dam.*

Show that, for $h \in [0, h_{\max}]$ and $v \in [-h^2/2, V_{\max} - h^2/2]$,

$$\begin{aligned} A(h) &= h; \\ V(h) &= h^2/2; \text{ and} \\ H(h, v) &= \sqrt{h^2 + 2v}. \end{aligned}$$

To test that your function `height(h, hmax, v, ftn)` works, define

```
A <- function(h) return(h)
```

then calculate `height(h, hmax = 4, v, ftn = A)` for the following values of h and v:

h	0	2	4	1	1
v	1	1	1	0.1	−0.1

22.1.2 Tracking height over time

Suppose that $h(t)$ is the level of the dam at the start of day t, and that $v(t)$ is the volume of rain falling into the catchment during day t, for $t = 1, \ldots, n$. Also let α be the volume of water taken from the dam for use per day, and let $\beta A(h(t))$ be the volume of water lost due to evaporation during day t. Then the level of water in the dam at the start of day $t + 1$ is given by

$$h(t+1) = H(h(t), v(t) - \alpha - \beta A(h(t))).$$

Further suppose that $h_{\text{max}} = 10$, $\alpha = 1$, $\beta = 0.05$, and $A(h)$ has the form

$$A(h) = \begin{cases} 100h^2 & \text{for } 0 \le h \le 2; \\ 400(h-1) & \text{for } 2 \le h. \end{cases}$$

The file `catchment.txt` (in the `spuRs` archive) gives $v(t)$ for $n = 100$ consecutive days. Write a program that reads this file then, for a given value of $h(1)$, calculates $h(2), \ldots, h(n+1)$. Plot your output for the cases $h(1) = 1$ and $h(1) = 5$, as in Figures 22.3 and 22.4, respectively.

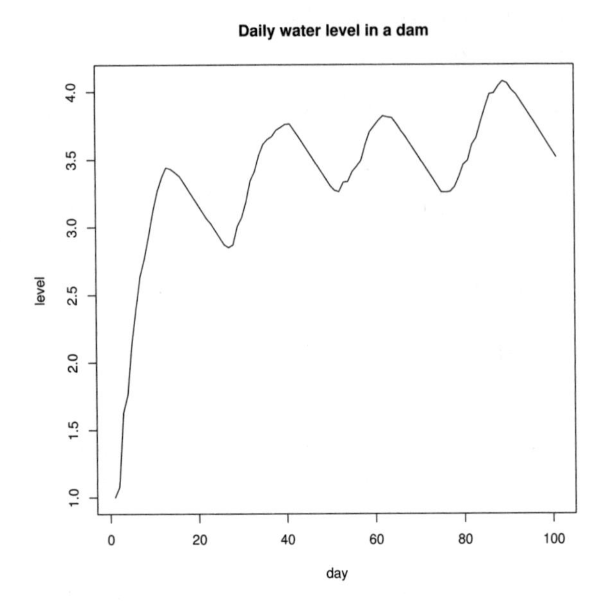

Figure 22.3 *Simulated time trace of water level for dam,* $h(1) = 1$.

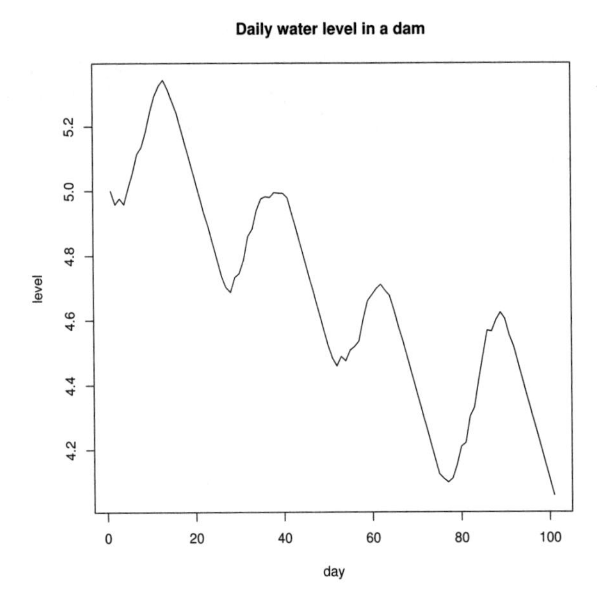

Figure 22.4 *Simulated time trace of water level for dam,* $h(1) = 5$.

22.2 Roulette

At the Crown Casino in Melbourne, Australia, some roulette wheels have 18 slots coloured red, 18 slots coloured black, and 1 slot (numbered 0) coloured green. The red and black slots are also numbered from 1 to 36. (Note that some of the roulette wheels also have a double zero, also coloured green, which nearly doubles the house percentage.)

You can play various 'games' or 'systems' in roulette. Four possible games are:

- A. Betting on Red

  This game involves just one bet. You bet \$1 on red. If the ball lands on red you win \$1, otherwise you lose.

- B. Betting on a Number

  This game involves just one bet. You bet \$1 on a particular number, say 17; if the ball lands on that number you win \$35, otherwise you lose.

- C. Martingale System

  In this game you start by betting \$1 on red. If you lose, you double your previous bet; if you win, you bet \$1 again. You continue to play until you have won \$10, or the bet exceeds \$100.

- D. Labouchere System

  In this game you start with the list of numbers (1, 2, 3, 4). You bet the sum of the first and last numbers on red (initially \$5). If you win you delete the first and last numbers from the list (so if you win your first bet it becomes (2,3)), otherwise you add the sum to the end of your list (so if you lose your first bet it becomes (1, 2, 3, 4, 5)). You repeat this process until your list is empty, or the bet exceeds \$100. If only one number is left on the list, you bet that number.

Different games offer different playing experiences, for example some allow you to win more often than you lose, some let you play longer, some cost more to play, and some risk greater losses. The aim of this assignment is to compare the four games above using the following criteria:

1. The expected winnings per game;
2. The proportion of games you win;
3. The expected playing time per game, measured by the number of bets made;
4. The maximum amount you can lose;
5. The maximum amount you can win.

22.2.1 Simulation

For each game write a function (with no inputs) that plays the game once and returns a vector of length two consisting of the amount won/lost and how many bets were made. Then write a program that estimates 1, 2, and 3, by simulating 100,000 repetitions of each game. Note that a game is won if you make money and lost if you lose money.

22.2.2 Verification

For games A and B, check your estimates for 1 and 2 by calculating the exact answers. What is the percentage error in your estimates for 100,000 repetitions?

For each game, work out the exact answers for 4 and 5. Of course, if this is not close to the answer given by your simulation, then you should suspect that either your calculation or your program is erroneous.

22.2.3 Variation

Repeat the simulation experiment of Part 22.2.1 five times. Report the minimum and maximum values for 1, 2, and 3 in a table as follows:

Game	Exp. winnings min–max	Prop. wins min–max	Exp. play time min–max
A			
B			
C			
D			

Modify your program from Part 22.2.1 so that in addition to estimating the expected winnings, expected proportion of wins, and expected playing time, it also estimates the *standard deviation* of each of these values. (You may use the built-in function `sd(x)` to do this.) For a single run, consisting of 100,000 repetitions of each game, report your results in a table as follows:

Game	Winnings mean, std dev	Prop. wins mean, std dev	Play time mean, std dev
A			
B			
C			
D			

For which game is the amount won most variable?

For which game is the expected playing time most variable?

22.3 Buffon's needle and cross

The following question was first considered by George Louis Leclerc, later Comte de Buffon, in 1733:

> *'If a thin, straight needle of length l is thrown at random onto the middle of a horizontal table ruled with parallel lines a distance $d \geq l$ apart, so that the needle lies entirely on the table, what is the probability that no line will be crossed by the needle?'*

The answer depends on π^{-1} and so simulation of this experiment offers a way of estimating π^{-1}. We will look at the complementary probability that the needle actually intersects with a ruled line on the table; call this a crossing.

22.3.1 Theoretical analysis

We can think of the position of the needle as being determined by two random variables:

Y : the perpendicular distance of the centre of the needle from the nearest line on the table and

X : the angle that the top half of the needle makes with a ray through its centre, parallel to the table lines and extending in a positive direction.

See Figure 22.5 for a sketch.

For the position of the needle to be random, we require Y to be $U(0, d/2)$ and X to be $U(0, \pi)$. We then define the sample space Ω of all possible outcomes or positions of the needle as $\Omega = [0, \pi] \times [0, d/2]$.

1. Identify the inequality that X and Y must satisfy if the needle is to cross a ruled table line. Draw a picture of the sample space Ω and use your inequality to shade that part of it that corresponds to a crossing. We will refer to this region as the crossing region C.
2. As the needle is thrown at random, the probability of falling in any region R in Ω can be calculated as the ratio of the area of R, denoted $|R|$, to the total area of Ω. That is, $2|R|/(\pi d)$. Using integration, find the area of C and hence confirm that the probability of a crossing is $2l/(\pi d)$.

22.3.2 Simulation estimates

Let T_1 be the number of crossings in n tosses of the needle, then $E_1 = T_1 d/(nl)$ is an unbiased estimator of $2/\pi$. Write a program to simulate E_1 using $n =$ 100,000 needle tosses.

Calculate the variance of E_1 and thus suggest the best needle length l to use, subject to the restriction $l \leq d$.

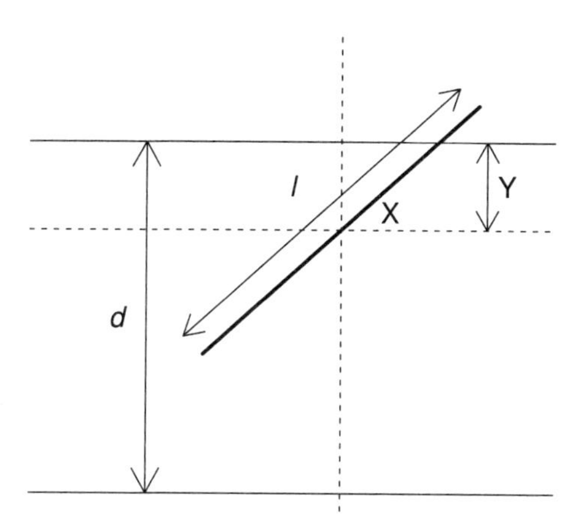

Figure 22.5 *Sketch of Buffon's needle.*

22.3.3 Buffon's cross

An extension of the Buffon needle problem is to think of throwing a cross made up of two equal length needles joined at right angles at their centres. We will assume that the needle lengths $l = d$. The cross can intersect the ruled lines 0, 1, or 2 times.

1. If the position of the first needle (and hence the cross) is specified by (X, Y) as above, show that the second needle crosses the ruled lines if:

$$
\begin{aligned}
Y &\le \frac{l}{2}\cos(X), &\quad \text{for } 0 < X < \frac{\pi}{2};\\
Y &\le \frac{-l}{2}\cos(X), &\quad \text{for } \frac{\pi}{2} < X < \pi.
\end{aligned}
$$

2. Write a program to estimate the probabilities of 0, 1, or 2 crossings, using n =50,000 simulated tosses of the cross.
3. You can think of the cross as just a convenient way of throwing two needles at once. So if T_2 represents the total crossings in n tosses of the cross, then $E_2 = T_2/2n$ should be another unbiased estimator of $2/\pi$.

   Write E_2 as $\sum_{i=1}^{n} Z_i/n$, where $Z_i \in \{0, 1, 2\}$ is the number of crossings on the i-th toss. We can estimate the variance of E_2 using S_Z^2/n, where $S_Z^2 = \sum_{i=1}^{n}(Z_i - \overline{Z})^2/(n-1)$ is the sample variance. Compare your answer with the theoretical variance of E_1 when n =100,000. Is it smaller, larger, or about the same? (This is an example of antithetic sampling.)

22.4 Insurance risk

This is a simplified version of two common problems faced by insurance companies: calculating the probability that they go bust and estimating how much money they will make.

Suppose that an insurance company has current assets of \$1,000,000. They have $n = 1{,}000$ customers who each pay an annual premium of \$5,500, paid at the start of each year. Based on previous experience, it is estimated that the probability of a customer making a claim is $p = 0.1$ per year, independently of previous claims and other customers. The size X of a claim varies, and is believed to have the following density, with $\alpha = 3$ and $\beta = 100{,}000$,

$$f(x) = \begin{cases} \dfrac{\alpha\beta^\alpha}{(x+\beta)^{\alpha+1}} & \text{for } x \geq 0, \\ 0 & \text{for } x < 0. \end{cases}$$

(Such an X is said to have a Pareto distribution, and in the real world is not an uncommon model for the size of an insurance claim.)

We consider the fortunes of the insurance company over a five-year period. Let $Z(t)$ be the company's assets at the end of year t, so

$$\begin{aligned} Z(0) &= 1{,}000{,}000, \\ Z(t) &= \begin{cases} \max\{Z(t-1) + \text{premiums} - \text{claims},\ 0\} & \text{if } Z(t-1) > 0, \\ 0 & \text{if } Z(t-1) = 0. \end{cases} \end{aligned}$$

Note that if $Z(t)$ falls below 0 then it stays there. That is, if the company goes bust then it stops trading.

22.4.1 Simulating X

Let X be the size of a typical claim as above. Calculate the cdf F_X, $\mathbb{E}X$, and $\operatorname{Var} X$.

Using the inversion method, write a subroutine to simulate X.

Use simulation to estimate the pdf of X and compare your estimate to the true pdf. Your answer should include a plot like Figure 22.6.

22.4.2 Simulating Z

Write a function to simulate the assets of the company over five years, then use it to plot the assets as a graph like Figure 22.7.

Using your function, estimate:

1. The probability that the company goes bust, and
2. The expected assets at the end five years.

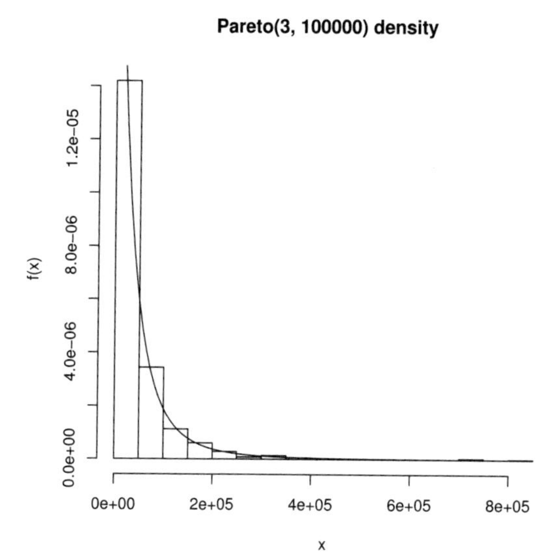

Figure 22.6 *Simulated and true pdf for insurance risk example.*

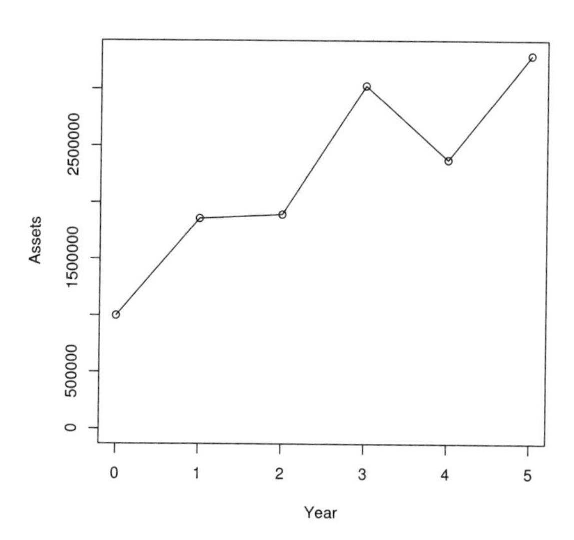

Figure 22.7 *Simulated assets for insurance risk example.*

22.4.3 Profit taking

Suppose now that the company takes profits at the end of each year. That is, if $Z(t) > 1{,}000{,}000$ then $Z(t)-$ 1,000,000 is paid out to the shareholders. If $Z(t) \leq 1{,}000{,}000$ then the shareholders get nothing that year.

Using this new scheme, estimate

1. The probability of going bust.
2. The expected assets at the end of five years, and
3. The expected total profits taken over the five years.

Compare these answers with your answers for Part 22.4.2 and comment.

22.5 Squash

A game of squash is played by two people: player 1 and player 2. The *game* consists of a sequence of *points*. If player i serves and wins the point, then his/her score increases by 1 and he/she retains the serve (for $i = 1$ or 2). If player i serves and loses the point, then the serve is transferred to the other player and the scores stay the same.

The winner is the first person to get 9 points, unless the score reaches 8 all first. If the score reaches 8 all then play continues until one player is 2 points ahead of the other, in which case he/she is the winner.

The object of this assignment is to simulate a game of squash and estimate the probability that player 1 wins. Define

$$\begin{aligned}
a &= \mathbb{P}(\text{ player 1 wins a point} \mid \text{player 1 serves }) \\
b &= \mathbb{P}(\text{ player 1 wins a point} \mid \text{player 2 serves }) \\
x &= \text{number of points won by player 1} \\
y &= \text{number of points won by player 2} \\
z &= \begin{cases} 1 & \text{if player 1 has the serve} \\ 2 & \text{if player 2 has the serve} \end{cases}
\end{aligned}$$

We will assume that player 1 serves first.

22.5.1 Status of the game

Write a function `status` that takes inputs x and y and returns one of the following text strings:

`"unfinished"` if the game has not yet finished;

`"player 1 win"` if player 1 has won the game;

`"player 2 win"` if player 2 has won the game;

`"impossible"` if x and y are impossible scores.

You may assume that the inputs x and y are integers.

When you have written your function, load or type the function `status.test` below.

```
# Program spuRs/resources/scripts/status.test.r

status.test <- function(s.ftn) {
  x.vec <- (-1):11
  y.vec <- (-1):11
  plot(x.vec, y.vec, type = "n", xlab = "x", ylab = "y")
  for (x in x.vec) {
    for (y in y.vec) {
```

```
      s <- s.ftn(x, y)
      if (s == "impossible") text(x, y, "X", col = "red")
      else if (s == "unfinished") text(x, y, "?", col = "blue")
      else if (s == "player 1 win") text(x, y, "1", col = "green")
      else if (s == "player 2 win") text(x, y, "2", col = "green")
    }
  }
  return(invisible(NULL))
}
```

Executing the expression `status.test(status)` should give you the output presented in Figure 22.8.

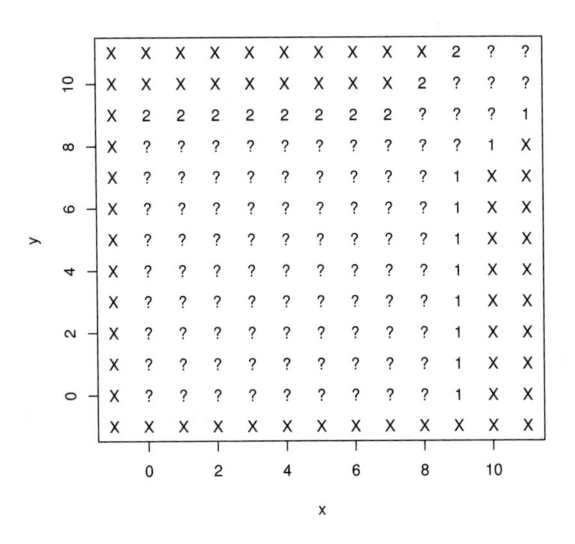

Figure 22.8 *Squash game status.*

22.5.2 Simulating a game

The vector *state* = (x, y, z) describes the current state of the game. Write a function `play_point` that takes inputs *state*, *a* and *b*, simulates the play of a single point, then returns an updated vector *state* representing the new state of the game.

Now code up the function `play_game` exactly as follows.

```
# Program spuRs/resources/scripts/play_game.r
```

```
play_game <- function(a, b) {
    state <- c(0, 0, 1)
    while (status(state[1], state[2]) == "unfinished") {
        # show(state)
        state <- play_point(state, a, b)
    }
    if (status(state[1], state[2]) == "player 1 win") {
        return(TRUE)
    } else {
        return(FALSE)
    }
}
```

Provided your functions `status` and `play_point` work properly, function `play_game` simulates a single game of squash and returns `TRUE` if player 1 wins and `FALSE` otherwise.

We define $p(a, b) = \mathbb{P}(\text{ player 1 wins the game} \mid \text{player 1 serves first })$. By simulating n squash games, estimate $p(0.55, 0.45)$ for $n = 2^k$ and $k = 1, 2, \ldots, 12$, then plot the results, as per Figure 22.9.

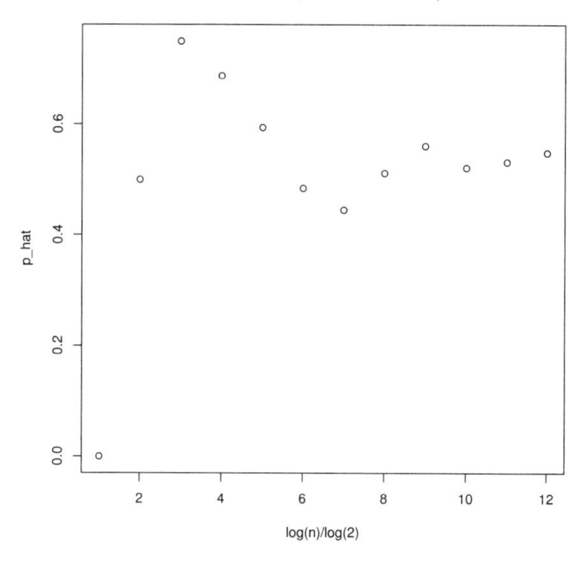

Figure 22.9 *Squash game simulations.*

Is $p(0.55, 0.45) = 0.5$? Explain your answer briefly?

Note that your code should specify a seed for the random number generator, so that you can reproduce your results exactly, if required.

22.5.3 Probability of winning

Let $X_1, \ldots, X_n$ be an iid sample of Bernoulli(p) random variables. We use $\hat{p} = \overline{X}$ to estimate p. Show that $\operatorname{Var}\hat{p} = p(1-p)/n$.

The standard deviation is the square root of the variance. What value of n will guarantee that the standard deviation of $\hat{p}$ is ≤ 0.01 for *any* value of p?

Using the value of n calculated above, reproduce the following table, which estimates $p(a, b)$ for different values of a and b.

`estimated p(a, b) for various a and b`

	b=0.1	b=0.2	b=0.3	b=0.4	b=0.5	b=0.6	b=0.7	b=0.8	b=0.9
a=0.1	0	0	0	0	0	0	0.01	0.05	0.51
a=0.2	0	0	0	0	0.01	0.04	0.16	0.51	0.96
a=0.3	0	0	0.01	0.03	0.09	0.25	0.51	0.85	1
a=0.4	0.01	0.01	0.05	0.12	0.28	0.53	0.79	0.97	1
a=0.5	0.02	0.06	0.15	0.32	0.53	0.76	0.94	0.99	1
a=0.6	0.06	0.19	0.35	0.55	0.76	0.9	0.98	1	1
a=0.7	0.18	0.36	0.56	0.75	0.9	0.97	1	1	1
a=0.8	0.35	0.59	0.79	0.9	0.97	0.99	1	1	1
a=0.9	0.66	0.82	0.92	0.98	0.99	1	1	1	1

Briefly explain the pattern of values you observe. Note that your numbers will be slightly different, as they are simulation estimates.

Make sure that your code specifies a seed for the random number generator, so that you can reproduce your results exactly, if required.

22.5.4 Length of a game

Modify the function `play_game` so that it returns the number of points played in the game (rather than the winning status of player 1).

Using your modified function, reproduce the following table, which estimates the expected number of points played in a game, for different values of a and b. Use the same value for n as above.

`average length of game for various a and b`

	b=0.1	b=0.2	b=0.3	b=0.4	b=0.5	b=0.6	b=0.7	b=0.8	b=0.9
a=0.1	12.23	14.87	18.29	22.90	28.81	38.71	54.38	85.37	151.84
a=0.2	12.54	15.38	19.06	23.69	30.52	40.21	54.54	74.18	84.84
a=0.3	12.82	15.92	19.63	24.86	31.46	39.78	48.48	53.86	53.11
a=0.4	13.26	16.44	20.43	25.37	30.52	35.53	39.41	39.18	37.49
a=0.5	13.83	17.22	20.94	24.58	28.09	30.28	29.91	29.12	27.79
a=0.6	14.46	17.51	20.48	22.79	23.87	24.07	23.54	22.59	21.57

```
a=0.7 |  14.81  17.06  18.91  19.46  19.88  19.10  18.48  17.72  17.04
a=0.8 |  14.36  15.63  16.14  15.91  15.32  14.93  14.50  14.00  13.87
a=0.9 |  12.39  12.90  12.61  12.23  11.94  11.70  11.34  11.22  11.09
```

Briefly explain the pattern of values you observe. Note that your numbers will be slightly different, as they are simulation estimates.

Make sure that your code specifies a seed for the random number generator, so that you can reproduce your results exactly, if required.

22.6 Stock prices

A popular model for stock prices is *Geometric Brownian Motion.* Let $S(i)$ be the stock price at the close of trading on day i (we take today as day 0), then using a Geometric Brownian Motion model we assume that

$$S(i+1) = S(i)\exp(\mu - \tfrac{1}{2}\sigma^2 + \sqrt{\sigma^2}Z(i+1))$$

where $Z(1), Z(2), \ldots,$ are iid $N(0,1)$ random variables. The parameter μ is called the drift and σ^2 is known as the volatility.

In practice both μ and σ^2 have to be estimated from the previous behaviour of the stock price.

22.6.1 *Simulating S*

Write a program that takes as input μ, σ^2, $S(0)$, and t, then simulates $S(1), \ldots, S(t)$ and plots them as a graph.

In your report include sample plots for at least two values of μ and two values of σ^2, and describe qualitatively what happens as μ increases/decreases and as σ^2 increases/decreases.

22.6.2 *Estimating $\mathbb{E}S(t)$*

Fix $S(0) = 1$ then show that $\log S(t) \sim N(\alpha, \boldsymbol{\beta}^2)$ for some α and $\boldsymbol{\beta}^2$, and find α and $\boldsymbol{\beta}^2$.

Unfortunately, $\mathbb{E}S(t) = \mathbb{E}\exp(\log S(t)) \neq \exp(\mathbb{E}\log S(t)) = \exp(\alpha)$. It turns out that $\mathbb{E}S(t)$ can be calculated exactly (the answer is $\exp(\mu t)$), but the calculation is rather difficult. Instead we will estimate $\mathbb{E}S(t)$ using simulation.

Write a program that takes as input μ, σ^2, and t, simulates $S(\mathrm{t})$ a number of times (at least 10,000) and then estimates $\mathbb{E}S(\mathrm{t})$ and $\mathbb{P}(S(\mathrm{t}) > S(0))$ and gives a 95% confidence interval for each estimate.

Use your program to complete the following table

μ σ^2	0.05 0.0025	0.01 0.0025	0.01 0.01
Estimate of $\mathbb{E}S(100)$ 95% CI for $\mathbb{E}S(100)$ Estimate of $\mathbb{P}(S(100) > S(0))$ 95% CI for $\mathbb{P}(S(100) > S(0))$			

22.6.3 Down-and-out call option

A *Down-and-Out Call Option* is a financial instrument that is sold alongside shares in our stock of interest. The option is determined by its *strike price* K, *time to maturity* t, and *barrier price* B. A single option gives you the right to buy a single share at time t for price K, provided the share price stayed above B.

Let $V(t)$ be the value of our option at maturity, then

$$V(t) = \begin{cases} S(t) - K & \text{if } S(t) \geq K \text{ and } \min_{0 \leq i \leq t} S(i) > B, \\ 0 & \text{if } S(t) < K \text{ or } \min_{0 \leq i \leq t} S(i) \leq B. \end{cases}$$

Options are used by companies to reduce the risks caused by changing prices. For example, a steel producer knows that it will need large quantities of iron ore 12 months in the future. Rather than buy the iron ore now it can buy options, which give a guaranteed price at which to buy the ore in the future.

Assuming $S(0) = 1$, write a program that asks for μ, σ^2, K, t, and B, then simulates $V(t)$ a number of times (at least 10,000) and estimates the cumulative distribution function of $V(t)$. Note that the cdf of $V(t)$ will have a jump at 0, but be continuous otherwise (it is an example of a mixed distribution).

Hence or otherwise, for $\mu = 0.01$, $\sigma \in \{0.0025, 0.005, 0.01\}$, $K = 2$, $t = 100$, and $B = 0.2$, estimate $P(V(100) > 0)$. What can you say about the distribution of $V(t)$ as σ^2 increases/decreases?

An important question (but one you do not have to answer) is what should we pay for the option now? Merton and Scholes won the 1997 Nobel Prize in Economics for their answer to this question (in the special case $B = 0$).

Glossary of R commands

Workspace and help

`getwd()`	get working directory
`setwd(dir)`	set working directory to `dir`
`help(topic) ?topic`	get help on `topic`
`help.search("keyword")`	search for help
`help.start()`	HTML help interface
`demo()`	list available demos
`save(..., file) load(file)`	save and load objects
`savehistory(f) loadhistory(f)`	save and load command history
`source(file)`	execute commands from `file`
`list.files(dir) dir(dir)`	list files in directory `dir`
`q()`	quit R

Objects

`mode(x)`	mode of `x`
`ls() objects()`	list existing objects
`rm(x) rm(list = ls())`	remove object `x` or all objects
`exists(x)`	test if object `x` already exists
`as.numeric(x) as.list(x) ...`	coerce mode of object `x`
`is.numeric(x) is.na(x) ...`	test mode of object `x`
`identical(x1, x2)`	test if objects are identical
`return(invisible(x))`	return invisible copy (doesn't print)

Packages

`install.packages(name)`	download and install package `name`
`download.packages(name, dir)`	download package `name` into `dir`
`library(name) require(name)`	load package `name`
`data(name)`	load dataset `name`
`.libPaths(dir)`	add directory `dir` to library paths
`sessionInfo()`	list loaded packages

Flow and control and function definition

```
if (logical_expression) expression_1 else expression_2
for (x in vector) expression
while (logical_expression) expression
name <- function(input_1, ...) {expression_1; ...; return(output)}
```

`stop(message)`	cease processing and print `message`
`browser()`	stop to inspect objects for debugging
`system.time(expression)`	report runtime for `expression`

Mathematical and logical operators and functions

`+ - * / ^ %% %/%`	algebraic operators
`< > <= >= == !=`	comparison operators
`& | !`	logical operators (and, or, not)
`&& ||`	and/or evaluated progressively from left
`xor(A, B)`	exclusive or (`A` or `B` but not both)
`ifelse(condition, x, y)`	choose `x` or `y` elementwise
`sin(x) cos(x) tan(x)`	sine, cosine, and tangent
`asin(x) acos(x) atan(x)`	inverse sine, cosine, and tangent
`exp(x) log(x)`	exponential and logarithm base e
`sqrt(x)`	square root
`abs(x)`	absolute value
`pi`	3.1415926...
`ceiling(x)`	smallest integer `>= x`
`floor(x)`	largest integer `<= x`
`all.equal(x, y)`	almost equal
`round(x, k)`	round `x` to `k` digits
`deriv(expression, vars)`	symbolic differentiation

Vectors

`x[i]`	select subvector using index vector
`x[logical] subset(x, subset)`	select subvector using logical vector
`c(...)`	combine vectors
`seq(from, to, by) from:to`	generate an arithmetic sequence
`rep(x, times)`	generate repeated values
`length(x)`	length of `x`
`which(x)`	indices of `TRUE` elements of `x`
`sum(...)`	sum over vector(s)
`prod(...)`	product over vector(s)
`cumsum(x) cumprod(x)`	cumulative sum and product
`min(...) max(...)`	minimum and maximum
`sort(x)`	sort a vector
`mean(x)`	sample mean

`var(x) sd(x)`	sample variance and standard deviation
`order(x)`	rank order of elements of `x`

Matrices

`matrix(data, nrow, ncol, byrow)`	create a matrix
`rbind(...) cbind(...)`	combine rows or columns
`diag(x)`	create a diagonal matrix
`%*%`	matrix multiplication
`nrow(A) ncol(A)`	number of rows and columns
`colMeans(A) colSums(A)`	column means or sums
`dim(A)`	dimensions of `x`
`det(A)`	determinant
`t(A)`	transpose
`solve(A, b)`	solution of `A x == b`
`solve(A)`	matrix inverse
`array(data, dim)`	create multidimensional array

Dataframes, factors and lists

`data.frame(...)`	create a dataframe
`str(x)`	summarise structure of `x`
`names(x)`	names of `x`
`dim(x)`	number of rows and columns of `x`
`attach(x)`	copy dataframe objects into workspace
`detach(x)`	delete dataframe objects from workspace
`factor(x)`	create a factor
`levels(x)`	list levels of factor `x`
`list(...)`	create a list
`unlist(x)`	flatten list `x` into a vector
`apply(x, i, f, ...)`	apply `f` over index `i` of array `x`
`sapply(x, f, ...)`	apply `f` to `x` and return a vector
`lapply(x, f, ...)`	apply `f` to `x` and return a list
`tapply(x, i, f, ...)`	apply `f` to subvectors of `x` given by levels of factor `i`
`mapply(f, ...)`	apply `f` to multiple arguments

Input and output

`scan(file, what, n, sep, skip)`	read from a file (or keyboard)
`read.table(file)`	read file in table format into dataframe
`read.csv(file)`	read comma separated data
`read.delim(file)`	read tab-delimited data into dataframe
`readline(prompt)`	read a line of text from the keyboard
`show(object)`	display `object` on screen

`head(object)`	list first few lines of `object`
`tail(object)`	list last few lines of `object`
`print(object)`	print `object`
`options(digits = x)`	display `x` digits in output
`cat(..., file)`	concatenate and write
`format(x, digits, nsmall, width)`	format `x` for output
`paste(..., sep = " ")`	paste strings together
`write(x, file, append = FALSE)`	write to a file (or screen)
`sink(file)`	redirect output to a file
`dump("x", file)`	write text representation of `x`
`write.table(x, file)`	write datafame `x` to a file

Plotting

`plot(x, y)`	plot `y` against `x`
`type = "?"`	determine the type:
`"p", "l", "b"`	for points, lines, or both
`"c"`	for the lines part alone of `"b"`
`"o"`	for both lines and points overplotted
`"h"`	for vertical lines (histogram like plot)
`"s", "S"`	for step function, across/up or reverse
`"n"`	no data plotted, only axes
`main = "title"`	provides plot title
`xlim = c(a,b)`	set lower and upper limits of x-axis
`ylim = c(a,b)`	set lower and upper limits of y-axis
`xlab ="?" ylab = "?"`	provide label for x-axis or y-axis
`pch = k`	set shape of points (k from 1 to 25)
`lwd = ?`	set line width, default 1
`col = "?"`	set line and point colour
`colour() or color()`	list R colours
`lines(x, y)`	add lines to plot
`abline(h) abline(v)`	draw horizontal and vertical lines
`points(x, y)`	add points to plot
`text(x, y, labels)`	place text on plot
`curve(f, from, to)`	plot `f`
`par(?)`	set graphical parameters:
`mfrow = c(nr, nc)`	create grid of plots with `nr` rows and `nc` columns filled by row (`mfcol` fills by col)
`oma = c(b, l, t, r)`	create outer margin around all plots
`mar = c(b, l, t, r)`	create margin around each plot
`las = 1`	make y-axis labels horizontal
`pty = "s"`	force the plot shape to be square
`cex = x`	magnify symbols and text by a factor `x`
`bty = "?"`	determine box type drawn around plot

Random numbers and probability distributions

`ddist(x, p1, ...)`	$\mathbb{P}(X = x)$ or $f(x)$
`pdist(q, p1, ...)`	$\mathbb{P}(X \leq q)$
`qdist(p, p1, ...)`	p-th quantile, equivalently $100p$%-point
`rdist(n, p1, ...)`	pseudo-random numbers
`dist p1, ...`	distribution and parameters:
`unif min = 0 max = 1`	uniform
`binom size prob`	binomial
`geom prob`	geometric
`hyper m n k`	hypergeometric
`nbinom size prob`	negative binomial
`pois lambda`	Poisson
`exp rate`	exponential
`chisq df`	chi square
`gamma shape rate`	gamma
`norm mean sd`	normal
`t df`	t distribution
`weibull shape scale`	Weibull
`set.seed(seed)`	set position in pseudo-random sequence
`.Random.seed`	state of the random number generator
`RNGkind()`	which random number generator?
`sample(x, n, replace = TRUE)`	sample of size `n` from `x`

Programs and functions developed in the text

add_event 398
antithetic sampling 366
ascent 210
Australian rules football 96
binom.sim 336
bisection 180
bp 384
bp_grid 386
change 108
compound 38
control variates 373
curve fitting 219
discrete_queue 292
err 80
expex 248
f3 216
fibonacci 37
fitDistances 413
fixedpoint 171
fixedpoint_show 182
forest_fire 389
gamma.sim 345
gsection 207
hit_miss 357
hit_miss2 358
importance sampling 371
inventory_2stage_sim 402
inventory_sim 399
Kew rainfall 303
life 82
linesearch 211
maxheads 263
mc.integral 360
moverings 84
mySum 145
n_choose_r 66
newton 215
newtonraphson 176
newton_gamma 203
newtonraphson_show 183
next.gen 125
nfact1 34
nfact2 74
pension 35
Phi 191
powers 50
ppoint 295
predprey 46
prime 73
primedensity 72
primesieve 76
print.transectHolder 415
quad1 30
quad2 32
quad2b 54
quad3 65
quadrature 197
quartiles1 53
rejecttriangle 342
Rosenbrock 222
scoping 134
seed-test 410
simpson 194
simpson_n 190
simpson_test 192
simulate.transectHolder 415
SIR_grid 382
SIRsim 379
sum of normals 295
swap 67
threexplus1 39
threexplus1array 45

transectHolder 414
trapezoid 189
trapTransect 139
tree growth 102
truncated normal 306
wmean 67

Index

`.C`, 145
`.Machine`, 152
`.Random.seed`, 333
`.libPaths`, 8
`:`, 15
`<-`, 12
`[ ]`
  dataframes, 90
  vectors, 21
`[[ ]]`
  dataframes, 90
  lists, 95
`{ }`, 31
3D-graphics, 123

adaptive quadrature for integration, 194
arguments, 13, 63
  optional, 70
arithmetic, 11
`array`, 25
assignments, 20
attaching dataframes, 93
attribute, 92
`axis`, 114

Bayes' theorem, 234
Bayesian probability, 228
Bernoulli distribution, 268
binomial distribution, 268, 276
  normal approximation to, 310
  simulating from, 334
bisection method of root-finding, 178
`box`, 114
Box-Muller algorithm, 347
`bquote`, 117
branching processes, 383
`browser`, 76
building a plot by components, 113

`cat`, 33, 41, 55
catastrophic cancellation, 155
Cauchy distribution, 249
cdf, *see* cumulative distribution function
Central Limit Theorem, 292, 309, 315, 345
Chebyshev's Inequality, 258
χ^2 distribution, 297
class, 137
coercion, 20, 25, 50, 87, 337
`colMeans`, 366
combination, 229
commenting, 30
compiled code, 144
`complete.cases`, 93
conditional execution, 31
conditional probability, 230
confidence interval
  for a proportion, 317
  for a small sample, 319
  Monte-Carlo, 321
confidence intervals, 314
congruential generators, 332
constructor, 139, 414
continuous random variables, 243
continuous review inventory model, 391
contraction, 395
control-variate estimation, 372
correlation, 257
countable, 228
counting operations, 157
counting probability, 229
covariance, 256
CRAN, 3, 129, 318
cumulative distribution function, 242
`curve`, 57, 116

`data.frame`, 91
dataframes, 88
  attaching, 93

debugging, 41, 134
dev.off, 118
df, *see* cumulative distribution function
differentiation, 217
dim, 92
dir, 52
discrete random variables, 242
distribution function, 242
do.call, 414
double precision, 152
download.packages, 8
dump, 56

e, 152
ecdf, 246
elements, 15
else, 31
empirical distribution function, 245
environment, 132
  parent environment, 132
equality, 21
estimator, 258
eval, 134
event, 227
example
  $\Phi(z)$, 190
  $\sin(x) - x$ near 0, 155
  n choose r, 65
  n factorial, 34, 74
  accuracy of an opinion poll, 319
  Al Stage tree growth, 102
  Australian rules football, 96
  Buffon's needle and cross, 364
  Cavendish's experiments, 246
  column sums of a matrix, 159
  compound interest, 38
  convergence of Simpson's rule, 192
  density of primes, 71
  discrete simulation of a queue, 289
  disjoint events, 232
  dreaded lurgy, 276
  exponential limit, 18
  factorial, 34
  Fibonacci numbers, 37
  file input, 52
  finding the root of
    $f(x) = \log(x) - \exp(-x)$, 171
  fitting curves, 219
  forest fire, 387
  gamma function, 251
  indigenous deaths in custody, 231
  insurance risk, 311
  inventory, 390
  Kew rainfall, 303
  life tables, 231
  lighting a Barbeque, 271
  loan repayments, 168
  mean and variance, 17
  meta-analysis of opinion polls, 322
  normal percentage points, 293
  numerical calculation of the mean, 248
  optimising
    $\sin(x^2/2 - y^2/4)\cos(2x - \exp(y))$, 215
  pension value, 34
  prostate cancer screening, 234
  quality control, 274
  radioactive decay, 283
  range reduction, 156
  redimensioning an array, 35
  roots of a quadratic, 30, 31, 53, 64
  rounding error, 21
  sampling a manufacturing line, 269
  seed dispersal, 138, 141, 405
  sieve of Eratosthenes, 74
  simple numerical integration, 17
  simulating a binomial random variable, 334
  simulating a gamma random variable, 344
  simulating a uniform random variable, 338
  simulating an exponential random variable, 338
  standard normal central probability, 370, 372
  standard normal tail probability, 369
  summing a vector, 33
  the Chevalier de Meré, 232
  time to the next disaster, 285
  transforming a continuous random variable, 252
  truncated normal, 249, 306
  two-up, 272
  Upper Flat Creek data, 89, 99, 109, 119, 123
  VCE students pass rates, 233

Winsorised mean, 66
expectation, 246
expectation of a transformed random variable, 253
exponential distribution, 283
simulating from, 338
expression, 133
`expression`, 116, 134
expressions, 19, 32
extracting columns from dataframes, 90

factors, 85
`fitdistr`, 413
fixed-point iteration for root-finding, 168
flattening a list, 96
`for`, 33
`format`, 50
formatting output, 50
frame, 132
frequentist probability, 228
functions, 13, 63

gamma distribution, 288
normal approximation to, 314
simulating from, 344
generic functions, 137
geometric distribution, 270
`getAnywhere`, 137
`getwd`, 4
global environment, 133
golden-section method for optimisation, 204, 210, 217
good programming habits, 42
graphics, 56
graphics device, 111

hazard function, 282
help, 5
Hessian, 208
histogram, 245
history, 25
hit-and-miss method, 355

`if`, 31
`ifelse`, 39, 113
iid, *see* independent and identically distributed
importance sampling, 367
independence of events, 232
independence of random variables, 245
independent and identically distributed, 245
infinite range random variable, 249
input from a file, 51
input from the keyboard, 53
`install.packages`, 6, 8, 129
`installed.packages`, 129
integers, representation, 151
interval estimation, 309, 315
inverse transformation method, 338

`lapply`, 100, 414
lattice graphics, 119
law of the random variable, 242
Law of Total Probability, 233, 234
`legend`, 114
length differences, resolving, 16
`library`, 7, 8, 128
line search, 210
`list.files`, 7, 52
lists, 94
`load`, 25
local search, 201
logical expressions, 20
looping, 33, 36
loss function, 219
`ls`, 25, 135

Markov's Inequality, 258
mathematical typesetting, 114
matrices, 23
matrix inversion, 23
maximum likelihood, 304
memory, 160
method of moments, 304
`methods`, 137, 140
minimum (parallel), 39
missing data, 18
mode, 91, 94
model building, 331
Monte-Carlo integration, 358
multivariate optimisation, 207

`NA`, 18
`na.omit`, 93

names, 92
NaN, 153
negative binomial distribution, 273
  normal approximation to, 314
new, 142
Newton method for optimisation, 202
Newton's method for multivariate optimisation, 213
Newton-Raphson method of root-finding, 173
normal distribution, 292
  simulating from, 345
NULL, 19

object-oriented programming
  S3 classes, 137, 412
  S4 classes, 141
optim, 218, 219
optimize, 218
output to a file, 55

package, 6, 127
package construction, 130
package management, 127
package.skeleton, 130
panel, 119
par, 57, 111
paste, 116
pdf, *see* probability density function
pdf, portable document format, 118
permutation, 229
plot, 110
pmf, *see* probability mass function
point estimates, 303
Poisson distribution, 274
  normal approximation to, 312
Poisson process, 287
  merging and thinning, 288
print.trellis, 121
probability axioms, 228
probability density function, 243
probability mass function, 242
program duration, 156
program flow, 39, 67
programming efficiency, 161
prompt, 3, 11
pseudo-code, 41
pseudo-random numbers, 331

qq-plot, 417
quartile, 52

random sample, 245
random seed, 332
random variables, 241
  simulating, 333, 338
read.table, 88
readline, 53
real numbers, representation, 152
recover, 135
recursive programming, 74
rejection method, 339, 342
  exponential envelope, 344, 346
renewal process, 392
return, 64
rm, 25, 30
RNGkind, 333
root bracketing, 178, 204
rounding error, 21, 24
roundoff error, 154
rv, *see* random variables

sample proportion, 259
sample space, 227
sample standard deviation, 261
sample variance, 260
sapply, 71, 100, 418
save.image, 25
scan, 51
scope, 68
script writing, 5
search path, 133
secant method of root-finding, 176
seed rain, 405
sessionInfo, 127
set.seed, 333
setClass, 141
setMethod, 141
setwd, 4
show, 30
significant digits, 12, 154
Simpson's rule for integration, 189
simulating sequences of independent trials, 336
SIR, *see* Susceptible, Infected and Removed
slotNames, 142

`source`, 29
special characters, 49
spuRs archive, 6
standard deviation, 256
standardisation, 309
steepest ascent method for
  optimisation, 209
`stop`, 76
stopping criteria, 201
`str`, 98
string, 49
Student's t distribution, 297, 319
subset, 21
sums of random variables, 254
survivor function, 282
Susceptible, Infected and Removed, 378
`system.time`, 156

`table`, 86
`tapply`, 99
Taylor expansion, 155–157, 181, 184,
  185, 213, 275, 310, 371, 373
text input and ouput, 49
transects, 405
transformations, 251
trapezoidal rule for integration, 187
trellis graphics, 119
two-stage inventory system, 400

uniform distribution, 282
  simulating from, 331, 338

variables, 12
variance, 256
vector programming, 38, 51, 70, 98, 158
vectors, 15

Weak Law of Large Numbers, 257, 276
Weibull distribution, 284
`which`, 21
`while`, 36
working directory, 4
workspace, 25
`write`, 55
`write.table`, 93